Ergebnisse der Mathematik und ihrer Grenzgebiete

3. Folge · Band 26
A Series of Modern Surveys in Mathematics

Seppo Rickman

Quasiregular Mappings

Springer-Verlag
Berlin Heidelberg New York
London Paris Tokyo
Hong Kong Barcelona
Budapest

Seppo Rickman

Department of Mathematics
University of Helsinki
Hallituskatu 15
SF-00100 Helsinki, Finland

Mathematics Subject Classification (1991): 30C65, 30C75, 30D35, 32A22, 32H25, 32H30

With 5 Figures

ISBN 3-540-56648-1 Springer-Verlag Berlin Heidelberg New York
ISBN 0-387-56648-1 Springer-Verlag New York Berlin Heidelberg

Library of Congress Cataloging-in-Publication Data.
Rickman, S. Quasiregular mappings / Seppo Rickman.
p. cm. – (Ergebnisse der Mathematik und ihrer Grenzgebiete; 3. Folge, Bd. 26)
Includes bibliographical references and index.
ISBN 3-540-56648-1 (Berlin: acid-free): DM 128.00. – ISBN 0-387-56648-1 (New York: acid-free)
1. Quasiconformal mappings. I. Title. II. Series.   QA360.R53   1993   515'.93–dc20
93-4824   CIP

© Springer-Verlag Berlin Heidelberg 1993
Printed in the United States of America

Typesetting: Camera ready copy from the author using a Springer TeX macro package
41/3140 - 5 4 3 2 1 0 – Printed on acid-free paper

To Eini

Preface

This monograph grew out from lecture series which I gave at the University of Helsinki in 1980, at the University of Minnesota in 1980 and 1983, and in Bonn at the Max–Planck–Institute in 1986. A great deal of the basic material originates from a seminar at the University of Helsinki during the academic year 1968–69 organized by Olli Martio, Jussi Väisälä, and myself. It was a suggestion of Frederick Gehring in the early 80's that I should publish the content of my courses on quasiregular mappings in book form. At that time the defect relation, which is one of the main results in the book, was not known in its sharp form. The proof of its sharpness has just recently been completed. This matter delayed considerably the finishing of the text.

During the preparation I have received much help from many people. I acknowledge my sincerest gratitude to Bruce Palka, who read most of the text and made numerous suggestions for improvements, especially to the language. I am also greatly indebted to Matti Vuorinen for careful reading of the manuscript and for valuable comments.

During various stages of the work I have had useful discussions with Juha Heinonen, Ilkka Holopainen, Tadeusz Iwaniec, Tero Kilpeläinen, Peter Lindqvist, Olli Martio, Pierre Pansu, and Jussi Väisälä. To all of them I would like to express my gratitude. Particular thanks are due to Jouni Kankaanpää for writing the manuscript in TEX with great care. He has also given valuable advice in the presentation of the text.

I am grateful for the financial support from Suomen Kulttuurirahasto, the Academy of Finland, and the Finnish Society of Sciences and Letters. Finally, I thank the editors for accepting my text in Springer-Verlag's well–known series.

Helsinki, Finland
May, 1993

Seppo Rickman

Contents

Introduction

Quasiregular mappings are defined in the same way as quasiconformal mappings, absent the homeomorphism requirement. They were first introduced and studied by Yu.G. Reshetnyak in a series of articles that began to appear in 1966. His discoveries were enhanced and furthered by the group O. Martio, S. Rickman, and J. Väisälä a few years later. The theory of quasiregular mappings gives a natural and beautiful generalization of the geometric aspects of the theory of (complex) analytic functions in the plane to Euclidean n-space $\mathbb{R}^n$, or more generally, to Riemannian n-manifolds. Quasiregular mappings are interesting not only because of the results obtained about them, but also because of the many new ideas generated in the course of the development of their theory. In addition, a part of classical complex function theory is enriched by its exposure to a different point of view.

Branched coverings often furnish examples of quasiregular mappings. In a general sense all quasiregular mappings are, in fact, topologically of said type. This is contained in one of Reshetnyak's main results. It states that a nonconstant quasiregular mapping, defined as a continuous element in the Sobolev class with locally L^n-integrable first order weak derivatives and with a uniform dilatation bound, is always discrete and open. In dimension $n \geq 3$ quasiregular mappings without branching exhibit considerable rigidity. This is expressed in a concrete way by a theorem of V.A. Zorich [Zo1] which asserts that locally homeomorphic quasiregular mappings of $\mathbb{R}^n$ into itself are homeomorphisms – and thus quasiconformal – when $n \geq 3$.

Quasiregular mappings have strong connections to certain partial differential equations that arise in potential theoretic sides to the theory. To harmonic functions in the plane correspond extremals of certain variational integrals which generalize the Dirichlet integral and which are invariant under a conformal change of metric. The Euler–Lagrange equations for these integrals are second order quasilinear partial differential equations. Launched in 1938 by C.B. Morrey [Mor1] the theory of such equations was further developed by J. Moser, L. Nirenberg, J. Serrin, and others. In the treatment of quasiregular mappings by Reshetnyak the elements of potential theory played an essential role. In recent years substantial advances in this nonlinear potential theory, with particular emphasis on applications to quasiregular theory, have been made on the one hand by Martio and his students [GLM1], [HKM], on the other by B. Bojarski and T. Iwaniec [BI2] (see also [IM], [I4]).

Quasiregular mappings in dimension two are of relatively little independent interest, at least from a purely function theoretic standpoint. This is due to the fact that in the Euclidean case each such mapping takes the form $g \circ h$, where h is quasiconformal and g analytic. Thus analytic functions are precisely the 2-dimensional quasiregular mappings with dilatation coefficient 1; i.e., the 1-quasiregular mappings. On the other hand, in dimension $n \geq 3$ nonconstant 1-quasiregular mappings turn out in the Euclidean setting to be restrictions of Möbius transformations. This is the content of a generalized Liouville theorem [G4], [Re3]. Therefore it is essential to allow dilatation in order to get a fruitful theory for $n \geq 3$. Branching in case $n \geq 3$ can occur only if the dilatation coefficient K exceeds a bound $K_n > 1$. Furthermore, when $K \to 1$, there is a degree of stability in the generalized Liouville theorem.

One of the most interesting problems in the theory of quasiregular mappings has been the existence of a Picard–type theorem on omitted values. As is well known, E. Picard proved in 1879 that an analytic function in $\mathbb{R}^2$ which omits two values in $\mathbb{R}^2$ must be constant. In 1967 Zorich [Zo1] posed the question whether a theorem of this type could be true for quasiregular mappings in dimension $n \geq 3$. Such a theorem was proved in [R5] in the following form: a quasiregular mapping of $\mathbb{R}^n$ into itself is constant if it omits at least q points, where q is an integer that depends only on n and the dilatation coefficient. This theorem is known to be qualitatively best possible in dimension three; namely, it is proved in [R11] that for each positive integer p there exists a nonconstant quasiregular mapping of $\mathbb{R}^3$ into itself omitting p points in $\mathbb{R}^3$. The case $p = 1$ is fairly straightforward and was already treated by Zorich in [Zo1]. The Picard–type theorem is a solution to a particular case of the general problem of characterizing Riemannian n-manifolds M and N for which there exists a nonconstant quasiregular mapping of M into N. At the present time very little is known about this subject (see IV.2.29).

Results in general value distribution, going far beyond Picard–type theorems, have been established for quasiregular mappings. For example, a defect relation in the spirit of Ahlfors' theory of covering surfaces [A2] was first proved in [R7]. An improved result can be found in [R16]. The latter is qualitatively best possible in dimension three in the sense that an arbitrary sequence of defect numbers subject to the constraints in the defect relation can be asymptotically realized by a quasiregular mapping of $\mathbb{R}^3$ into $\overline{\mathbb{R}}^3 = \mathbb{R}^3 \cup \{\infty\}$.

The purpose of this monograph is first to lay a solid foundation for the theory of quasiregular mappings from a geometric point of departure and then to proceed in a reasonably direct way to deeper results like Picard–type theorems and value distribution theory. We restrict ourselves for the most part to mappings of domains in $\mathbb{R}^n$ or $\overline{\mathbb{R}}^n$. Reshetnyak's result on discreteness and openness is used from the outset, but its proof is postponed until Chapter VI.

Chapter I contains basic material that is focused primarily on the definition of quasiregular mappings. Among the most powerful tools for obtaining geometric information in the theory of quasiregular mappings has been the method of extremal length. For present purposes it centers around specific inequalities for moduli of path families due to E.A. Poletskiĭ and J. Väisälä. The method is presented in Chapter II. Some immediate applications, such as distortion theorems, are given in Chapter III.

In Chapter IV we prove the Picard–type theorem and consider some related questions. The details of the proof of the sharpness of these results (for $n = 3$) are overly complicated, so are not included. The interested reader is referred to [R11]. Chapter V deals with the general question of value distribution. The main result is the defect relation, of which a complete proof is given. On the other hand, the proof that prescribed defects are realizable is omitted. For it the reader is referred to [R16].

Prior to Chapter VI the dominant theme in arguments is extremal length. In Chapter VI we introduce the method of variational integrals and employ it to prove the Reshetnyak result on discreteness mentioned above.

For Chapter VII we have chosen some theorems that illustrate the boundary behavior of quasiregular mappings. Both main methods, extremal length and variational integrals, turn out to be useful in this regard. In many cases the results on quasiregular mappings follow as corollaries from corresponding statements for extremals for variational integrals.

Two monographs on quasiregular mappings have been published earlier. The first is [Re11] by Reshetnyak. It largely contains material from his original articles on the subject dating from the late 1960's, the emphasis being on variational integrals and analytical tools. In contrast with the present book, rather little is said about the more geometric part of the theory. The second monograph is [Vu12] by M. Vuorinen. The stress there is placed upon distortion results and conformal invariants, so there is little overlap with the present volume. It also contains a lot of material on quasiconformal mappings and an extensive bibliography.

The early studies by the group Martio, Rickman, and Väisälä are published in [MRV1]–[MRV3]. Another presentation of the foundations of the theory of quasiregular mappings is available in the paper by Bojarski and Iwaniec [BI2]. Like Reshetnyak, they base the theory on variational integrals, but their methods are more direct and considerably simpler than those in [Re11]. Associations of quasiregular mappings with isoperimetric inequalities on Riemannian manifolds are investigated by M. Gromov in [Gro2, Chapitre 6]. As additional surveys on quasiregular mappings we refer to [V7], [R13], and [R14].

It is assumed that the reader is familiar with real analysis and with topology in Euclidean spaces. Some knowledge of quasiconformal mappings is also recommended. The standard reference is Väisälä's monograph [V4]. When convenient, we will cite [V4] rather than repeat a proof. Apart from these deviations our presentation remains essentially selfcontained.

Chapter I. Basic Properties of Quasiregular Mappings

We shall define quasiregular mappings analytically following Yu.G. Reshetnyak. At an early stage we take full advantage of Reshetnyak's important discovery that a nonconstant quasiregular mapping is discrete and open, but put off the proof to Chapter VI. This way we are able to have a coherent geometric treatment without introducing an excessive amount of machinery at the outset. Section 1 on ACL^p mappings contains fairly standard preliminary results. Discrete open mappings are considered in Section 4 as a separate topic, mostly without proofs. The material of the first chapter is primarily concerned with various aspects of the definition of quasiregularity.

1. ACL^p Mappings

1.1. Notation. The standard orthonormal basis vectors in the Euclidean space $\mathbb{R}^n$ are denoted by $e_1, \ldots, e_n$. For $x, y \in \mathbb{R}^n$ the inner product is $x \cdot y$ and norm $|x| = (x \cdot x)^{1/2}$. The open ball and sphere with center x and radius $r > 0$ are written $B^n(x, r)$ and $S^{n-1}(x, r)$. We sometimes omit the superscript here and also write $B^n(r) = B^n(0, r)$, $S^{n-1}(r) = S^{n-1}(0, r)$, $B^n = B^n(1)$, $S^{n-1} = S^{n-1}(1)$. The diameter of a set $A \subset \mathbb{R}^n$ is denoted by $d(A)$ and the line segment connecting the points x and y by $[x, y]$. The Lebesgue measure in $\mathbb{R}^n$ is denoted by m or m_n. We also use the notation $|A|$ for the Lebesgue measure of a measurable set $A \subset \mathbb{R}^n$ and occasionally write dx instead of dm. The set of integers is denoted by $\mathbb{Z}$, the set of complex numbers by $\mathbb{C}$, and we define $\mathbb{N} = \{ z \in \mathbb{Z} : z > 0 \}$.

The stereographic projection $\pi \colon \mathbb{R}^n \to S^n(e_{n+1}/2, 1/2)$ of the one–point compactification $\overline{\mathbb{R}}^n = \mathbb{R}^n \cup \{\infty\}$ of $\mathbb{R}^n$ is defined by

$$\pi(x) = e_{n+1} + \frac{x - e_{n+1}}{|x - e_{n+1}|^2} \,, \quad x \in \mathbb{R}^n \, ; \quad \pi(\infty) = e_{n+1} \,.$$

We give $\overline{\mathbb{R}}^n$ the spherical metric $d\sigma^2$ which makes π an isometry. In addition to the spherical distance $\sigma(x, y)$ we sometimes also employ the chordal distance $q(x, y) = |\pi(x) - \pi(y)|$ for $x, y \in \overline{\mathbb{R}}^n$. If not otherwise stated, for subsets of $\overline{\mathbb{R}}^n$ we perform all topological operations with respect to $\overline{\mathbb{R}}^n$. The open ball $\{ y \in \overline{\mathbb{R}}^n : \sigma(x, y) < r \}$ and the sphere $\{ y \in \overline{\mathbb{R}}^n : \sigma(x, y) = r \}$, $0 < r < 1$, are denoted by $D(x, r)$ and $\Sigma(x, r)$ respectively.

For $\beta \geq 0$, $t > 0$, and $A \subset \mathbb{R}^n$ we set

$$\mathcal{H}_t^\beta(A) = \inf \sum \lambda_\beta d(A_i)^\beta \, ,$$

where the infimum is taken over all countable coverings $\{A_i : i = 1, 2, \ldots\}$ of A with $d(A_i) < t$ and λ_β is the constant

$$\lambda_\beta = \frac{\Gamma(\frac{1}{2})^\beta}{2^\beta \Gamma(\frac{1}{2}\beta + 1)} \, .$$

The normalized β-dimensional Hausdorff outer measure of A is

$$\mathcal{H}^{\beta*}(A) = \lim_{t \to 0} \mathcal{H}_t^\beta(A) \, ,$$

and $\mathcal{H}^\beta$ is the corresponding measure. The measure $\mathcal{H}^n$ coincides with the Lebesgue measure in $\mathbb{R}^n$. Let $h \colon [0, \infty[\rightarrow [0, \infty[$ be a continuous and increasing function with $h(0) = 0$ and $h(t) > 0$ for $t > 0$. If we above replace $\lambda_\beta d(A_i)^\beta$ by $h(d(A_i)/2)$, we obtain the h-Hausdorff measure Λ_h. The measure $m(B^n)$ of the unit ball is denoted by Ω_n and $\mathcal{H}^{n-1}(S^{n-1})$ by ω_{n-1}. For any set A we let χ_A be the characteristic function of A.

Let Q be a closed n-interval $\{x \in \mathbb{R}^n : a_i \leq x_i \leq b_i, i = 1, \ldots, n\}$. A mapping $f \colon Q \to \mathbb{R}^m$ is called ACL (*absolutely continuous on lines*) if f is continuous and if f is absolutely continuous on almost every line segment in Q parallel to the coordinate axes; more precisely, $x_1 \mapsto f(x_1, x_2, \ldots, x_n)$ is absolutely continuous for m_{n-1} almost every $(x_2, \ldots, x_n)$ in $[a_2, b_2] \times \ldots \times [a_n, b_n]$ and similarly for other coordinates. In the following U will be an open set in $\mathbb{R}^n$. A mapping $f \colon U \to \mathbb{R}^m$ is ACL if $f|Q$ is ACL for every closed n-interval $Q \subset U$. Such a mapping has partial derivatives $D_i f(x)$ a.e. in U and if $A_i \subset U$ is the subset where $D_i f(x)$ exists, $D_i f$ is a Borel function in A_i [V4, 26.4]. If, in addition, $p \geq 1$ and the partial derivatives are locally L^p-integrable, f is said to be ACL^p or in ACL$^p(U)$.

There is a close connection between ACL^p mappings and Sobolev spaces. Let U be open in $\mathbb{R}^n$ and $u \in L^1_{\text{loc}}(U)$ be a real valued function. We say that $v \in L^1_{\text{loc}}(U)$ is the ith *weak partial derivative* of u if

$$\int_U \varphi v \, dm = - \int_U u D_i \varphi \, dm$$

for all $\varphi \in \mathcal{C}_0^1(U)$. Note that by approximation it is equivalent to require this for all $\varphi \in \mathcal{C}_0^\infty(U)$. If v exists, it is also denoted by $D_i u$. The *Sobolev space* $W_p^1(U)$, $1 \leq p < \infty$, consists of all real valued functions u in $L^p(U)$ with weak first order partial derivatives which are themselves in $L^p(U)$. By $W_{p,\text{loc}}^1(U)$ we denote functions which locally belong to W_p^1. By considering component functions we extend these definitions to $\mathbb{R}^m$-valued mappings without separate notation.

1.2. Proposition. *A mapping* $f: U \to \mathbb{R}^m$ *is* ACL^p *if and only if it is continuous and belongs to* $W_{p,\mathrm{loc}}^1(U)$. *In this case the weak and ordinary partial derivatives coincide a.e.*

Proof. We may assume $m = 1$. Let f be ACL^p. Let $\varphi \in C_0^1(U)$ and extend φ and f to $\mathbb{R}^n$ by $\varphi|\mathbb{R}^n \setminus U = f|\mathbb{R}^n \setminus U = 0$. When necessary, C_0^k functions are always thought to be extended this way. Then φf is absolutely continuous on compact subintervals of almost all lines parallel to the x_i-axis. If L is such a line, we get by integration

$$0 = \int_L D_i(\varphi f)\, dm_1 = \int_L D_i \varphi f\, dm_1 + \int_L \varphi D_i f\, dm_1 \ .$$

Fubini's theorem then gives

$$\int_U \varphi D_i f\, dm = - \int_U D_i \varphi f\, dm \ ,$$

which shows that the partial derivative $D_i f$ is also the ith weak derivative of f. By the definition of ACL^p $D_i f$ belongs to $L_{\mathrm{loc}}^p(U)$ and, being continuous, so does f. Hence $f \in W_{p,\mathrm{loc}}^1(U)$.

Next let $f \in W_{p,\mathrm{loc}}^1(U) \cap C(U)$. Let v be the first weak partial derivative of f. Fix a closed n-interval $Q = I \times J$ in U, where $I = [a, b]$ is an interval and J an $(n-1)$-interval. For $x \in Q$ we write $x = (t, y) \in I \times J$. Let $V = B^{n-1}(\eta, \rho)$ be an $(n-1)$-ball in J and let $C = [a, c]$, $a \leq c \leq b$. Consider test functions $\varphi \in C^1$ of the form $\varphi(x) = h(t)g(y)$ with support in $C \times V$. Then

$$\int_{C \times V} v(x)h(t)g(y)\, dx = - \int_{C \times V} f(x)h'(t)g(y)\, dx \ .$$

Letting g run through a sequence g_k such that $0 \leq g_k \leq 1$ and $g_k \to 1$ in V, we get first

$$\int_{C \times V} v(x)h(t)\, dx = - \int_{C \times V} f(x)h'(t)\, dx \ ,$$

after which Fubini's theorem gives

$$\frac{1}{|V|} \int_V \left(\int_C v(t, y)h(t)\, dt \right) dy = - \frac{1}{|V|} \int_V \left(\int_C f(t, y)h'(t)\, dt \right) dy \ .$$

By letting $\rho \to 0$ we see using Lebesgue's theorem that for almost every $y \in J$

$$\int_a^c v(t, y)h(t)\, dt = - \int_a^c f(t, y)h'(t)\, dt \ .$$

Letting h run through a sequence h_k with $h_k|[a+1/k, c-1/k] = 1$ and $|h_k'| \leq 2k$ we infer from the continuity of f that

$$(1.3) \qquad \int_a^c v(t,y)\, dt = f(c,y) - f(a,y)$$

for all $y \in J \setminus E_c$, where $E_c \subset J$ is a set of zero $(n-1)$-dimensional measure. Letting c run through rational values we get by the continuity of f that (1.3) holds outside an $(n-1)$-dimensional null set E for all c. Hence $t \mapsto f(t,y)$ is absolutely continuous and v is the first partial derivative of f a.e. in I for $y \in J \setminus E$. Repeating this for other coordinates we conclude that f is ACL. Since $v \in L_{\text{loc}}^p(U)$, f is ACL^p.   □

A mapping $f: U \to \mathbb{R}^m$ in $W_p^1(U)$ is normed by

$$\|f\|_{1,p,U} = \|f\|_{1,p} = \|f\|_p + \||Df|\|_p .$$

Here $\|\ \ \|_p$ is the norm in L^p and Df is defined in terms of the weak partial derivatives $D_i f_j$, thus $Df(x): \mathbb{R}^n \to \mathbb{R}^m$ is the linear mapping for which $Df(x)e_i = \sum_j D_i f_j(x)e_j$. The mapping $Df(x)$ is called the *formal derivative*. For the norm of $Df(x)$ we take the operator norm

$$|Df(x)| = \sup_{|h|=1} |Df(x)h| .$$

Often we replace $Df(x)$ by $f'(x)$. With respect to the standard basis $Df(x)$ will also mean the corresponding matrix, and this convention will be applied to any linear mapping $\mathbb{R}^n \to \mathbb{R}^n$. The Jacobian determinant $\det Df(x)$ is written $J_f(x)$.

The space W_p^1 is a Banach space, which is proved as follows. We may assume $m = 1$. If (f_k) is a Cauchy sequence in W_p^1, then (f_k) and $(D_i f_k)$ are Cauchy sequences in L^p. Let $f = \lim f_k$ and $g_i = \lim D_i f_k$ be the limits in L^p. Then

$$\int_U \varphi g_i\, dm = - \int_U D_i \varphi f\, dm , \quad \varphi \in \mathcal{C}_0^1(U) ,$$

so g_i is the ith weak derivative of f.

We shall often use smoothing operations on functions. Fix a nonnegative $\psi \in \mathcal{C}_0^\infty(\mathbb{R}^n)$ with spt $\psi = \overline{B}^n$ and

$$\int_{B^n} \psi\, dm = 1 .$$

For $\varepsilon > 0$ and any h in $L_{\text{loc}}^1(\mathbb{R}^n)$ we define the regularization h_ε of h to be the convolution $\psi_\varepsilon * h$ where $\psi_\varepsilon(x) = \varepsilon^{-n}\psi(x/n)$. With such regularizations one proves that $W_p^1(U) \cap \mathcal{C}^\infty(U)$ is dense in $W_p^1(U)$ (see [St, p. 122]). Closely connected with this density result is the following proposition for ACL^p mappings. For the proof we refer to [V4, 27.7]. For future reference we mention that the closure of $\mathcal{C}_0^\infty(U)$ in $W_p^1(U)$ is denoted by $W_{p,0}^1(U)$.

1.4. Proposition. *Let* $f: U \to \mathbb{R}^m$ *be* ACL^p *. Then there is a sequence of mappings* $f_k: U \to \mathbb{R}^m$ *such that the following hold:*

(1) $f_k \in C^\infty$ *.*

(2) $f_k \to f$ *locally uniformly in* U *.*

(3) *For each compact* $F \subset U$ *and for* $1 \le i \le n$ *,* $D_i f_k \to D_i f$ *in* $L^p(F)$ *.*

(4) *If the support of* f *is a compact subset of* U *, then the same is true for* f_k *.*

The ACL^p property remains invariant under coordinate changes. This is included in Proposition 1.11. We need the concept of the adjunct of a linear mapping $\mathbb{R}^n \to \mathbb{R}^n$.

Let $\bigwedge_k \mathbb{R}^n$ and $\bigwedge^k \mathbb{R}^n$ be the spaces of k-vectors and k-covectors or k-forms of $\mathbb{R}^n$. For $\xi \in \bigwedge_k \mathbb{R}^n$ and $\alpha \in \bigwedge^k \mathbb{R}^n$ we write $\langle \alpha, \xi \rangle = \alpha(\xi)$. Let $\omega_1, \ldots, \omega_n$ be the base of 1-forms dual to the standard basis $e_1, \ldots, e_n$, and let E be the n-vector $e_1 \wedge \ldots \wedge e_n$ and Ω the n-form $\omega_1 \wedge \ldots \wedge \omega_n$. We then have the isomorphisms $D_k: \bigwedge_k \mathbb{R}^n \to \bigwedge^{n-k} \mathbb{R}^n$ and $D^k: \bigwedge^k \mathbb{R}^n \to \bigwedge_{n-k} \mathbb{R}^n$ given by (see [F, 1.52])

$$\langle D_k \eta, \xi \rangle = \langle \Omega, \xi \wedge \eta \rangle , \quad \eta \in \bigwedge_k \mathbb{R}^n , \quad \xi \in \bigwedge_{n-k} \mathbb{R}^n ,$$
$$\langle \alpha, D^k \beta \rangle = \langle \alpha \wedge \beta, E \rangle , \quad \alpha \in \bigwedge^{n-k} \mathbb{R}^n , \quad \beta \in \bigwedge^k \mathbb{R}^n ,$$

and $D_p^{-1} = D^{n-p}$ holds.

Let $A: \mathbb{R}^n \to \mathbb{R}^n$ be linear. The *adjunct* $\mathrm{ad}\, A$ of A is the linear mapping

(1.5) $$\mathrm{ad}\, A = D^{n-1} \circ \bigwedge^{n-1} A \circ D_1: \mathbb{R}^n \to \mathbb{R}^n,$$

where $\bigwedge^{n-1} A: \bigwedge^{n-1} \mathbb{R}^n \to \bigwedge^{n-1} \mathbb{R}^n$ is the induced mapping. If $A = (a_{ij})$ as a matrix, the matrix coordinates for $\mathrm{ad}\, A$ are

$$(\mathrm{ad}\, A)_{ji} = \det A'_{ij},$$

where A'_{ij} is the matrix formed from A by replacing the coordinate a_{ij} by 1 and all others in the row i and column j by 0 . Thus $\det A'_{ij}$ is a cofactor of A .

We have (see [Gre, p. 58])

(1.6) $$A \, \mathrm{ad}\, A = \det A \cdot I ,$$

where I is the identity. If A is invertible, $\mathrm{ad}\, A = \det A \cdot A^{-1}$.

If $g: U \to \mathbb{R}^n$ is C^1 , we write $\mathrm{ad}\, g'$ for the mapping of U with $(\mathrm{ad}\, g')(x) = \mathrm{ad}\, g'(x)$.

1.7. Lemma. *Let* $g: U \to \mathbb{R}^n$ *be in* C^2. *Then each of the vector fields* ad $g'e_j$ *is divergence free, i.e.,*

$$(1.8) \qquad\qquad \operatorname{div} \operatorname{ad} g'e_j = 0 \,, \quad j = 1, \ldots, n \,.$$

Proof. In coordinates (1.8) becomes

$$
\begin{aligned}
0 &= \sum_{i=1}^{n} D_i(\operatorname{ad} g')_{ij} \\
&= \sum_{i=1}^{n} \sum_{\substack{k_1,\ldots,k_n \\ k_i=j}} \varepsilon_{k_1 \cdots k_n} D_i \left(D_1 g_{k_1} \cdots D_{i-1} g_{k_{i-1}} D_{i+1} g_{k_{i+1}} \cdots D_n g_{k_n} \right) \\
&= \sum_{i=1}^{n} \sum_{s=1}^{n} \sum_{m \neq i} \sum_{\substack{k_1,\ldots,k_n \\ k_i=j \\ k_m=s}} \varepsilon_{k_1 \cdots k_n} \left[D_i D_m g_s \prod_{\substack{\nu \neq m \\ \nu \neq i}} D_\nu g_{k_\nu} \right] \,.
\end{aligned}
$$

The expression in brackets is symmetric in m and i. An interchange of m and i causes an interchange of k_m and k_i, and then $\varepsilon_{k_1 \cdots k_n}$ changes sign. Therefore the sum vanishes. $\qquad\qquad\qquad\qquad\qquad\qquad\qquad\qquad\qquad$ □

Let $v: U \to \mathbb{R}^n$ be in $L^1_{\text{loc}}(U)$. We say that v is divergence free if it satisfies the corresponding condition weakly, i.e., if

$$(1.9) \qquad\qquad \int_U v \cdot \nabla\varphi \, dm = 0$$

for all $\varphi \in C_0^\infty(U)$, where $\nabla\varphi$ is the gradient of φ. If $v \in \text{ACL}^1(U)$, then (1.9) is equivalent to $\operatorname{div} v = 0$ a.e.

1.10. Remark. Under the definition (1.9) the assertion in 1.7 remains true for mappings $g: U \to \mathbb{R}^n$ that are ACL^{n-1}. In this case $\operatorname{ad} g' \in L^1_{\text{loc}}(U)$ and if $v = \operatorname{ad} g'e_j$, (1.9) follows by approximation.

1.11. Proposition. *Let* $f: U \to \mathbb{R}^m$ *be* ACL^p, $g: V \to U$ *a* C^2 *diffeomorphism, and* $h: fU \to \mathbb{R}^q$ *a Lipschitz mapping. Then* $h \circ f$ *and* $f \circ g$ *are* ACL^p. *Moreover,*

$$(1.12) \qquad\qquad D_i(f \circ g) = \sum_{j=1}^{n} (D_j f \circ g) D_i g_j \quad \text{a.e.}$$

Proof. The mapping $h \circ f$ is clearly ACL. If L is the Lipschitz constant of h, we have $|D_i(h \circ f)| \leq L|D_i f|$ a.e. which shows that $h \circ f$ is also ACL^p.

For the rest of the proposition we may assume that $m = 1$ and $J_g > 0$. Let $\varphi \in C_0^1(U)$. Then $\varphi \circ g$ represents an arbitrary test function in V, and by 1.2 and 1.7 we obtain

$$\int_V \sum_j (D_j f \circ g) D_i g_j (\varphi \circ g) \, dm = \int_V \sum_j (D_j f \circ g)(D_i g_j / J_g)(\varphi \circ g) J_g \, dm$$

$$= \int_U \sum_j D_j f (\operatorname{ad} g^{-1'})_{ji} \varphi \, dm = - \int_U \sum_j f D_j \big((\operatorname{ad} g^{-1'})_{ji} \varphi\big) \, dm$$

$$= - \int_U f \sum_j \big(D_j (\operatorname{ad} g^{-1'})_{ji} \varphi + (\operatorname{ad} g^{-1'})_{ji} D_j \varphi\big) \, dm$$

$$= - \int_U f \sum_j (\operatorname{ad} g^{-1'})_{ji} D_j \varphi \, dm$$

$$= - \int_V (f \circ g) \sum_j (D_i g_j / J_g)(D_j \varphi \circ g) J_g \, dm$$

$$= - \int_V (f \circ g) D_i (\varphi \circ g) \, dm \ .$$

Hence $f \circ g$ has weak derivatives given by (1.12). We used above a form of a transformation formula for integrals which is a special case of 4.14(c).

For a compact $F \subset V$ we get by (1.12)

$$\int_F |D_i (f \circ g)|^p dm \le M \int_F \Big|\sum_j D_j f \circ g\Big|^p J_g \, dm = M \int_{gF} \Big|\sum_j D_j f\Big|^p dm < \infty$$

where

$$M = \sup_{1 \le j \le n} \sup_F J_g^{-1} |D_i g_j|^p < \infty \ .$$

Thus $f \circ g$ is ACL^p by 1.2. $\square$

1.13. Notes. The proofs of 1.2 and 1.11 are n-dimensional versions of [A5, pp. 28–30]. Lemma 1.7 is taken from [BI2].

2. Quasiregular Mappings

We shall now define quasiregular mappings in the Euclidean case as follows.

2.1. Definition. A mapping $f \colon G \to \mathbb{R}^n$, $n \ge 2$, of a domain G in $\mathbb{R}^n$ is *quasiregular* (abbreviated qr) if
  (1) f is ACL^n, and
  (2) there exists K, $1 \le K < \infty$, such that

(2.2) $|f'(x)|^n \leq KJ_f(x)$ a.e.

The smallest K in (2.2) is the *outer dilatation* $K_O(f)$ of f. If f is quasiregular, then it is also true that

(2.3) $J_f(x) \leq K'\ell(f'(x))^n$ a.e.

for some $K' \geq 1$ where $\ell(f'(x)) = \inf_{|h|=1} |f'(x)h|$. The smallest $K' \geq 1$ in (2.3) is the *inner dilatation* $K_I(f)$ of f. The *(maximal) dilatation* is $K(f) = \max(K_O(f), K_I(f))$ and a qr mapping is called K-*quasiregular* if $K(f) \leq K$. The relationships $K_O(f) \leq K_I(f)^{n-1}$ and $K_I(f) \leq K_O(f)^{n-1}$ hold as is easily seen by linear algebra. Thus $K_O(f) = K_I(f)$ for $n = 2$. A quasiregular homeomorphism $f: G \to fG$ is called *quasiconformal* (abbreviated qc).

2.4 Theorem [Re2]. *A qr mapping is differentiable a.e.*

We shall use 2.4, but we postpone its proof until VI.4.

The definition of quasiregularity extends easily to the setting of Riemannian n-manifolds. Let M and N be oriented connected $\mathcal{C}^\infty$ Riemannian n-manifolds, $n \geq 2$, and $f: M \to N$ continuous. In fact, we formally use here the same definition for quasiregularity of f as in 2.1. The ACL^n (or more generally the ACL^p) property is well defined for such f by Proposition 1.11. The condition (2) in 2.1 is imposed on the derivative on tangent spaces, and 2.4 is used. In this work Riemannian n-manifolds are always oriented, connected, and $\mathcal{C}^\infty$.

If N here is $\overline{\mathbb{R}}^n = \mathbb{R}^n \cup \{\infty\}$ equipped with the spherical metric (thus $\overline{\mathbb{R}}^n$ is isometric via stereographic projection with the sphere $S^n(1/2)$ in $\mathbb{R}^{n+1}$) and M is a domain in $\overline{\mathbb{R}}^n$, we call a qr mapping $f: M \to N$ also *quasimeromorphic* (abbreviated qm).

Unless otherwise stated, we shall in this monograph consider quasiregularity in the Euclidean case. Except for the quasimeromorphic case we shall only occasionally refer to results on manifolds.

In the case $n = 2$ we make the following observations. All (complex) analytic functions are clearly 1-qr. The converse is also true and amounts essentially to Weyl's lemma (see [A3, p. 45] or [Mor2, p. 42]). A qr mapping f for $n = 2$ is always of the form $f = g \circ h$ where h is qc and g analytic [LV2, p. 247].

If $n \geq 3$, there is the following strong rigidity result for 1-qr mappings. It is known as the generalized Liouville's theorem.

2.5. Theorem. *For $n \geq 3$ every 1-qr mapping is a restriction of a Möbius transformation or a constant.*

We shall not prove 2.5. In this form it was proved by Yu.G. Reshetnyak [Re3], and it follows as well from the corresponding result for 1-qc mappings by F.W. Gehring [G4, p. 388]. Both these proofs rely on regularity results for elliptic partial differential equations. More recently, B. Bojarski and T.

Iwaniec [BI1] have given a direct proof depending only on basic properties of Sobolev spaces. See too [IM]. More remarks on the history of 2.5 can be found in [V4, 5.8 and 13.7].

Because of 2.5 it is essential in the case $n \geq 3$ to allow $K(f) > 1$ in order to get a sufficiently rich class of mappings. On the other hand, if the dilatation is not uniformly bounded, the beautiful global properties that are counterparts of the theory of analytic functions in the plane are lost.

Quasiregular mappings exhibit also another kind of rigidity when $n \geq 3$. It was anticipated already in 1938 by M.A. Lavrentiev [La] and proved in 1967 by V.A. Zorich [Zo1] that a locally homeomorphic qr mapping $f: \mathbb{R}^n \to \mathbb{R}^n$ for $n \geq 3$ must in fact be homeomorphic and hence qc. We shall get this as a corollary of a more general result in III.3.4.

It is essential not to restrict oneself to overly smooth mappings in the definition of quasiregularity in the case $n \geq 3$. This is so because too smooth nonconstant qr mappings are locally homeomorphic for $n \geq 3$ and hence rigid in the above sense. This was observed by P.T. Church [C] for qr mappings in the class C^n. An improvement of his result is obtained as follows. If $f: G \to \mathbb{R}^n$ is in C^k, $k \geq 1$, and $f'(x) = 0$ in a Borel set $E \subset G$, then by [F, 3.4.3] the (n/k)-dimensional Hausdorff measure $\mathcal{H}^{n/k}(fE)$ of fE is zero. Let B_f be the *branch set*, i.e.,

$$B_f = \{\, x \in G : f \text{ is not locally homeomorphic at } x \,\} \,.$$

If f is qr and C^1, $f'(x) = 0$ in B_f. If we now assume that f is nonconstant and $B_f \neq \emptyset$, take $E = B_f$, and use the result that $\mathcal{H}^{n-2}(fB_f) > 0$ from III.5.3, we get a contradiction for $k = 3$ if $n = 3$ and for $k = 2$ if $n \geq 4$. It is not known whether a nonconstant qr C^1 mapping must have $B_f = \emptyset$ for $n \geq 3$.

The ACL^n requirement in the definition of quasiregularity is natural in the sense that we then have that a locally uniformly convergent sequence of K-qr mappings has a K-qr limit (Theorem VI.8.6). It is true that we can replace condition (1) in 2.1 by $f \in W^1_{n,\mathrm{loc}}(G)$ and remain essentially in the same class of mappings, namely, f then coincides a.e. with a qr mapping. This is Theorem VII.3.9. By a recent result by T. Iwaniec [I4] it is in fact enough here to asume $f \in W^1_{p,\mathrm{loc}}$ where $p = p(n, K) < n$. Slightly earlier he established such a result for even dimensions together with G. Martin in [IM]. For a discussion of this, see VII.1.19.

2.6. Notes. Yu.G. Reshetnyak introduced qr mappings in [Re2] with Definition 2.1 but with the term "mappings with bounded distortion". 2-dimensional qr mappings have earlier been called "quasiconformal functions" [LV2]. Reshetnyak also uses dilatation coefficients different from ours. The use of $K_O(f)$ and $K_I(f)$ is common in qc theory and these coefficients are convenient in connection with inequalities of moduli of path families. Although the systematic study of qr mappings began in 1966 with the papers [Re1]–[Re10] by Reshetnyak and a few years later by O. Martio, S. Rickman, and J. Väisälä in [MRV1]–[MRV3], qr mappings appear in the literature earlier: in 1960 E.D. Callender [Ca] established the Hölder continuity of qr mappings.

3. Examples

In this section we shall present a few simple examples of qr mappings. Our purpose is to illustrate certain properties that are typical of such mappings.

3.1. Winding mapping. Let k be a positive integer and let $f : \mathbb{R}^3 \to \mathbb{R}^3$ be the mapping $(r, \varphi, x_3) \mapsto (r, k\varphi, x_3)$ in cylindrical coordinates. Then f is clearly qr with $K_O(f) = k^2$ and $K_I(f) = k$. The branch set B_f is the x_3-axis. The n-dimensional version is $(r, \varphi, y) \mapsto (r, k\varphi, y)$, again in cylindrical coordinates with $y \in \mathbb{R}^{n-2}$.

If we denote by $i(x, f)$ the infimum of $\sup_y \operatorname{card} f^{-1}(y) \cap U$ when U runs through the neighborhoods of x, we observe that $i(x, f) = k$ for $x \in B_f$ and $i(x, f) = 1$ otherwise. The number $i(x, f)$ is the *local (topological) index* at x. We see that the dilatation tends to ∞ together with k. This is typical for general qr mappings when $n \geq 3$. In fact, if the local index exceeds a certain value depending on n and $K_I(g)$ at all points of a continuum, then g is constant (see III.5.8).

This simple example f is interesting for a second reason. For $k = 2$ we have $K_I(f) = 2$. It is an open problem whether there exists a nonconstant qr mapping g with $K_I(g) < 2$ and with nonempty branch set.

Topologically f (the same holds for the example in 3.2) – or, to be more precise, its extension $f_0 : \overline{\mathbb{R}}^n \to \overline{\mathbb{R}}^n$ – belongs to a group of examples, namely *branched covers* $g : M \to N$ between compact manifolds in the PL category (see for example [Hil]). Indeed, all such mappings are qr when M and N are provided with some natural Riemannian structures. In this connection, note a difference in terminology: our branch set B_g is in the literature often called the singular set and gB_g is called the branch set.

3.2. Counterparts to power mappings. Here we shall give an example in three–space which is in many respects a counterpart to the power mapping $f_m : z \mapsto z^m$ in the plane (with complex notation). The sectors $f_m^{-1}H$, where H is the upper half plane, will correspond to infinite cones with triangular bases. A detailed construction goes as follows.

Let Q be the cube $\{ x \in \mathbb{R}^3 : -1 \leq x_i \leq 1, i = 1, 2, 3 \}$ and let k be a positive integer. We divide each face in ∂Q into congruent squares by the planes $x_i = j/2^k$, $j = -(2^k - 1), -(2^k - 2), \ldots, 2^k - 2, 2^k - 1$, $i = 1, 2, 3$. Each such square A is further divided into four congruent closed triangles by line segments connecting the center of A to the vertices of A. If T is such a triangle, we let C_T be the cone $\{ ty : y \in T, t \geq 0 \}$. In Fig. 1 one such cone is exhibited. In a face of Q we have for any pair T, T' of triangles a well defined mapping $T \to T'$ obtained by repeated reflections in the sides of the triangles. This mapping extends naturally also to pairs T, T' with T and T' in different faces of Q. Fix one triangle meeting the positive x_3-axis, T_0 say, and let $\varphi_T \colon T \to T_0$ be the mapping described above. Let $\psi_T \colon C_T \to C_{T_0}$ be the mapping $\psi_T(ty) = t\varphi_T(y)$. Let y_0 be the barycenter of T_0 and let $B^3(y_0, \rho)$ be the maximal ball not touching the sides of T_0. We map the round cone $D = \{ tz : z \in \overline{B}^3(y_0, \rho), t \geq 0 \}$ onto C_{T_0} by $\psi_0(tz) = th(z)$, $z \in D \cap \partial Q$, $t \geq 0$, where $h \colon D \cap \partial Q \to T_0$ is the radial stretch mapping centered at y_0. We can map $\operatorname{int} D$ K_0-quasiconformally onto the half space $H = \{ x \in \mathbb{R}^3 : x_3 > 0 \}$, K_0 not depending on ρ. In fact, the cone $V = \{ x : x \cdot e_3 > |x| \cos \alpha \}$ congruent to $\operatorname{int} D$ is mapped K_0-quasiconformally onto H by $g \colon (t, \varphi, \vartheta) \mapsto (t^{\pi/2\alpha}, \varphi, \pi\vartheta/2\alpha)$ in spherical coordinates, $0 < t$, $0 \leq \varphi < 2\pi$, $0 \leq \vartheta < \alpha$.

The final mapping $f \colon \mathbb{R}^3 \to \mathbb{R}^3$ is now given by

$$f|C_T = u_T \circ g \circ \omega \circ \psi_0^{-1} \circ \psi_T \, ,$$

where u_T is the identity if ψ_T is sense–preserving and the reflection in ∂H otherwise, and ω is an orthogonal mapping sending D onto V. The mapping f is K-qr, with K independent of k. However, the local index $i(0, f)$ is $3 \cdot 2^{2k+4}$, and tends to ∞ with k. Note also that $|f(x)|$ is of the order $|x|^{\pi/2\alpha}$ and $1/\alpha^2 \sim i(0, f)$. The branch set of f consists of all edges of the cones C_T. The construction of f extends to any dimension n in a rather straight–forward manner (see [MRV3, 4.9]).

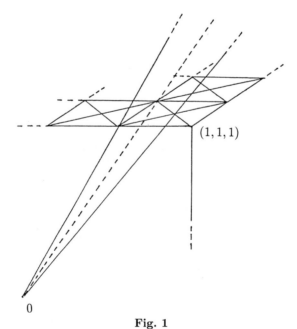

$(1, 1, 1)$

0

Fig. 1

3.3. Zorich's mapping. In [Zo1] V.A. Zorich gave the following example of a qr mapping of $\mathbb{R}^3$ with range $\mathbb{R}^3 \setminus \{0\}$. It is an analogue of the exponential function in the plane. We divide $\mathbb{R}^3$ into congruent infinite cylinders C by the planes $x_1 = i$ and $x_2 = j$, $i, j \in \mathbb{Z}$. Fix one such (closed) cylinder C_0. We map $\operatorname{int} C_0$ onto the half space $H = \{x \in \mathbb{R}^3 : x_3 > 0\}$ quasiconformally as follows. First we map $\operatorname{int} C_0$ quasiconformally onto the round cylinder $V = \{x \in \mathbb{R}^3 : x_1^2 + x_2^2 < 1\}$ by stretching in planes $\mathbb{R}^2 \times \{x_3\}$ and by translation (cf. 3.2) and then map V quasiconformally onto H by the mapping $(r, \varphi, x_3) \mapsto (t, \varphi, \vartheta)$ with $t = e^{x_3}$, $\vartheta = \pi r/2$, where cylindrical and spherical coordinates, respectively, are used. We extend the mapping obtained by repeated reflection in the faces of the cylinders C and in ∂H in the range (cf. 3.2). As a result we obtain a qr mapping $f: \mathbb{R}^3 \to \mathbb{R}^3 \setminus \{0\}$. It is doubly periodic with periods $2e_1$ and $2e_2$ and it grows by the rule $|f(x)| = e^{x_3}$. The branch set consists of the edges of the cylinders C.

The n-dimensional version is obtained by taking cylinders of the form $\{x \in \mathbb{R}^n : x = y + te_n, y \in Q, t \in \mathbb{R}^1\}$, where Q runs through the set of $(n-1)$-cubes into which $\mathbb{R}^{n-1}$ is partitioned by the planes $x_k = i$, $k = 1, \ldots, n-1$, $i \in \mathbb{Z}$.

3.4. Automorphic qr mappings. Let Γ be a discrete Möbius group acting on a domain D in $\overline{\mathbb{R}}^n$. We assume that the elements in Γ are all sense–preserving. A qm mapping $f: D \to \overline{\mathbb{R}}^n$ is said to be automorphic with respect to Γ if $f \circ g = f$ for all $g \in \Gamma$.

The examples in 3.1 and 3.3 are clearly automorphic with respect to nontrivial Möbius groups. In 3.1 we can take Γ to be the group generated by the rotation $(r, \varphi, x_3) \mapsto (r, \varphi + 2\pi/k, x_3)$. In 3.3 we can for example take the group generated by the translations $x \mapsto x + 2e_1$, $x \mapsto x + 2e_2$, and by the rotation $(r, \varphi, x_3) \mapsto (r, \varphi + \pi, x_3)$. Note that this group consists of the elements obtained by an even number of reflections in the planes containing some face of one cylinder C.

Let Q be the cube $[0,1]^n$ in $\mathbb{R}^n$. We send $\operatorname{int} Q$ by radial stretching and translation quasiconformally by a mapping f_0 onto the unit ball B^n. By repeated reflection in the $(n-1)$-planes containing the faces of Q and reflection in ∂B^n we extend f_0 to a qm mapping $f: \mathbb{R}^n \to \overline{\mathbb{R}}^n$. This mapping is n-periodic and it serves as a counterpart for the Weierstrass p-function. The mapping f is automorphic with respect to the Möbius group consisting of all elements obtained by an even number of reflections in the planes containing some face of Q.

In [Sr1] U. Srebro presented examples in dimension three which can be regarded as counterparts for the elliptic modular function. One of these is constructed as follows. Let

$$D = \{x \in H : x_1 > -1/2, \ x_1 - 1 < \sqrt{3}\,x_2 < 1 - x_1, \ |x| > 1\}$$

and let Γ' be the Möbius group generated by reflections in the faces of D. Because the dihedral angles of D are all $\pi/3$, the group Γ' is a discrete

Möbius group acting on H. The domain D is a fundamental domain for Γ'. Let Γ be the subgroup of sense–preserving elements of Γ'. There exists a qc mapping $f_0\colon D \to H$ by the constructions in [GV1, 10.4] (cf. 3.3). We normalize so that f_0 keeps on the boundary the vertices a_1, a_2, a_3, ∞ of D fixed. Next extend f_0 by repeated reflection in the faces of D and ∂H to obtain a qr mapping $f\colon H \to \mathbb{R}^3 \smallsetminus \{a_1, a_2, a_3\}$ (cf. 3.3). Then f is automorphic with respect to Γ. A similar example, where D is replaced by a domain with an arbitrary number of cusps, is given in [R10, 3.5].

Automorphic qm mappings have been further studied by O. Martio and U. Srebro in [MS1]–[MS3] and by P. Tukia in [T]. It is known that given a discrete Möbius group Γ acting on B^n with finite hyperbolic volume, there exists a nonconstant qm mapping of B^n which is automorphic with respect to Γ [MS3]. In [T] the same statement is proved for groups without torsion, even if they have infinite volume. Tukia's theorem follows also from a more general result by K. Peltonen [Pel]. Recently Srebro exhibited an example of a discrete Möbius group Γ acting on B^3 such that there is no qm mapping automorphic with respect to Γ.

3.5. Note. There exist many examples of qr mappings of a more sophisticated nature in the literature. Each of these serves a specific purpose. We shall discuss some of them at the appropriate places in the text.

4. Discrete Open Mappings

Let X and Y be topological spaces. A mapping $f\colon X \to Y$ is *discrete* if $f^{-1}(y)$ is discrete for all $y \in Y$ and f is *open* if it takes open sets onto open sets. One of Reshetnyak's main results in the theory of qr mappings is the following [Re2], [Re4]:

4.1. Theorem. *A nonconstant qr mapping is discrete and open.*

As mentioned earlier we shall postpone the proof of this important theorem to VI.5. In this section we shall review results on discrete open mappings relevant to qr theory, mostly without proofs.

4.2. Degree of a mapping. Let $f\colon S^n \to S^n$, $n \geq 1$, be continuous. The *degree* $\mu(f) \in \mathbb{Z}$ of f is defined by $f_*(\alpha) = \mu(f)\alpha$ where $f_*\colon H_n(S^n) \to H_n(S^n)$ is the induced mapping on the singular n-dimensional homology group $H_n(S^n) \approx \mathbb{Z}$.

Let $f\colon A \to S^n$ be a continuous mapping of a set $A \subset S^n$, let $U \subset A$ be open in S^n with $\overline{U} \subset A$, and let $y \in S^n$ be a point with $y \notin f\partial U$. Such a point y is called (f, U)-*admissible*. In this situation we shall define the *local degree* or *topological index* $\mu(y, f, U) \in \mathbb{Z}$ of f at y with respect to U as follows. Consider the following sequence of induced mappings on homology of pairs:

$$H_n(S^n) \xrightarrow{j_*} H_n(S^n, S^n \setminus (U \cap f^{-1}(y))) \xleftarrow{e_*} H_n(U, U \setminus f^{-1}(y))$$
$$\xrightarrow{f_{1*}} H_n(S^n, S^n \setminus \{y\}) \xleftarrow{k_*} H_n(S^n) .$$

Here j, e, and k are inclusions and f_1 is defined by f. Then e_* is an isomorphism by excision and k_* is an isomorphism because $S^n \setminus \{y\}$ is homologically trivial. We get a homomorphism $h = k_*^{-1} f_{1*} e_*^{-1} j_*\colon H_n(S^n) \to H_n(S^n)$ and we define $\mu(y, f, U)$ by $h(\alpha) = \mu(y, f, U)\alpha$. If $f\colon S^n \to S^n$ and if $f^{-1}(y) \subset U$, then $\mu(y, f, U) = \mu(f)$.

4.3. Remark. Another way to define the local degree is to use cohomology groups with compact supports, see [RR], [V3]. This method also applies directly to mappings $f\colon M \to N$ where M and N are connected oriented n-manifolds [V3].

The following proposition summarizes well–known properties of the local degree. We shall omit the proof (see [RR]).

4.4. Proposition. *The local degree satisfies the following conditions:*

D$_1$. *The mapping $y \mapsto \mu(y, f, U)$ is constant in each component of $S^n \setminus f\partial U$ and $\mu(y, f, U) = 0$ if $y \notin f\overline{U}$.*

D$_2$. *If f is injective, $|\mu(y, f, U)| = 1$ for $y \in fU$.*

D$_3$. *If f is an inclusion, $\mu(y, f, U) = 1$ for $y \in fU$.*

D$_4$. *If $U_1, \ldots, U_k$ are disjoint open sets and if $U \cap f^{-1}(y) \subset \bigcup_i U_i \subset U$, then*

$$\mu(y, f, U) = \sum_{i=1}^{k} \mu(y, f, U_i) \quad \text{(if defined)} .$$

D$_5$. *Let f and g be homotopic via a homotopy h_t, $t \in [0,1]$, $h_0 = f$, $h_1 = g$. Suppose further that y is (h_t, U)-admissible for all $t \in [0,1]$. Then $\mu(y, f, U) = \mu(y, g, U)$.*

D$_6$. *Let $T\colon \mathbb{R}^n \to \mathbb{R}^n$ be a linear bijection and $a \in \mathbb{R}^n$. If $g\colon \overline{\mathbb{R}}^n \to \overline{\mathbb{R}}^n$ is the obvious extension of $T + a$ to $\overline{\mathbb{R}}^n$, then $\mu(g) = \det T/|\det T|$.*

In what follows we let $n \geq 2$. Let G be a domain in $\overline{\mathbb{R}}^n$ and $f\colon G \to \overline{\mathbb{R}}^n$ continuous. We say that f is *sense-preserving* (*weakly sense-preserving*) if $\mu(y, f, D) > 0$ ($\mu(y, f, D) \geq 0$) for all domains $D \subset\subset G$ and $y \in fD \setminus f\partial D$. Similarly we define *sense-reversing* mappings by opposite inequalities.

4.5. Theorem [Re2]. *A nonconstant qr mapping is sense–preserving.*

We shall prove this in VI.5.2 in connection with the proof of Theorem 4.1.

We make the convention that henceforth the term " f is *discrete open*" includes the assumption that f is also continuous.

Let $f: G \to \overline{\mathbb{R}}^n$ be discrete open and let $x \in G$. There exists a neighborhood V of x such that $\overline{V} \cap f^{-1}(f(x)) = \{x\}$. If V' is another neighborhood with this property, D_4 implies $\mu(f(x), f, V) = \mu(f(x), f, V \cap V') = \mu(f(x), f, V')$ and $\mu(f(x), f, V)$ is thus independent of the choice of V. We write $i(x, f) = \mu(f(x), f, V)$ and call $i(x, f)$ the *local (topological) index* of f at x. This concept was mentioned in 3.1 in connection with examples of qr mappings. That the definition in 3.1 coincides with the present one follows from 4.10(4) below. By D_2, $|i(x, f)| = 1$ outside the branch set B_f, and by D_1, $x \mapsto i(x, f)$ is locally constant in the open set $G \setminus B_f$.

The following result was first proved by A.V. Chernavskiĭ [Ch1], [Ch2]. For a shorter proof we refer the reader to J. Väisälä [V3]. By $\dim A$ we denote the *topological dimension* of a set A (see [HW]).

4.6. Theorem. *A discrete open mapping f satisfies*

$$\dim B_f = \dim f B_f \le n - 2 .$$

P.T. Church and J.G. Timourian have for $n = 5$ given an example in [CT] which shows that there can be strict inequality in 4.6 for $B_f \ne \emptyset$. We can now infer from 4.6, and the fact that B_f is closed in G, that $G \setminus B_f$ is a domain, hence either $i(x, f) = 1$ for all $x \in G \setminus B_f$ or $i(x, f) = -1$ for all $x \in G \setminus B_f$. In the first case f is sense–preserving. For if $y \in fD \setminus f\partial D$, we get using $\dim f B_f \le n - 2$ a point $z \notin fB_f$ in the same component of $\overline{\mathbb{R}}^n \setminus f\partial D$ as y, and by D_1 and D_4 we conclude

$$\mu(y, f, D) = \mu(z, f, D) = \sum_{x \in D \cap f^{-1}(z)} i(x, f) > 0 .$$

In the second case f is sense–reversing.

A domain $D \subset\subset G$ is called a *normal domain* for f if $f\partial D = \partial fD$. The openness implies that $\partial fD \subset f\partial D$ always holds. The condition $f\partial D \subset \partial fD$ means that $f|D: D \to fD$ is a closed map, i.e., fE is closed in fD whenever $E \subset D$ is closed in D. If $x \in G$ and U is a normal domain such that $U \cap f^{-1}(f(x)) = \{x\}$, then U is called a *normal neighborhood* of x.

If D is a normal domain of f, then $fD \cap f\partial D = \emptyset$ and hence $\mu(y, f, D)$ is constant for $y \in fD$. This constant is denoted by $\mu(f, D)$.

4.7. Lemma. *If $U \subset \overline{\mathbb{R}}^n$ is a domain and D is a component of $f^{-1}U$ such that $D \subset\subset G$, then D is a normal domain and $fD = U \subset\subset fG$.*

Proof. Let $y \in f\partial D$ and let $x \in \partial D$ be such that $y = f(x)$. Since D is a component of $f^{-1}U$, we have $x \notin f^{-1}U$ and thus $y \notin U \supset fD$. Hence $y \in \overline{fD} \setminus fD = \overline{fD} \setminus fD = \partial fD$ and D is a normal domain. Furthermore, $f\partial D \cap U = \emptyset$. Thus $fD = U \cap f\overline{D}$ is both open and closed in U, whence $fD = U$. From $f\overline{D} = \overline{U}$ we get $U \subset\subset fG$. $\qquad\square$

4.8. Lemma. *Let D be a normal domain of f. If E is a continuum in fD, then f maps every component of $D \cap f^{-1}E$ onto E. Furthermore, if $F \subset fD$ is compact, then $D \cap f^{-1}F$ is compact.*

Proof. If $E \subset fD$ is compact, then $A = D \cap f^{-1}E = \overline{D} \cap f^{-1}E$ is compact. For any open $U \subset D$ we have $f(U \cap A) = fU \cap fA$, which shows that $f|A: A \to E$ is an open map. If E is a continuum, every component of A is mapped onto E by [Wh, (7.5), p. 148]. □

4.9. Lemma. *For $x \in G$ and $s > 0$ let $U(x, f, s)$ be the x-component of $f^{-1}B^n(f(x), s)$. Then there exists $s_x > 0$ such that $U(x, f, s)$ is a normal neighborhood of x and $fU(x, f, s) = B^n(f(x), s)$ for $0 < s \le s_x$. Moreover, the diameter $d(U(x, f, s))$ tends to 0 as $s \to 0$.*

Proof. We choose a neighborhood W of x such that $W \subset\subset G$ and $\overline{W} \cap f^{-1}f(x) = \{x\}$. Then $U(x, f, s) \subset W$ for $0 < s \le s_x = d(f(x), f\partial W)$ and $U(x, f, s)$ is a normal neighborhood with $fU(x, f, s) = B^n(f(x), s)$ by 4.7. Since W can be chosen arbitrarily small, $d(U(x, f, s)) \to 0$ as $s \to 0$. □

For $A \subset G$ and $y \in \mathbb{R}^n$ we write $N(y, f, A) = \operatorname{card} f^{-1}(y) \cap A$ and $N(f, A) = \sup_y N(y, f, A)$.

4.10. Proposition. *Let $f: G \to \overline{\mathbb{R}}^n$ be sense–preserving, discrete, and open. Then*

(1) *If $U \subset\subset G$, then $N(y, f, U) \le \mu(y, f, U)$ for all $y \notin f\partial U$ and $N(y, f, U) = \mu(y, f, U)$ for $y \notin f\big(\partial U \cup (U \cap B_f)\big)$.*
(2) *If D is a normal domain, $N(f, D) = \mu(f, D)$.*
(3) *If $A \subset G$ is compact, $N(f, A) < \infty$.*
(4) *Every $x \in G$ has a neighborhood V such that, if U is a neighborhood of x and if $U \subset V$, then $N(f, U) = i(x, f)$.*
(5) *$x \in B_f$ if and only if $i(x, f) \ge 2$.*
(6) *The mapping $x \mapsto i(x, f)$ is upper semicontinuous.*

Proof. (1) Let $y \notin f\partial U$ and let $U \cap f^{-1}(y) = \{x_1, \ldots, x_k\}$. Then

$$\mu(y, f, U) = \sum_{j=1}^{k} i(x_j, f).$$

Since $i(x_j, f) \ge 1$, $\mu(y, f, U) \ge k = N(y, f, U)$. If $y \notin f\big(\partial U \cup (U \cap B_f)\big)$, $i(x_j, f) = 1$ for all j and $\mu(y, f, U) = N(y, f, U)$.

(2) Using $\dim fB_f \le n - 2$ we find a point $y \in fD \smallsetminus fB_f$. By (1), $\mu(f, D) = \mu(y, f, D) = N(y, f, D) \le N(f, D)$ and also $N(z, f, D) \le \mu(f, D)$ for all $z \notin f\partial D$. Hence $\mu(f, D) = N(f, D)$.

(3) By 4.9 A can be covered by a finite number of normal domains $D_1, \ldots, D_k$. Then (2) gives

$$N(f, A) \le \sum_{i=1}^{k} N(f, D_i) = \sum_{i=1}^{k} \mu(f, D_i) < \infty .$$

(4) By 4.9 x has a normal neighborhood V. If $U \subset V$ is a neighborhood of x, there is a normal neighborhood V_1 of x such that $V_1 \subset U$. Then (2) implies $i(x, f) = N(f, V_1) \le N(f, U) \le N(f, V) = i(x, f)$.

(5) follows from (4).

(6) If U is a normal neighborhood of x, we have $i(x, f) = \mu(f(x), f, U) = \mu(f(z), f, U) \ge i(z, f)$ for $z \in U$. $\square$

4.11. Lemma. Let $f \colon G \to \mathbb{R}^n$ be sense–preserving, discrete, and open, and let f be differentiable at x_0. Then $J_f(x_0) \ge 0$. If $x_0 \in B_f$, $J_f(x_0) = 0$. If A is a Borel set in G and if f is differentiable a.e. in A, then

$$(4.12) \qquad \int_A J_f \, dm \le \int_{\mathbb{R}^n} N(y, f, A) \, dy .$$

Proof. If $J_f(x_0) \ne 0$, D_5 and D_6 imply $i(x_0, f) = \operatorname{sgn} J_f(x_0)$. Then necessarily $i(x_0, f) = 1$ and $x_0 \notin B_f$ by 4.10(5). To prove (4.12) we express $A \smallsetminus B_f$ as a union of disjoint Borel sets $A_1, A_2, \dots$ such that each A_i is contained in a domain D_i where f is injective. Since (4.12) is well known for homeomorphisms and since $J_f(x) = 0$ a.e. in $B_f \cap A$, we obtain

$$\int_A J_f \, dm = \int_{A \smallsetminus B_f} J_f \, dm = \sum_i \int_{A_i} J_f \, dm \le \sum_i \int_{\mathbb{R}^n} N(y, f, A_i) \, dy$$

$$\le \int_{\mathbb{R}^n} N(y, f, A) \, dy . \quad \square$$

4.13. Lemma. Let $f \colon G \to \mathbb{R}^n$ be a nonconstant qr mapping and suppose that every point in $G \smallsetminus B_f$ has a neighborhood U such that $K_O(f|U) \le a$ and $K_I(f|U) \le b$. Then $K_O(f) \le a$ and $K_I(f) \le b$.

Proof. By 2.4, 4.1, 4.5, and 4.11, $J_f(x) = 0$ a.e. in B_f. Hence also $f'(x) = 0$ a.e. in B_f and the lemma follows. $\square$

By appealing to the discreteness and openness we are now able to prove the following result for quasiregular mappings.

4.14. Proposition. Let $f \colon G \to \mathbb{R}^n$ be qr. Then the following holds:

(a) f satisfies the condition (N), that is, $m(E) = 0$ implies $m(fE) = 0$ for all $E \subset G$. Furthermore, if $E \subset G$ is measurable, so is fE.

(b) $m(fB_f) = 0$.

(c) The transformation formula

$$\int_E (h \circ f) J_f \, dm = \int_{\mathbb{R}^n} h(y) N(y, f, E) \, dy$$

holds for any measurable $h: \mathbb{R}^n \to [0, \infty]$ *and all measurable* $E \subset G$.

Proof. We may assume that f is nonconstant and hence, by 4.1 and 4.5, sense–preserving, discrete, and open.

(a) If we assume that f is injective, we have by 2.4 and (4.12) that

$$(4.15) \qquad \int_E J_f \, dm \le m(fE)$$

for all Borel sets $E \subset G$. For the general case we claim that

$$(4.16) \qquad m(fE) \le \int_E J_f \, dm$$

for all Borel sets $E \subset G$.

Let Q be a closed n-interval in G and let $\varepsilon > 0$. Take a closed n-interval Q' in G with $Q \subset \text{int } Q'$ and $m(Q' \setminus Q) < \varepsilon$. Suppose $m(f\partial Q_0) > 0$ for all n-intervals Q_0 with $Q \subset Q_0 \subset Q'$. Then there exists $\delta > 0$ and a sequence of n-intervals Q_i, $Q \subset Q_i \subset Q'$, with disjoint boundaries such that $m(f\partial Q_i) \ge \delta$ for all i. By 4.10(3), $N(f, Q') < \infty$ and hence

$$\infty = \sum_i m(f\partial Q_i) = \int_{\mathbb{R}^n} \sum_i \chi_{f\partial Q_i} dm \le N(f, Q') m(fQ') < \infty$$

gives a contradiction. It follows that there exists an n-interval $Q_0 \supset Q$ with $m(f\partial Q_0) = 0$ and $m(Q_0 \setminus Q) < \varepsilon$. By using this and the fact that J_f is locally integrable we find for each $\varepsilon_1 > 0$ a sequence of closed n-intervals $Q_i \subset G$ such that $m(f\partial Q_i) = 0$, $E \subset \bigcup_i Q_i$, and

$$\sum_i \int_{Q_i} J_f \le \int_E J_f + \varepsilon_1 .$$

Since $m(fE) \le \sum_i m(fQ_i)$, it suffices to show (4.16) for each Q_i. Fix i and for simplicity write $Q = Q_i$.

By 1.4 we get a sequence $f_k : G \to \mathbb{R}^n$ of C^∞ mappings such that $f_k \to f$ locally uniformly and $D_i f_{k_j} \to D_i f_j$ locally in L^n. Then also $J_{f_k} \to J_f$ locally in L^1. We write $\chi = \chi_{fQ}$ and $\chi_k = \chi_{f_k Q}$. We want to show $\chi_k \to \chi$ a.e. Let $y \in fQ \setminus f\partial Q$. Then for all $k \ge k_0$ for some k_0 we have that the homotopy $h_t(x) = t f_k(x) + (1 - t) f(x)$ satisfies the condition in D_5 for the point y and the set $U = \text{int } Q$. Then by D_5 and 4.5 we get $\mu(y, f_k, \text{int } Q) = \mu(y, f, \text{int } Q) > 0$, and by D_1, $y \in f_k Q$ for $k \ge k_0$. Thus $\chi_k(y) \to \chi(y)$. If $y \notin fQ$, D_1 clearly implies $\chi_k(y) \to \chi(y)$. Since $m(f\partial Q) = 0$, $\chi_k \to \chi$ a.e. Now Fatou's lemma gives

$$m(fQ) = \int_{\mathbb{R}^n} \chi\, dm \leq \liminf_{k \to \infty} \int_{\mathbb{R}^n} \chi_k dm = \liminf_{k \to \infty} m(f_k Q)$$

$$\leq \liminf_{k \to \infty} \int_Q |J_{f_k}|\, dm = \int_Q J_f\, dm \, ,$$

and (4.16) follows. Here the latter inequality is elementary calculus. Using a Borel cover we see that (4.16) implies (N). The last statement follows from (N) together with Borel covers. More precisely, if E is any measurable set in G, let $F \supset E$ be a Borel cover in G. Then the measurable set fF differs from fE by a zero set because of (N), and thus fE is measurable. We also observe that similar reasoning gives (4.15) in the injective case and (4.16) for all measurable sets $E \subset G$.

(b) The proof follows from 2.4, 4.11, and (4.16).

(c) We first assume that h is a simple Borel function of the form

$$h = \sum_{j=1}^m a_j \chi_{B_j} \geq 0 \, .$$

Since $m(fB_f) = 0$ and $J_f(x) = 0$ for almost every x in B_f, we may assume $E \cap B_f = \emptyset$. Let $E_1, E_2, \ldots$ be a measurable partition of E such that each E_k is contained in a domain $D_k \subset G \setminus B_f$ for which $f|D_k$ is injective. Each fE_k is measurable by (a) and we get

$$\int_E (h \circ f) J_f\, dm = \sum_{j,k} a_j \int_{E_k \cap f^{-1}B_j} J_f\, dm = \sum_{j,k} a_j m(fE_k \cap B_j)$$

$$= \int_{\mathbb{R}^n} \sum_j a_j \chi_{B_j} \sum_k \chi_{fE_k}\, dm = \int_{\mathbb{R}^n} h(y) N(y, f, E)\, dy \, .$$

Here we also used (4.15) and (4.16) for measurable sets.

Given a measurable $h \geq 0$ we approximate it by an increasing sequence (h_i) of nonnegative simple Borel functions such that $h_i \to h$ a.e. Outside the set $\{x \in G : J_f(x) = 0\}$ $h_i \circ f \to h \circ f$ a.e. For if not, we can find a measurable set A in some E_k with the properties

$$0 < \int_A J_f\, dm = m(fA)$$

and $h_i \circ f(x) \nrightarrow h \circ f(x)$, $x \in A$. But this means $h_i(y) \nrightarrow h(y)$ for $y \in fA$ which then contradicts $h_i \to h$ a.e. Assertion (c) now follows from the monotone convergence theorem. $\qquad\square$

4.17. Notes. The part 4.7–4.13 is from [MRV1] in a shortened form. The condition (N) for qr mappings was first proved by Yu.G. Reshetnyak in [Re2], without using discreteness in the proof. Earlier reference for the transformation formula in 4.14(c) for the qr theory has been [RR, Theorem 3, p. 364].

That the heavy machinery in [RR] can be avoided here by using discreteness was shown by M. Pesonen. We have followed the plan of proof from [Pe3] in establishing 4.14. Another proof for the transformation formula 4.14(c) is given in [BI2]. For the condition (N), see also [MZ].

Chapter II. Inequalities for Moduli of Path Families

In this chapter we shall derive the important inequalities for moduli of path families which form the basis for the geometric part of qr theory. The main reference for path families in connection with qc theory is the book [V4] by J. Väisälä. When convenient, we shall refer to [V4] rather than repeat proofs. The main results, namely Poletskiĭ's and Väisälä's inequalities, involve a somewhat technical measure theoretic step concerning path families whose modulus is neglible. We shall call this step Poletskiĭ's lemma. In qc theory a corresponding result is known as Fuglede's theorem. Inequalities for path families are more general and more effective than inequalities for capacities of condensers, which historically came first [MRV1]. Here we shall obtain the capacity inequalities as corollaries in Section 10.

1. Modulus of a Path Family

We shall for the most part adopt the notation in [V4]. A path in a topological space X is a continuous mapping $\alpha\colon \Delta \to X$, where Δ is a (possibly unbounded) interval in $\mathbb{R}^1$. The path α is called closed or open according as Δ is compact or open. The locus $|\alpha|$ of α is the image set $\alpha\Delta$.

We shall here deal with paths in $\mathbb{R}^n$. Let $\alpha\colon [a,b] \to \mathbb{R}^n$ be a closed path. If α is rectifiable, i.e. the length $\ell(\alpha)$ of α is finite, we define the *length function* $s_\alpha\colon [a,b] \to \mathbb{R}^1$ by $s_\alpha(t) = \ell(\alpha|[a,t])$. If a path β is obtained from α by a change of parameter, then $\ell(\beta) = \ell(\alpha)$. If α is rectifiable, there is a unique path $\alpha^0\colon [0, \ell(\alpha)] \to \mathbb{R}^n$ obtained from α by an increasing change of parameter such that $s_{\alpha^0}(t) = t$ and $\alpha = \alpha^0 \circ s_\alpha$ [V4, 2.4]. The path α^0 is called the *normal representation* of α.

A path α in $\mathbb{R}^n$ is locally rectifiable if each closed subpath of α is rectifiable. We write $\ell(\alpha) = \sup \ell(\beta)$, where the supremum is taken over all closed subpaths β of α. If $\ell(\alpha) < \infty$, α is rectifiable.

Let $A \subset \mathbb{R}^n$ be a Borel set and $\rho\colon A \to [0, \infty]$ a Borel function. If $\alpha\colon [a,b] \to A$ is a rectifiable path, the line integral of ρ along α is defined by

$$\int_\alpha \rho\, ds = \int_0^{\ell(\alpha)} \rho(\alpha^0(t))\, dt\,.$$

We also use the notation

$$\int\limits_\alpha \rho \, ds = \int\limits_\alpha \rho(x)|dx| \, .$$

If α is absolutely continuous, we have [V4, 4.1]

$$\int\limits_\alpha \rho \, ds = \int\limits_a^b \rho(\alpha(t))|\alpha'(t)| \, dt \, .$$

If $\alpha\colon \Delta \to A$ is a locally rectifiable path, we define

$$\int\limits_\alpha \rho \, ds = \sup_\beta \int\limits_\beta \rho \, ds \, ,$$

where the supremum is taken over all closed subpaths of α.

Let Γ be a family of paths in $\mathbb{R}^n$, $n \geq 2$. We let $\mathcal{F}(\Gamma)$ be the set of all Borel functions $\rho\colon \mathbb{R}^n \to [0, \infty]$ such that

$$\int\limits_\gamma \rho \, ds \geq 1$$

for every locally rectifiable path $\gamma \in \Gamma$. The functions in $\mathcal{F}(\Gamma)$ are called *admissible* for Γ. For $1 \leq p < \infty$ we define

(1.1) $$\mathsf{M}_p(\Gamma) = \inf_{\rho\in\mathcal{F}(\Gamma)} \int\limits_{\mathbb{R}^n} \rho^p \, dm$$

and call $\mathsf{M}_p(\Gamma)$ the *p-modulus* of Γ. If $\mathcal{F}(\Gamma) = \emptyset$ (which happens only if Γ contains constant paths), we set $\mathsf{M}_p(\Gamma) = \infty$. The term *n-modulus* is usually abbreviated to *modulus* and is denoted by $\mathsf{M}(\Gamma)$. The *extremal length* of Γ is $\mathsf{M}(\Gamma)^{1/(1-n)}$.

If $A \subset \mathbb{R}^n$ is a Borel set such that

$$\int\limits_\gamma \chi_{\mathbb{R}^n \smallsetminus A} \, ds = 0$$

for all locally rectifiable γ in Γ, then $\mathcal{F}(\Gamma)$ can be replaced by the set of restrictions $\rho|A$, $\rho \in \mathcal{F}(\Gamma)$, and integration in (1.1) can be taken over A.

If $0 < \mathsf{M}_p(\Gamma) < \infty$, then the definition (1.1) can clearly be given in the alternate form

(1.2) $$\mathsf{M}_p(\Gamma) = \inf_\varphi \frac{1}{L(\Gamma, \varphi)^p} \int\limits_{\mathbb{R}^n} \varphi^p \, dm \, .$$

Here

(1.3)
$$L(\Gamma, \varphi) = \inf_{\gamma} \int_{\gamma} \varphi \, ds$$

where $\varphi \colon \mathbb{R}^n \to [0, \infty]$ is a Borel function and the infimum is taken over all locally rectifiable γ in Γ. The infimum in (1.2) is taken over all φ with $L(\Gamma, \varphi) > 0$.

Unless otherwise stated, in what follows all path families lie in $\mathbb{R}^n$. It is clear from the definition that $M_p(\Gamma) = M_p(T\Gamma)$ for any Euclidean motion T and for any path family Γ. Here $T\Gamma = \{T \circ \gamma : \gamma \in \Gamma\}$. Let Γ_1 and Γ_2 be families of paths. We say that Γ_2 is *minorized* by Γ_1 and write $\Gamma_1 < \Gamma_2$ if every $\gamma \in \Gamma_2$ has a subpath which belongs to Γ_1. If $\Gamma_1 < \Gamma_2$, $\mathcal{F}(\Gamma_1) \subset \mathcal{F}(\Gamma_2)$ and thus $M_p(\Gamma_1) \geq M_p(\Gamma_2)$. If $\Gamma_1 \supset \Gamma_2$, $\Gamma_1 < \Gamma_2$.

The path families $\Gamma_1, \Gamma_2, \ldots$ are called *separate* if there exist disjoint Borel sets E_i in $\mathbb{R}^n$ such that

(1.4)
$$\int_{\gamma} \chi_{\mathbb{R}^n \setminus E_i} \, ds = 0$$

for all locally rectifiable $\gamma \in \Gamma_i$, $i = 1, 2, \ldots$.

1.5. Proposition. *Let* $\Gamma, \Gamma_1, \Gamma_2, \ldots$ *be a sequence of path families. Then*

(1) $M_p(\bigcup_i \Gamma_i) \leq \sum_i M_p(\Gamma_i)$.

(2) *If* $\Gamma_1, \Gamma_2, \ldots$ *are separate and* $\Gamma < \Gamma_i$ *for all* i, *then*

$$M_p(\Gamma) \geq \sum_i M_p(\Gamma_i).$$

Equality holds if $\Gamma = \bigcup_i \Gamma_i$.

(3) *If* $\Gamma_1, \Gamma_2, \ldots$ *are separate and* $\Gamma_i < \Gamma$ *for all* i, *then*

$$\frac{1}{M_p(\Gamma)^{1/(p-1)}} \geq \sum_i \frac{1}{M_p(\Gamma_i)^{1/(p-1)}}, \quad p > 1.$$

Proof. (1) For $\varepsilon > 0$ take $\rho_i \in \mathcal{F}(\Gamma_i)$ such that

$$\int_{\mathbb{R}^n} \rho_i^p \, dm \leq M_p(\Gamma_i) + \frac{\varepsilon}{2^i}.$$

Then $\rho = (\sum_i \rho_i^p)^{1/p} \in \mathcal{F}(\bigcup_i \Gamma_i)$ and

$$M_p\left(\bigcup_i \Gamma_i\right) \leq \int_{\mathbb{R}^n} \rho^p \, dm = \sum_i \int_{\mathbb{R}^n} \rho_i^p \, dm \leq \sum_i M_p(\Gamma_i) + \varepsilon.$$

(2) If $\rho \in \mathcal{F}(\Gamma)$, $\rho_i = \chi_{E_i} \rho \in \mathcal{F}(\Gamma_i)$ where E_i is as in (1.4). Hence

$$\sum_i \mathsf{M}_p(\Gamma_i) \le \sum_i \int_{\mathbb{R}^n} \rho_i^p \, dm = \sum_i \int_{E_i} \rho^p \, dm \le \int_{\mathbb{R}^n} \rho^p \, dm \, .$$

(3) Since $\Gamma_i < \Gamma$ for all i, we may assume $0 < \mathsf{M}_p(\Gamma) < \infty$ and $0 < \mathsf{M}_p(\Gamma_i) < \infty$ for all i. For $\varepsilon > 0$ there exist admissible functions ρ_i for the families Γ_i such that $\rho_i | \mathbb{R}^n \smallsetminus E_i = 0$, $L(\Gamma_i, \rho_i) = 1$ with the notation in (1.3), and

(1.6)
$$\sum_i t_i \ge \sum_i \frac{1}{\mathsf{M}_p(\Gamma_i)^{1/(p-1)}} - \varepsilon$$

where

$$t_i = \left(\int_{E_i} \rho_i^p \, dm \right)^{1/(1-p)} \, .$$

Set $\varphi = \sum_i t_i \rho_i$. Since the families Γ_i are separate and $\Gamma_i < \Gamma$ for all i, we obtain

$$L(\Gamma, \varphi) \ge \sum_i t_i \, .$$

We may assume the right hand side to be positive. Furthermore,

$$\int_{\mathbb{R}^n} \varphi^p \, dm = \sum_i t_i^p \int_{E_i} \rho_i^p \, dm = \sum_i t_i \, ,$$

hence

$$\frac{1}{\mathsf{M}_p(\Gamma)} \ge \frac{L(\Gamma, \varphi)^p}{\int_{\mathbb{R}^n} \varphi^p \, dm} \ge \left(\sum_i t_i \right)^{p-1}$$

which together with (1.6) gives (3). $\square$

1.7. Examples. If $E, F, D \subset \mathbb{R}^n$, we denote by $\Delta(E, F; D)$ the family of closed paths γ which join E and F in D, that is, if $\gamma \colon [a, b] \to \mathbb{R}^n$, then one of the end points $\gamma(a)$, $\gamma(b)$ belongs to E and the other to F, and $\gamma(t) \in D$ for $a < t < b$. For the calculation of the moduli in (a) and (b), see [V4, pp. 21–23].

(a) *The cylinder.* Let E be a Borel set in $\mathbb{R}^{n-1}$ and $h > 0$. Set $D = \{x \in \mathbb{R}^n : (x_1, \ldots, x_{n-1}) \in E, \ 0 < x_n < h\}$ and $F = E + h e_n$. If $\Gamma = \Delta(E, F; D)$, then $\mathsf{M}_p(\Gamma) = m_{n-1}(E)/h^{p-1} = m(D)/h^p$. If Γ_0 is the subfamily of Γ consisting of all paths $\gamma_y \colon [0, h] \to \mathbb{R}^n$, $\gamma_y(t) = y + t e_n$, $y \in E$, then $\mathsf{M}_p(\Gamma_0) = \mathsf{M}_p(\Gamma)$.

(b) *The spherical ring.* Let A be a spherical ring $B^n(b) \smallsetminus \overline{B}^n(a)$, $0 < a < b$, and let $\Gamma = \Delta\bigl(S^{n-1}(a), S^{n-1}(b); A\bigr)$. Then

$$\mathsf{M}_p(\Gamma) = \begin{cases} \omega_{n-1} \left(\displaystyle\int_a^b r^{q-1} dr \right)^{1-p} , \quad q = \dfrac{p-n}{p-1} & \text{if } p > 1, \\[2ex] \omega_{n-1} a^{n-1} & \text{if } p = 1, \end{cases}$$

where ω_{n-1} is the $(n-1)$-measure of the unit sphere S^{n-1}. In particular,

$$\mathsf{M}(\Gamma) = \frac{\omega_{n-1}}{\left(\log \frac{b}{a}\right)^{n-1}} \, .$$

If $\overline{B}^n(a)$ degenerates to a point, we get $\mathsf{M}(\Gamma) = 0$. From this it follows also that if the paths of a family Γ_0 are nonconstant and pass through a fixed point, then $\mathsf{M}(\Gamma_0) = 0$. Let E be a Borel set in S^{n-1} and let Γ_E be the family of all paths $\gamma_y \colon [a, b] \to \mathbb{R}^n$, $\gamma_y(t) = ty$, $y \in E$. Then by the argument in [V4, 7.5] we obtain

$$\mathsf{M}(\Gamma_E) = \frac{\mathcal{H}^{n-1}(E)}{\left(\log \frac{b}{a}\right)^{n-1}} \, .$$

The definition of the p-modulus $\mathsf{M}_p(\Gamma)$ can be extended to a family of paths Γ in a Riemannian n-manifold in the obvious way and properties such as those in 1.5 remain true there. Let $f \colon M \to N$ be a conformal mapping between two oriented Riemannian n-manifolds, that is, f is a $\mathcal{C}^1$ diffeomorphism and $|f'(x)|^n = J_f(x)$ for all $x \in M$. Let Γ be a family of paths in M and let ρ' be admissible for the image family $f\Gamma = \{ f \circ \gamma : \gamma \in \Gamma \}$. If $\rho \colon M \to [0, \infty]$ is defined by $\rho(x) = \rho'(f(x))|f'(x)|$, then

$$\int_\gamma \rho \, ds = \int_{f \circ \gamma} \rho' \, ds$$

for every locally rectifiable $\gamma \in \Gamma$ and thus $\rho \in \mathcal{F}(\Gamma)$. This gives

$$\mathsf{M}(\Gamma) \le \int_M \rho^n \, dm = \int_M (\rho' \circ f)^n J_f \, dm = \int_N \rho'^n \, dm \, ,$$

hence $\mathsf{M}(\Gamma) \le \mathsf{M}(f\Gamma)$. Since f^{-1} is also conformal, we see that the n-modulus is a *conformal invariant*, i.e., $\mathsf{M}(f\Gamma) = \mathsf{M}(\Gamma)$.

Let us apply the above to the stereographic projection $\pi \colon S_0 \setminus \{e_{n+1}\} \to \mathbb{R}^n$ where S_0 is the sphere $S^n(e_{n+1}/2, 1/2)$ in $\mathbb{R}^{n+1}$. Let Γ be a family of nonconstant paths in $\overline{\mathbb{R}}^n$ and let Γ' be the subfamily of paths $\gamma \in \Gamma$ with $\infty \in |\gamma|$. The remark in 1.7(b) concerning paths passing through a fixed point applies obviously also to manifolds, hence the modulus of Γ' in $\overline{\mathbb{R}}^n$ (equipped with the spherical metric) is zero and $\mathsf{M}(\Gamma) = \mathsf{M}(\Gamma \setminus \Gamma')$. Since the stereographic projection π is conformal, $\mathsf{M}(\Gamma)$ coincides with the modulus of $\Gamma \setminus \Gamma'$ in the Euclidean metric of $\mathbb{R}^n$. It follows that we can formally use the definition (1.1) also for path families in $\overline{\mathbb{R}}^n$ (cf. [V4, 6.1]).

1.8. Spherical caps. Let C be a cap in a sphere $N = S^{n-1}(x, r)$, let E and F be nonempty disjoint subsets of $\overline{C}$, and let $\Gamma = \Delta(E, F; C)$. Regarding N as a Riemannian $(n-1)$-manifold with metric induced from $\mathbb{R}^n$ we denote the n-modulus of Γ on N by $\mathsf{M}_n^N(\Gamma)$. By definition,

$$M_n^N(\Gamma) = \inf_{\rho \in \mathcal{F}(\Gamma)} \int_N \rho^n d\mathcal{H}^{n-1} .$$

In [V4, 10.2] there is given a specific constant $b(n) > 0$ depending only on n such that

(1.9) $$M_n^N(\Gamma) \geq \frac{b(n)}{r}$$

holds. If the cap $\overline{C}$ is the whole sphere N, then

(1.10) $$M_n^N(\Gamma) \geq \frac{c(n)}{r}$$

for the constant $c(n) = 2^n b(n)$ and equality holds if $E = \{u\}$ and $F = \{v\}$ where u and v are antipodal points in N.

The proof of (1.9) in [V4] is based on the idea in the proof of an oscillation lemma by F.W. Gehring which is the special case $\overline{C} = N$ of Lemma 1.11 below. He gave the proof for $n = 3$ in [G4, Lemma 1]. For the n-dimensional version, see [Mo, p. 69]. If u is a real valued function in a set A, we write

$$\mathrm{osc}(u, A) = \sup_A u - \inf_A u$$

for the oscillation of u in A.

1.11. Lemma. Let u be a C^1 function on a cap C of an $(n-1)$-sphere N of radius r. Then

$$\mathrm{osc}(u, C)^n \leq A_n r \int_C |\nabla u|^n d\mathcal{H}^{n-1}$$

where A_n depends only on n. Here the gradient is taken with respect to N.

On the other hand, 1.11 is an easy consequence of (1.9), which is seen as follows. Let $x, y \in C$ be such that $a = |u(x) - u(y)| > 0$ and let Γ be the family $\Delta(\{x\}, \{y\}; C)$. Then $\rho = |\nabla u|/a \in \mathcal{F}(\Gamma)$ and hence by (1.9)

$$\frac{1}{a^n} \int_C |\nabla u|^n d\mathcal{H}^{n-1} \geq \frac{b(n)}{r} .$$

It follows that we can take $A_n = b(n)^{-1}$. In the case $\overline{C} = N$ we can take $c(n)^{-1}$ for the constant.

1.12. Example. Let $0 < a < b$, let E and F be disjoint sets in $\mathbb{R}^n$ which meet the sphere $S^{n-1}(t)$ for all t in a measurable set $T \subset [a, b]$, and let $A = B^n(b) \setminus \overline{B}^n(a) \subset D \subset \mathbb{R}^n$. Set $\Gamma = \Delta(E, F; D)$. Suppose $\rho \in \mathcal{F}(\Gamma)$. For $t \in T$ write $S_t = S^{n-1}(t)$, $\Gamma_t = \Delta(E \cap S_t, F \cap S_t; S_t)$, and $\rho_t = \rho|S_t$. Then $\rho_t \in \mathcal{F}(\Gamma_t)$ and by (1.10) we get

$$\int \rho^n \, dm \geq \int_T \left(\int_{S_t} \rho^n \, d\mathcal{H}^{n-1} \right) dt \geq c(n) \int_T \frac{dt}{t} \, ,$$

hence

(1.13)
$$\mathsf{M}(\Gamma) \geq c(n) \int_T \frac{dt}{t} \, .$$

In particular, if E and F meet $S^{n-1}(t)$ for $a < t < b$,

(1.14)
$$\mathsf{M}(\Gamma) \geq c(n) \log \frac{b}{a} \, .$$

There is equality in (1.14) if $D = A$ and if E and F are the components of $A \cap L$ where L is a line through the origin [V4, 10.12].

1.15. Remarks. 1. We have observed that the p-modulus is invariant under Euclidean motion and that the modulus is a conformal invariant. The fact that the modulus is quasi–invariant under qc mappings, is contained in the inequalities 2.4 and 8.1. In [V4] this quasi–invariance is taken as a definition of qc mappings. The equivalence of that definition with our definition is given in [V4, 32.3].

2. If Γ is a family of paths in $\mathbb{R}^n$ and if Γ_r is the subfamily of all rectifiable paths in Γ, then $\mathsf{M}(\Gamma \setminus \Gamma_r) = 0$ and $\mathsf{M}(\Gamma) = \mathsf{M}(\Gamma_r)$ [V4, 6.10]. It should be noted, however, that these are not true in general for Riemannian manifolds.

3. If $E, F, D \subset \mathbb{R}^n$, let $\Delta_0(E, F; D)$ be the family of open paths γ such that $|\gamma| \subset D$ and $\overline{|\gamma|} \cap E \neq \emptyset \neq \overline{|\gamma|} \cap F$. Then by [V4, 7.10], $\mathsf{M}\big(\Delta(E, F; D)\big) = \mathsf{M}\big(\Delta_0(E, F; D)\big)$. A similar remark holds also for half–open paths.

1.16. Notes. The concept of extremal length was first published by L.V. Ahlfors and A. Beurling [AB] in 1950 in the 2-dimensional case. The p-modulus in $\mathbb{R}^n$ was introduced and studied by B. Fuglede [Fu] in 1957, along with several generalizations of the concept. We refer to [Vu12, 5.72] for further comments on the history of the modulus.

2. The K_O-Inequality

In this section we shall prove an inequality which gives a lower bound for the modulus of the image of a path family under a qr mapping. The proof for the qr case differs from the qc case only in that we need the transformation formula in the more general form I.4.14(c).

If $f: U \to \mathbb{R}^m$ is a continuous mapping of an open set U in $\mathbb{R}^n$, we set

(2.1)
$$L(x, f) = \limsup_{h \to 0} \frac{|f(x + h) - f(x)|}{|h|} \, .$$

The mapping $x \mapsto L(x, f)$ is a Borel function. If f is differentiable at x, $L(x, f) = |f'(x)|$.

Let α be a rectifiable closed path in $\mathbb{R}^n$. We say that a mapping $g \colon |\alpha| \to \mathbb{R}^m$ is *absolutely continuous* on α if $g \circ \alpha^0$ is absolutely continuous on $[0, \ell(\alpha)]$. For the proofs of 2.2 and 2.3 we refer to [V4, 5.3] and [V4, 28.2], respectively.

2.2. Lemma. *Let* $f \colon U \to \mathbb{R}^m$ *be a continuous mapping of an open set* U *in* $\mathbb{R}^n$ *and let* $\alpha \colon \Delta \to U$ *be a locally rectifiable path such that* f *is absolutely continuous on every closed subpath of* α. *Then* $f \circ \alpha$ *is locally rectifiable. If* $\rho \colon |f \circ \alpha| \to [0, \infty]$ *is a Borel function, then*

$$\int\limits_{f \circ \alpha} \rho \, ds \leq \int\limits_{\alpha} \rho(f(x)) L(x, f) \, |dx| \, .$$

2.3. Fuglede's theorem. *Let* $f \colon U \to \mathbb{R}^m$ *be* ACL^p *and let* Γ *be the family of all locally rectifiable paths in* U *which have a closed subpath on which* f *is not absolutely continuous. Then* $\mathsf{M}_p(\Gamma) = 0$.

The K_O-inequality is the following.

2.4. Theorem. *Let* $f \colon G \to \mathbb{R}^n$ *be a nonconstant qr mapping. Let* $A \subset G$ *be a Borel set with* $N(f, A) < \infty$, *and let* Γ *be a family of paths in* A. *Then*

$$\mathsf{M}(\Gamma) \leq K_O(f) N(f, A) \mathsf{M}(f\Gamma) \, .$$

Proof. Let $\rho' \in \mathcal{F}(f\Gamma)$. Define a Borel function $\rho \colon \mathbb{R}^n \to [0, \infty]$ by setting $\rho(x) = \rho'(f(x)) L(x, f)$ for $x \in A$ and $\rho(x) = 0$ for $x \in \mathbb{R}^n \setminus A$. Let Γ_0 be the family of all locally rectifiable paths $\gamma \in \Gamma$ such that f is absolutely continuous on every closed subpath of γ. Then by 2.3, $\mathsf{M}(\Gamma \setminus \Gamma_0) = 0$, and hence $\mathsf{M}(\Gamma) = \mathsf{M}(\Gamma_0)$. By 2.2,

$$\int\limits_{\gamma} \rho \, ds \geq \int\limits_{f \circ \gamma} \rho' \, ds \geq 1$$

for all $\gamma \in \Gamma_0$. Thus $\rho \in \mathcal{F}(\Gamma_0)$ and

$$\mathsf{M}(\Gamma) = \mathsf{M}(\Gamma_0) \leq \int\limits_{\mathbb{R}^n} \rho^n \, dm = \int\limits_{A} \rho'(f(x))^n L(x, f)^n dx$$

$$\leq K_O(f) \int\limits_{A} \rho'(f(x))^n J_f(x) \, dx = K_O(f) \int\limits_{\mathbb{R}^n} \rho'(y)^n N(y, f, A) \, dy$$

$$\leq K_O(f) N(f, A) \int\limits_{\mathbb{R}^n} \rho'^n dm$$

by I.2.4 and I.4.14(c). $\qquad\square$

2.5. Remark. Theorem 2.4 is from [MRV1, 3.2]. Sometimes more useful than 2.4 is the inequality

$$(2.6) \qquad \mathsf{M}(\Gamma) \leq K_O(f) \int_{\mathbb{R}^n} \rho'(y)^n N(y, f, A) \, dy$$

obtained from the proof. It holds for any $\rho' \in \mathcal{F}(f\Gamma)$.

2.7. Corollary. *If* $f \colon G \to G'$ *is qc and* Γ *is a family of paths in* G, *then*

$$\mathsf{M}(\Gamma) \leq K_O(f)\mathsf{M}(f\Gamma) \, .$$

2.8. Example. Let $f \colon \mathbb{R}^2 \to \mathbb{R}^2$ be the exponential function $f(z) = e^z$ and let Γ be the family of paths $\gamma_u \colon [0, 1] \to \mathbb{R}^2$, $\gamma_u(t) = te_1 + ue_2$, $0 \leq u \leq k \, 2\pi$. Then $\mathsf{M}(\Gamma) = k \, 2\pi$ and $\mathsf{M}(f\Gamma) = 2\pi$. The family Γ lies in $A = [0, 1] \times [0, k \, 2\pi]$ and $N(f, A) = k$, hence 2.4 is sharp in this case.

3. Path Lifting

In this section we shall give a direct proof of path lifting for discrete open mappings. Recall that "discrete open" according to our convention means also continuous. We shall formulate the main result in terms of finite sequences of liftings, following [R1]. We need such a general result in the proof of Poletskiĭ's lemma in Section 5 and also in Chapters IV and V.

Throughout this section we let $f \colon G \to \mathbb{R}^n$ be a discrete, open, and sense–preserving mapping of a domain G in $\mathbb{R}^n$. Let $\beta \colon [a, b[\to \mathbb{R}^n$ be a path and let $x \in f^{-1}(\beta(a))$. A path $\alpha \colon [a, c[\to G$ is called a *maximal f-lifting of β starting at* x if (1) $\alpha(a) = x$, (2) $f \circ \alpha = \beta| \, [a, c[$, and (3) if $c < c' \leq b$, then there does not exist a path $\alpha' \colon [a, c'[\to G$ such that $\alpha = \alpha'| \, [a, c[$ and $f \circ \alpha' = \beta| \, [a, c'[$. We also say that α is an *f-lifting of β starting at* x if (1) and (2) are satisfied (even if $c < b$).

Let $x_1, \ldots, x_k$ be k different points of $f^{-1}(\beta(a))$ and let

$$(3.1) \qquad m = \sum_{i=1}^{k} i(x_i, f) \, .$$

We say that the sequence $\alpha_1, \ldots, \alpha_m$ is a *maximal sequence of f-liftings of β starting at the points* $x_1, \ldots, x_k$ if

(a) each α_j is a maximal f-lifting of β,
(b) $\mathrm{card}\{ j : \alpha_j(a) = x_i \} = i(x_i, f)$, $1 \leq i \leq k$,
(c) $\mathrm{card}\{ j : \alpha_j(t) = x \} \leq i(x, f)$ for all $x \in G$ and for all t.

Similarly we define a *maximal sequence of f-liftings of β terminating at* $x_1, \ldots, x_k$ if $\beta \colon]b, a] \to \mathbb{R}^n$. We also say that the paths $\alpha_1, \ldots, \alpha_m$ are *essentially separate* if (c) is satisfied. These definitions extend in an obvious way to the case where $[a, b[$ (or $]b, a]$) is replaced by a closed interval or a point. If γ is a path and if γ' is a restriction of γ to a subinterval or a point, we write $\gamma' \subset \gamma$.

3.2. Theorem. *Let* $f \colon G \to \mathbb{R}^n$ *be discrete, open, and sense–preserving, let* $\beta \colon [a, b[\to \mathbb{R}^n$ *(respectively,* $\beta \colon]b, a] \to \mathbb{R}^n$ *) be a path, and let* $x_1, \ldots, x_k$ *be distinct points in* $f^{-1}(\beta(a))$. *Then* β *has a maximal sequence of* f*-liftings starting (respectively, terminating) at* $x_1, \ldots, x_k$.

Proof. In the first step of the proof we show that the local existence of liftings implies the statement in the theorem. In the second step we prove the local existence of liftings.

Step 1. We assume that every $x \in G$ has a normal neighborhood U such that the following holds. Every path $\gamma \colon [d, e[\to fU$ with $\gamma(d) = f(x)$ has a maximal sequence of $f|U$-liftings $\tau_\mu \colon [d, e[\to U$, $\mu = 1, \ldots, r = i(x, f)$, starting at x, and similarly for a path $\gamma \colon]e, d] \to fU$. We claim that then β has a maximal sequence of f-liftings starting (or terminating) at $x_1, \ldots, x_k$.

We assume $\beta \colon [a, b[\to \mathbb{R}^n$. The other case is similar. Let m be defined by (3.1). Let A be the set of maximal sequences $\alpha = (\alpha_1, \ldots, \alpha_m)$ of f-liftings $\alpha_j \colon J_\alpha \to G$ of $\beta|J_\alpha$ starting at $x_1, \ldots, x_k$, where for each α the set J_α is a connected subset of $[a, b[$ with $a \in J_\alpha$. In A we define an ordering by setting $\alpha < \alpha'$ if $\alpha_j \subset \alpha'_j$ for all j and $J_\alpha \neq J_{\alpha'}$.

Clearly there exists $\alpha \in A$ with $J_\alpha = \{a\}$, hence $A \neq \emptyset$. Let $E \subset A$ be linearly ordered and nonempty. We define $\sigma = (\sigma_1, \ldots, \sigma_m)$ by

$$J_\sigma = \bigcup_{\alpha \in E} J_\alpha$$

and $\alpha_j \subset \sigma_j$ for all j. Then σ is an upper bound of E in A. The conditions in Zorn's lemma are thus satisfied and there is a maximal element $\alpha = (\alpha_1, \ldots, \alpha_m)$ in A.

We claim that for some j the path α_j is a maximal f-lifting of β starting at $\alpha_j(a)$. Suppose this is not true. Then for some $c \in [a, b[$, $J_\alpha = [a, c[$. Note that J_α cannot be of the form $[a, c[$ because each α_j could then be extended to c, giving a sequence α' in A with $\alpha < \alpha'$ and contradicting the maximality of α. Set $\{z_1, \ldots, z_l\} = \{\alpha_j(c) : 1 \leq j \leq m\}$ and let U_s be a normal neighborhood of z_s as in the assumption. Let $e \in]c, b[$ be such that $\beta|[c, e[\subset \cap fU_s$. By extending each path α_j via a suitably chosen member of the maximal sequence of $f|U$-liftings of $\beta|[c, e[$ starting at $\alpha_j(c)$ we obtain $\tilde{\alpha} \in A$ with $\tilde{\alpha} > \alpha$, a contradiction. (For a presentation of this with indices, see [R1].) We have proved that, for some j, α_j is a maximal f-lifting of β starting at $\alpha_j(a)$.

We may assume that for some $m_1 < m$, $\alpha_{m_1+1}, \ldots, \alpha_m$ are maximal f-liftings of β and $\alpha_1, \ldots \alpha_{m_1}$ are not. Then $J_\alpha = [a, c[$, $a < c < b$. We continue α_j for $1 \leq j \leq m_1$ to a path $\overline{\alpha}_j \colon [a, c] \to G$. We apply the above argument to $\beta_1 = \beta|[c, b[$ and the points of $\{z_1, \ldots, z_{l_1}\} = \{\overline{\alpha}_j(c) : 1 \leq j \leq m_1\} \subset f^{-1}(\beta(c))$ to obtain a maximal element $\alpha^1 = (\alpha_1^1, \ldots, \alpha_p^1)$ where $p \geq m_1$. We can continue each $\overline{\alpha}_j$, $1 \leq j \leq m_1$, by a suitably chosen member of α^1 to obtain a path $\hat{\alpha}_j \colon [a, c] \cup J_{\alpha^1} \to G$. Some of the paths $\hat{\alpha}_j$ are again maximal f-liftings of β. After a finite number of steps we obtain the required maximal sequence of f-liftings of β.

Step 2. We now show by induction on the local index that the assumption in step 1 is true.

For points $x \in G$ where $i(x, f) = 1$ the assertion is clear. Suppose it has been proved for points $x \in G$ with $1 \le i(x, f) < r$. Let now x be a point in G with $i(x, f) = r$. Let U be any normal neighborhood of x and let $\gamma \colon [d, e[\to fU$ be a path with $\gamma(d) = f(x)$. Set $F = \{ z \in U : i(z, f) = r \}$. Since $z \mapsto i(z, f)$ is upper semicontinuous, F is closed in U. Since U is a normal neighborhood of x, $f\overline{F} \cap U = fF$, hence fF is closed in fU. It follows that $[d, e[\setminus \gamma^{-1} fF$ is a countable union of open intervals $]a_\lambda, b_\lambda[$, $\lambda = 1, 2, \ldots$. For every λ we choose $c_\lambda \in]a_\lambda, b_\lambda[$ and set $\gamma_\lambda = \gamma | [c_\lambda, b_\lambda[$, $\gamma'_\lambda = \gamma |]a_\lambda, c_\lambda]$.

Since $F \subset B_f$ and F is closed in U, $U \setminus F$ is a domain and, by the induction hypothesis, the assumption in step 1 is satisfied for the mapping $g = f | (U \setminus F)$. Fix λ. It follows that γ_λ has a maximal sequence of g-liftings starting at the points of $\{ u_1, \ldots, u_\mu \} = (U \setminus F) \cap f^{-1}(\gamma_\lambda(c_\lambda)) = U \cap f^{-1}(\gamma_\lambda(c_\lambda))$. Since U is a normal neighborhood of x, we have

$$r = \sum_{i=1}^{\mu} i(u_i, f) \,.$$

We claim that the g-liftings in the maximal sequence are defined on the whole interval $[c_\lambda, b_\lambda[$. Suppose this is not true and that γ_λ has a maximal g-lifting $l \colon [c_\lambda, t_1[\to U \setminus F$ where $c_\lambda < t_1 < b_\lambda$. Let $\{ v_1, \ldots, v_\nu \} = (U \setminus F) \cap f^{-1}(\gamma_\lambda(t_1)) = U \cap f^{-1}(\gamma_\lambda(t_1))$ and let $V_1, \ldots, V_\nu \subset U \setminus F$ be disjoint normal neighborhoods of $v_1, \ldots, v_\nu$ respectively. For some $t_2 \in]c_\lambda, t_1[$, $\gamma_\lambda(t) \in \bigcap fV_i$ whenever $t_2 < t < t_1$. Since $f^{-1}(\bigcap fV_i) \cap U \subset \bigcup V_i$, it follows that for some i, $l(t) \in V_i$ whenever $t_2 < t < t_1$. Hence

$$v_i = \lim_{t \to t_1} l(t)$$

and l has an extension to a path $\bar{l} \colon [c_\lambda, t_1] \to U \setminus F$. Again by the induction hypothesis and step 1, $\gamma_\lambda | [t_1, b[$ has a maximal g-lifting $l_1 \colon [t_1, t_3[\to U \setminus F$ starting at v_i. Then the path consisting of l followed by l_1 is a g-lifting of γ_λ contradicting the maximality of l. We conclude that the g-liftings of γ_λ in the maximal sequence are defined on the whole interval $[c_\lambda, b_\lambda[$. A similar statement holds for the γ'_λ. Therefore there exist paths $\alpha_{\lambda q} \colon]a_\lambda, b_\lambda[\to U \setminus F$, $q = 1, \ldots, r$, such that

  (i) $f \circ \alpha_{\lambda q} = \gamma |]a_\lambda, b_\lambda[$,
  (ii) $\operatorname{card}\{ q : \alpha_{\lambda q}(c_\lambda) = u_i \} = i(u_i, f)$ for $1 \le i \le \mu$,
  (iii) $\operatorname{card}\{ q : \alpha_{\lambda q}(t) = z \} \le i(z, f)$ for all $z \in U \setminus F$ and $t \in]a_\lambda, b_\lambda[$.

If $t \in \gamma^{-1} fF$, there is exactly one point x_t in U such that $f(x_t) = \gamma(t)$. Then the mappings $\tau_q \colon [d, e[\to U$, $1 \le q \le r$, defined by

$$\tau_q(t) = \begin{cases} \alpha_{\lambda q}(t) & \text{if } t \in]a_\lambda, b_\lambda[, \\ x_t & \text{if } t \in \gamma^{-1} fF, \end{cases}$$

form a maximal sequence of $f|U$-liftings of γ starting at x. The proof for a path $\gamma \colon]e, d] \to fU$ is similar. The theorem is proved. $\square$

3.3. Corollary. *Let f and β be as in 3.2. If $x \in f^{-1}(\beta(a))$, then β has a maximal f-lifting starting at x.*

3.4. Corollary. *Let f be as in 3.2, D a normal domain for f, $\beta: [a, b[\to fD$ a path and $m = N(f, D)$. Then there exist paths $\alpha_j: [a, b[\to D$, $1 \le j \le m$, such that*

(1) $f \circ \alpha_j = \beta$,
(2) $\operatorname{card}\{ j : \alpha_j(t) = x \} = i(x, f)$ *for* $x \in D \cap f^{-1}\beta(t)$,
(3) $|\alpha_1| \cup \ldots \cup |\alpha_m| = D \cap f^{-1}|\beta|$.

Proof. Let $\{ x_1, \ldots, x_k \} = D \cap f^{-1}(\beta(a))$. Since D is a normal domain,

$$\sum_{i=1}^{k} i(x_i, f) = m .$$

By 3.2 there exists a maximal sequence $\alpha_1, \ldots, \alpha_m$ of f-liftings of β starting at $x_1, \ldots, x_k$. As in step 2 of the proof of 3.2 it follows that the paths α_j are defined on $[a, b[$. Since

$$\sum_{x \in D \cap f^{-1}(y)} i(x, f) = m$$

for every $y \in fD$, (2) follows from (c). The equality (3) follows from (2). □

3.5. Notes. This section is from [R1]. The assertion in Corollary 3.4 had been proved earlier by E.A. Poletskiĭ in [P1, Lemma 4] under a restrictive assumption. The idea of the proof of Theorem 3.2 is partially due to him. Corollary 3.3 is also proved in [MRV3, 3.12] for light open mappings. A mapping f is *light* if $f^{-1}(y)$ is totally disconnected for all y. The path lifting problem for light open mappings had been considered previously in [Sto1, p. 354], [Sto2, p. 109], [Wh, p. 186], [Fl, p. 574], and [MRV1, 2.7].

4. Linear Dilatations

Throughout this section $f: G \to \mathbb{R}^n$ is a discrete and open mapping of a domain G in $\mathbb{R}^n$. If $x \in G$, $0 < r < d(x, \partial G)$, and $0 < s < d(f(x), \partial fG)$, we set

$$l(x, r) = l(x, f, r) = \inf_{|y-x|=r} |f(y) - f(x)| ,$$

$$L(x, r) = L(x, f, r) = \sup_{|y-x|=r} |f(y) - f(x)| ,$$

$$l^*(x, s) = l^*(x, f, s) = \inf_{z \in \partial U(x, f, s)} |x - z| ,$$

$$L^*(x, s) = L^*(x, f, s) = \sup_{z \in \partial U(x, f, s)} |x - z| ,$$

Recall that $U(x, f, s)$ is the x-component of $f^{-1}B^n(f(x), s)$. In the present discussion we shorten $U(x, f, s)$ to $U(x, s)$. We call

$$H(x, f) = \limsup_{r \to 0} \frac{L(x, r)}{l(x, r)}$$

the *linear dilatation* of f at x and

$$H^*(x, f) = \limsup_{s \to 0} \frac{L^*(x, s)}{l^*(x, s)}$$

the *inverse linear dilatation* of f at x.

In this section we shall prove that $H(x, f)$ and $H^*(x, f)$ are locally bounded for a nonconstant qr mapping. The latter result will be applied for example in the proof of Poletskiĭ's lemma in Section 5. Both the linear dilatation and the inverse linear dilatation can be used to characterize quasiregularity, and we shall do this in Theorem 6.2 for the linear dilatation. It is true that $H^*(x, f)$ is actually bounded if f is qr, but the proof of this fact is postponed until III.4.6.

A ring in $\mathbb{R}^n$ is a domain $D \subset \mathbb{R}^n$ such that $\complement D = \mathbb{R}^n \smallsetminus D$ has exactly two components.

4.1. Lemma. *For every $x \in G$ there exists $\sigma_x > 0$ such that the following conditions are satisfied for $0 < s \le \sigma_x$:*

(1) $U(x, s)$ *is a normal neighborhood of x;*
(2) $U(x, s) = U(x, \sigma_x) \cap f^{-1}B^n(f(x), s)$;
(3) $\partial U(x, s) = U(x, \sigma_x) \cap f^{-1}S^{n-1}(f(x), s)$ *if $s < \sigma_x$;*
(4) $\complement U(x, s)$ *and $\complement \overline{U}(x, s)$ are connected;*
(5) $U(x, s) \smallsetminus \overline{U}(x, t)$ *is a ring for $0 < t < s \le \sigma_x$;*
(6) $l^*(x, L(x, r)) = L^*(x, l(x, r)) = r$ *for $0 < r < l^*(x, \sigma_x)$.*

Proof. By I.4.9 there exists $s_x > 0$ such that $U(x, s)$ is a normal neighborhood of x and $fU(x, s) = B^n(f(x), s)$ for $0 < s \le s_x$. Let $D = U(x, s_x)$ and choose $\sigma_x \in \,]0, s_x[$ such that $U_0 = U(x, \sigma_x) \subset B^n(x, \rho) \subset D$ for some $\rho > 0$. We may assume $s < \sigma_x$. Set $U = U(x, s)$ and $V = fU$.

Condition (1) is clear. Since $U_0 \cap f^{-1}(f(x)) = \{x\}$, (2) follows from I.4.7. To prove (3) let $z \in U_0 \cap f^{-1}\partial V$. Since f is open, every neighborhood of z meets $f^{-1}V$. By (2), this gives $z \in \overline{U}$. Since $z \notin U$, we get $U_0 \cap f^{-1}\partial V \subset \partial U$. The last statement of I.4.8 implies $\overline{U} \subset U_0$. With $f\partial U = \partial fU = \partial V$ we get $\partial U \subset U_0 \cap f^{-1}\partial V$ and (3) follows.

Since $U \subset B^n(x, \rho) \subset D$, there is exactly one component E of $\complement U$ which meets $\complement D$. We show that $E = \complement U$. Set $F = \complement U \smallsetminus E$. Since $U = D \cap f^{-1}V$, fF does not meet V. The openness of f implies $\partial fF \subset f\partial F \subset f\partial U = \partial V$ and the boundedness of fF gives $fF \subset \partial V$. Hence $\operatorname{int} F = \emptyset$. Setting $U_1 = \complement E$ we thus have $\overline{U}_1 = \overline{U}$. Hence $fU_1 \subset \operatorname{int} f\overline{U} = V$, which implies $fF \subset V$. It follows that $F = \emptyset$ and $\complement U$ is connected. If $y, z \in \complement \overline{U}$, there exists $s_1 \in \,]s, \sigma_x[$ such that $\mathbb{C}U(x, s_1) \subset \overline{\mathbb{C}U}$. Since $\mathbb{C}U(x, s_1)$ is connected, y and z can be connected in $\overline{\mathbb{C}U}$, and (4) follows.

Since $\overline{U}(x, t) \subset U(x, s)$, $A = U(x, s) \smallsetminus \overline{U}(x, t) \neq \emptyset$. Since $\mathbb{C}U(x, s)$ is connected, the components of the complement of A are $\overline{U}(x, t)$ and $\mathbb{C}U(x, s)$. Since $\mathbb{C}\overline{U}(x, t)$ is connected, it follows from the Phragmén–Brouwer theorem that A is connected, so (5) is proved.

To prove (6) set $l = l(x, r)$ and $L = L(x, r)$. Obviously $l \leq L < \sigma_x$. We prove $l^*(x, L) = r$. The proof for $L^*(x, l) = r$ is similar. Since $B^n(x, r) \subset U(x, L)$, $l^*(x, L) \geq r$. Choose $a \in S^{n-1}(x, r)$ such that $|f(a) - f(x)| = L$. By (2), $\partial U(x, L) = U_0 \cap f^{-1} S^{n-1}(f(x), L)$. Thus $a \in \partial U(x, r)$, which implies $l^*(x, L) \leq |a - x| = r$. $\qquad\square$

4.2. Lemma. *Let $x \in G$ and let σ_x be as in 4.1. For $0 < s < t \leq \sigma_x$ let $\Gamma(s, t)$ be the family of all paths that join $\partial U(x, s)$ and $\partial U(x, t)$ in $U(x, t) \smallsetminus \overline{U}(x, s)$. Suppose that there exist constants b and σ, $1 \leq b < \infty$, $0 < \sigma \leq \sigma_x$, such that $\mathsf{M}(\Gamma(s, t)) \leq b\,\mathsf{M}(f\Gamma(s, t))$ for $0 < s < t < \sigma$. Then $H(x, f) \leq C < \infty$, where C depends only on n and b.*

Proof. As in 4.1(6) we assume $0 < r < l^*(x, \sigma_x)$ and set $L = L(x, r)$, $l = l(x, r)$. We may assume $l < L$. By Example 1.7(b), $\mathsf{M}(f\Gamma(l, L)) \leq \omega_{n-1}/(\log(L/l))^{n-1}$. By 4.1(6) $\partial U(x, l)$ and $\partial U(x, L)$ meet $S^{n-1}(x, r)$. From this and the fact that $U(x, L) \smallsetminus \overline{U}(x, l)$ is a ring follows the estimate $\mathsf{M}(\Gamma(l, L)) \geq a_n > 0$, where a_n depends only on n [V4, 11.7]. Since $\mathsf{M}(\Gamma(l, L)) \leq b\,\mathsf{M}(f\Gamma(l, L))$, we get $L/l \leq C$ with $C = \exp\big((b\omega_{n-1}/a_n)^{1/(n-1)}\big)$, and the lemma follows. $\qquad\square$

4.3. Theorem. *Let $f : G \to \mathbf{R}^n$ be a nonconstant qr mapping. Then for every $x \in G$*

$$H(x, f) \leq C < \infty$$

where C depends only on n and the product $i(x, f)K_O(f)$.

Proof. By I.4.10(2) and 2.4 the conditions in 4.2 are satisfied with $b = i(x, f)K_O(f)$ and $\sigma = \sigma_x$. The result follows from 4.2. $\qquad\square$

4.4. Theorem. *Let $f : G \to \mathbf{R}^n$ be a nonconstant qr mapping. Then for every $x \in G$*

$$H^*(x, f) \leq C^* < \infty,$$

where C^* depends only on n and the product $i(x, f)K_O(f)$.

Proof. Let $x \in G$. Choose σ_x as in Lemma 4.1 and set $U_0 = U(x, \sigma_x)$. Let $t > 0$ be such that $L(x, t) < \sigma_x$, and let $r_0 > 0$ be such that $U(x, r) \subset B^n(x, t)$ if $0 < r \leq r_0$. Assume $0 < r \leq r_0$ and set $L^* = L^*(x, r)$, $l^* = l^*(x, r)$, $L = L(x, L^*)$, and $l = l(x, l^*)$. We choose a line J through $y = f(x)$. Let $A' = \{v : l < |v - y| < L\}$, let E' and F' be the components of $J \cap A'$, and set $E = U_0 \cap f^{-1}E'$, $F = U_0 \cap f^{-1}F'$ (Fig. 2). If E_0 is any component of E, $fE_0 = E'$ by I.4.8. By 4.1(3), $\partial U(x,l) = U_0 \cap f^{-1}S^{n-1}(y,l)$ and $\partial U(x,L) = U_0 \cap f^{-1}S^{n-1}(y,L)$. Hence E_0 meets both $\partial U(x,l)$ and $\partial U(x,L)$. Since $\partial U(x,l) \subset \overline{B}^n(x,l^*)$ and $\partial U(x,L) \subset \mathbb{C}B^n(x,L^*)$, we get thus $\emptyset \neq S^{n-1}(x,u) \cap E_0 \subset S^{n-1}(x,u) \cap E$, and similarly $S^{n-1}(x,u) \cap F \neq \emptyset$ for every u, $l^* < u < L^*$.

Set $A = \{z : l^* < |x - z| < L^*\}$, and let Γ be the family of paths which join E and F in A. Then by 1.12, $\mathsf{M}(\Gamma) \geq c(n)\log(L^*/l^*)$. Every path in $f\Gamma$ joins E' and F' in A', hence again by 1.12, $\mathsf{M}(f\Gamma) \leq c(n)\log(L/l)$. Owing to I.4.10(2) and 2.4 we thus get

$$c(n)\log\frac{L^*}{l^*} \leq \mathsf{M}(\Gamma) \leq i(x,f)K_O(f)\mathsf{M}(f\Gamma)$$

$$\leq i(x,f)K_O(f)c(n)\log\left(\frac{L}{r}\cdot\frac{r}{l}\right).$$

But $r = l(x,L^*) = L(x,l^*)$, so letting $r \to 0$ we obtain

(4.5) $$H^*(x,f) \leq H(x,f)^{2i(x,f)K_O(f)}.$$

An appeal to Theorem 4.3 completes the proof.   □

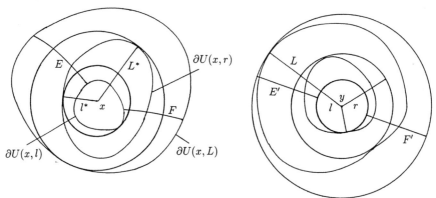

Fig. 2

4.6. Remarks. 1. Although $H^*(x,f)$ is uniformly bounded (III.4.6) $H(x,f)$ need not be. Let $g\colon \mathbb{R}^2 \to \mathbb{R}^2$ be the K-qr mapping defined by

$$g(z) = (x + iKy)^k$$

in complex notation $z = x + iy$, where $K > 1$ and k is a positive integer. Then $H(0,g) = K^k$ and $i(0,g) = k$. Replacing $z \mapsto z^k$ by an analytic function $\varphi\colon \mathbb{R}^2 \to \mathbb{R}^2$ with $\sup\{i(x,\varphi) : x \in \mathbb{R}^2\} = \infty$ we get a K-qr mapping $f\colon \mathbb{R}^2 \to \mathbb{R}^2$ with $\sup\{H(x,f) : x \in \mathbb{R}^2\} = \infty$. For dimensions $n > 2$ the corresponding result is obtained by the use of mappings like in I.3.2. See also [MRV3, 4.10].

2. It is true that both $H(x, f)$ and $H^*(x, f)$ have a bound independent of n. For $H(x, f)$ this fact is given in [Vu12, 10.27] and for $H^*(x, f)$ we apply (4.5).

In the next section we need the following semicontinuity result for l^* and L^*.

4.7. Lemma. *Let* $C \subset G$ *be compact. There exists* $t > 0$ *such that* $(x, s) \mapsto$ $l^*(x, s)$ *is upper semicontinuous and* $(x, s) \mapsto L^*(x, s)$ *is lower semicontinuous in the set* $C \times \,]0, t[$.

Proof. For $x \in C$ let $s_x > 0$ be such that $\overline{U}(x, s_x) \subset B^n(x, a)$ where $a = d(C, \partial G)$. We cover C by sets $U(x_i, s_{x_i}/2)$, $i = 1, \ldots, k$. If $x \in C$, $U(x, s_{x_j}/2) \subset U(x_j, s_{x_j})$ for some j. Hence $\overline{U}(x, s) \subset C + aB^n$ if $0 < s < t = \min(s_{x_1}/2, \ldots, s_{x_k}/2)$. Assume $(x_0, s_0) \in C \times \,]0, t[$ and $0 < \varepsilon < \min\bigl(l^*(x_0, s_{x_0}), d(U(x_0, s_0), \partial G)\bigr)$, and set $y_0 = f(x_0)$.

To prove the upper semicontinuity of $(x, s) \mapsto l^*(x, s)$ at $(x_0, s_0) \in C \times \,]0, t[$, we let $z \in \partial U(x_0, s_0)$ be such that $|x_0 - z| = l^*(x_0, s_0)$. The set $f B^n(z, \varepsilon/2)$ is a neighborhood of $f(z)$. There is therefore a point $v \in B^n(z, \varepsilon/2) \cap \complement \overline{U}(x_0, s_0)$ such that $v' = f(v) \notin \overline{B}^n(y_0, s_0)$. Set $\tau = (|v' - y_0| - s_0)/2$, and let $\delta \in \,]0, \varepsilon/2[$ be such that $|x - x_0| < \delta$ implies $|f(x) - y_0| < \tau$. If now $(x, s) \in C \times \,]0, t[$ is such that $|x - x_0| < \delta$ and $|s - s_0| < \tau$, then $v' \notin \overline{B}^n(f(x), s)$, whence $v \notin \overline{U}(x, s)$. This implies $l^*(x, s) < |v - z| + |z - x_0| + |x_0 - x| < l^*(x_0, s_0) + \varepsilon$.

To prove the lower semicontinuity of $(x, s) \mapsto L^*(x, s)$ at (x_0, s_0), let $z \in \partial U(x_0, s_0)$ be such that $|z - x_0| = L^*(x_0, s_0)$ and let $u \in U(x_0, s_0) \cap B^n(z, \varepsilon/2)$. There exists a continuum A in $U(x_0, s_0)$ such that $u \in A$ and $B^n(x_0, \varepsilon/2) \subset A$. Then $2\tau = d(fA, S^{n-1}(x_0, s_0))$ is positive. If δ and (x, s) are as in the first part of this proof, then we have $fA \subset B^n(f(x), s)$ and therefore $A \subset U(x, s)$. Hence $L^*(x, s) \geq |x - u| \geq L^*(x_0, s_0) - \varepsilon$. $\square$

4.8. Remarks. 1. In fact $(x, s) \mapsto l^*(x, s)$ is continuous in $C \times \,]0, t[$ (see [MRV1, 4.8]), but we shall not need this information. The winding mapping $f \colon \mathbb{R}^3 \to \mathbb{R}^3$, $f(r, \varphi, x_3) = (r, k\varphi, x_3)$, in I.3.1 has $x \mapsto L^*(x, s)$ discontinuous at points $x = (s, \varphi, x_3)$.

2. All material in this section is from [MRV1].

5. Poletskiĭ's Lemma

Poletskiĭ's lemma is a substitute for Fuglede's theorem 2.2 in the "inverse" direction. To state it we need some terminology.

Let $f \colon G \to \mathbb{R}^n$ be a discrete open mapping of a domain G in $\mathbb{R}^n$, let $\beta \colon I_0 \to \mathbb{R}^n$ be a closed rectifiable path, and let $\alpha \colon I \to G$ be a path such that $f \circ \alpha \subset \beta$. If the length function $s_\beta \colon I_0 \to [0, \ell(\beta)]$ is constant on some interval

$J \subset I$, β is also constant on J, and the discreteness of f implies that α too is constant on J. It follows that there is a unique mapping $\alpha^* \colon s_\beta I \to G$ such that $\alpha = \alpha^* \circ (s_\beta | I)$. Clearly α^* is continuous and $f \circ \alpha^* \subset \beta^0$. The path α^* is called the *f-representation of α (with respect to β)* if $\beta = f \circ \alpha$. Let $\beta = f \circ \alpha$. We say that f is *absolutely precontinuous on α* if α^* is absolutely continuous. If $f \colon G \to G'$ is homeomorphism onto a domain $G' \subset \mathbb{R}^n$, then f is absolutely precontinuous on α if and only if f^{-1} is absolutely continuous on $f \circ \alpha$.

5.1. Poletskiĭ's lemma. *Let $f \colon G \to \mathbb{R}^n$ be a nonconstant K-qr mapping. Let Γ be a family of paths γ in G such that $f \circ \gamma$ is locally rectifiable and there is a closed subpath α of γ on which f is not absolutely precontinuous. Then $\mathsf{M}(f\Gamma) = 0$.*

The proof of 5.1 splits into several parts. In this section we define – see (5.2) – a substitute for an inverse function and as a preliminary step we show that it is ACL.

Let $f \colon G \to \mathbb{R}^n$ be a nonconstant K-qr mapping, let U be a normal domain of f, and let $m = \mu(f, U)$. Recall the notation $\mu(f, U)$ from I.4. We define a mapping g_U of $V = fU$ into $\mathbb{R}^n$ by

$$(5.2) \qquad g_U(y) = \frac{1}{m} \sum_{x \in f^{-1}(y) \cap U} i(x, f)x \,.$$

5.3. Lemma. *The mapping g_U is ACL.*

Proof. We first show that g_U is continuous. For this let $g = mg_U$, $y_0 \in V$, and $\varepsilon > 0$. Set $\{x_1, \ldots, x_q\} = f^{-1}(y_0) \cap U$ and choose disjoint normal neighborhoods $U_j \subset U$ of the points x_j such that $U_j \subset B^n(x_j, \varepsilon/m^2)$. We claim that $z \in W = fU_1 \cap \cdots \cap fU_q$ implies $|g(z) - g(y_0)| < \varepsilon$. We have

$$(5.4) \qquad g(z) = \sum_{j=1}^{q} \sum_{x \in f^{-1}(z) \cap U_j} i(x, f)x \,.$$

Here $\sum \{ i(x, f) : x \in f^{-1}(z) \cap U_j \} = i(x_j, f)$, which yields

$$\left| \sum_{x \in f^{-1}(z) \cap U_j} i(x, f)x - i(x_j, f)x_j \right| < \frac{i(x_j, f)\varepsilon}{m^2} \le \frac{\varepsilon}{m} \,.$$

Our claim then follows from (5.4).

To show that g_U is ACL, we again choose for $y_0 \in V$ disjoint normal neighborhoods $U_j \subset U$ of the points x_j. Let Q be an open n-interval such that $\overline{Q} \subset \bigcap_j fU_j$. Write $Q = Q_0 \times J_0$ where Q_0 is an open $(n-1)$-interval in $\mathbb{R}^{n-1}$ and $J_0 = \,]a, b[\,\subset \mathbb{R}^1$. For each Borel set $A \subset Q_0$ put $\varphi(A) = m(U \cap f^{-1}(A \times J_0))$. Then φ is a bounded completely additive set function and by the Lebesgue–Banach theorem it therefore has a finite derivative $\varphi'(z)$ for almost every $z \in Q_0$. The existence of $\varphi'(z)$ means that the upper and lower derivatives $\overline{\varphi}'(z)$ and $\underline{\varphi}'(z)$ coincide. Here we use the definitions

$$\overline{\varphi}'(z) = \lim_{h \to 0} \sup_{d(B) < h} \frac{\varphi(B)}{m(B)},$$

$$\underline{\varphi}'(z) = \lim_{h \to 0} \inf_{d(B) < h} \frac{\varphi(B)}{m(B)},$$

where B runs through all open $(n-1)$-cubes and open $(n-1)$-balls such that $z \in B \subset Q_0$. Fix such a z, and set $J = \{z\} \times J_0$. By symmetry, it suffices to show that g_U is absolutely continuous on $\overline{J}$.

Let $h \colon \overline{J} \to U$ be a continuous mapping such that $f \circ h$ is the identity on $\overline{J}$. We will show that h is absolutely continuous. There exists by 4.4 a constant c such that $H^*(x, f) < c$ for $x \in U$. Let $\varepsilon > 0$ and let $F^i = [y_i, \overline{y}_i]$, $i = 1, \ldots, p$, be disjoint closed intervals of J. More precisely, $y_i = (z, t_i)$, $\overline{y}_i = (z, \overline{t}_i)$, and $a < t_1 < \overline{t}_1 < \ldots < t_p < \overline{t}_p < b$. Set $F = F^1 \cup \cdots \cup F^p$. Let k_0 be an integer such that $0 < 1/k_0 < d(F, \partial Q)$. For $k \geq k_0$ let F_k be the set of all points $y \in F$ such that $0 < r < 1/k$ implies $L^*(h(y), r) \leq c l^*(h(y), r)$. Then $F_k \subset F_{k+1}$ and $F = \bigcup \{ F_k : k \geq k_0 \}$. By choosing k_0 sufficiently large, we conclude from 4.7 that $y \mapsto L^*(h(y), r)$ is lower semicontinuous and the function $y \mapsto l^*(h(y), r)$ upper semicontinuous. Hence each F_k is compact.

Let $\eta, t > 0$ and fix $k \geq k_0$. By [V4, 31.1] we can find $\delta > 0$ such that for $0 < r < \delta$ there exists a covering of F_k by open segments $\Delta_1, \ldots, \Delta_l$ in J such that (1) $m_1(\Delta_s) = 2r$, (2) the center v_s of Δ_s belongs to F_k, (3) every point of F_k belongs to at most two different Δ_s, and (4) $lr < m_1(F_k) + \eta$. Choose $r > 0$ such that $r < \min(\delta, 1/k)$ and such that $|h(y) - h(y')| < t/(2c)$ whenever $|y - y'| \leq 2r$, $y, y' \in J$.

Set $x_s = h(v_s)$ and $W_s = U(x_s, r)$, $1 \leq s \leq l$. Since $v_s \in F_k$, we have $L_s^* \leq c l_s^*$ with $L_s^* = L^*(x_s, r)$, $l_s^* = l^*(x_s, r)$. On the other hand, $l_s^* \leq d(h\Delta_s) < t/(2c)$, which implies $d(W_s) \leq 2L_s^* < t$. Since $h\Delta_s \subset W_s$, this yields via Hölder's inequality

$$\mathcal{H}_t^1(hF_k)^n \leq \left(\sum_{s=1}^l d(W_s) \right)^n \leq 2^n c^n \left(\sum_{s=1}^l l_s^* \right)^n \leq 2^n c^n l^{n-1} \sum_{s=1}^l l_s^{*n}.$$

Since $lr \leq m_1(F_k) + \eta$ and since $\Omega_n l_s^{*n} \leq m(W_s)$, this implies

$$\mathcal{H}_t^1(hF_k)^n \leq 2^n c^n \Omega_n^{-1} r^{1-n} (m_1(F) + \eta)^{n-1} \sum_{s=1}^l m(W_s).$$

Set $B = B^{n-1}(z, r)$. Then each W_s is contained in $U \cap f^{-1}(B \times J_0)$, which gives $\sum m(W_s) \leq 2m(U \cap f^{-1}(B \times J_0)) = 2\varphi(B)$. Thus

$$\mathcal{H}_t^1(hF_k)^n \leq C(m_1(F) + \eta)^{n-1} \frac{\varphi(B)}{m_{n-1}(B)},$$

where $C = 2^{n+1}c^n \Omega_{n-1}/\Omega_n$. Letting first $r \to 0$, then $\eta \to 0$, then $t \to 0$, and lastly $k \to \infty$ we discover that

$$\left(\sum_{i=1}^{p} \mathcal{H}^1(hF^i)\right)^n = \mathcal{H}^1(hF)^n \le Cm_1(F)^{n-1}\varphi'(z).$$

Since $|h(y_i) - h(\bar{y}_i)| \le \mathcal{H}^1(hF^i)$, it follows by a simple limiting argument that h is absolutely continuous.

Let $\beta\colon \bar{J}_0 \to \bar{J}$ be the path $\beta(t) = (z, t)$. By 3.4 there exist m liftings $\alpha_j\colon \bar{J}_0 \to U$, $j = 1, \ldots, m$, of β such that if $x \in U \cap f^{-1}\bar{J}$ and $t = \beta^{-1}(f(x))$, then $\alpha_j(t) = x$ for exactly $i(x, f)$ paths α_j. By applying the above to $h = \alpha_j \circ \beta^{-1}$ we see that each map $\alpha_j \circ \beta^{-1}$ is absolutely continuous. Since

$$g_U(y) = \frac{1}{m}\sum_{j=1}^{m} \alpha_j \circ \beta^{-1}(y)$$

for every $y \in \bar{J}$, $g_U|\bar{J}$ is absolutely continuous. The lemma is proved. $\square$

6. Characterizations of Quasiregularity

It is convenient to interrupt the proof of Poletskiĭ's lemma and give a couple of characterizations of qr mappings. For this we need the theorem of Rademacher–Stepanov, the proof of which can be found in [V4, 29.1].

6.1. Theorem of Rademacher–Stepanov. *Let A be open in $\mathbb{R}^n$ and let $\varphi\colon A \to \mathbb{R}^m$ be a mapping with the following properties:*

(1) φ *is continuous.*
(2) *The partial derivatives of φ exist a.e.*
(3) $L(x, \varphi) < \infty$ *a.e. (see (2.1)).*

Then φ is differentiable a.e.

We first characterize qr mappings by means of the linear dilatation as follows.

6.2. Theorem. *A nonconstant mapping $f\colon G \to \mathbb{R}^n$ is qr if and only if it satisfies the following conditions:*

(1) f *is sense–preserving, discrete, and open.*
(2) $H(x, f)$ *is locally bounded in G.*
(3) *There exists $a < \infty$ such that $H(x, f) \le a$ for almost every $x \in G \setminus B_f$.*

Proof. If f is qr, then (1)–(3) follow from I.4.1, I.4.5, and 4.3. Suppose conversely that f satisfies (1)–(3).

Let $Q \subset\subset G$ be an open n-interval and suppose $Q = Q_0 \times J_0$ where Q_0 is an $(n-1)$-interval in $\mathbb{R}^{n-1}$ and $J_0 =]a, b[\subset \mathbb{R}^1$. For each Borel set $A \subset Q_0$ we set $\varphi(A) = m(f(A \times J_0))$. Then φ is a bounded nonnegative q-quasiadditive set function in Q_0, where $q = N(f, Q)$. This means that (a) $A \subset B$ implies $\varphi(A) \leq \varphi(B)$ and (b) if $A_1, \ldots, A_k$ are disjoint Borel sets in a Borel set $A \subset Q_0$, then

$$\sum_{i=1}^{k} \varphi(A_i) \leq q\,\varphi(A).$$

It follows that the upper derivative $\overline{\varphi}'(z)$ is finite for almost every $z \in Q_0$. The proof of this fact is completely analogous to that of the theorem of Lebesgue–Banach. We fix such a $z \in Q_0$ and set $J = \{z\} \times J_0$. By a simplified version of the argument in the proof of 5.3 (cf. also [V4, 31.2] and [MRV1, 4.11]) we obtain

$$(6.3) \qquad \mathcal{H}^1(fF)^n \leq q\,C_1\overline{\varphi}'(z)m_1(F)^{n-1}$$

for every compact $F \subset J$. Here $C_1 = 2^{n+1}c_1^n \Omega_{n-1}/\Omega_n$ with $c_1 = \sup\{H(x, f) : x \in Q\}$. Let F^i, $i = 1, \ldots, p$, be disjoint closed subintervals of J. We have $\sum d(fF^i) \leq \sum \mathcal{H}^1(fF^i) \leq q\mathcal{H}^1(f(\bigcup F^i))$, and if we apply (6.3) to $F = F^1 \cup \cdots \cup F^p$, we obtain

$$(6.4) \qquad \left(\sum_{i=1}^{p} d(fF^i)\right)^n \leq q^{n+1}C_1\overline{\varphi}'(z)\left(\sum_{i=1}^{p} m_1(F^i)\right)^{n-1}.$$

Hence f is ACL.

If for Borel sets $A \subset G$ we set $\mu_f(A) = m(fA)$, the upper derivative $\overline{\mu}_f'(x)$ is finite for almost every $x \in G$. Let x be such a point and let $H(x, f) < c$. There exists $r_0 > 0$ such that $L(x, r) \leq cl(x, r)$ and thus $\Omega_n L(x, r)^n \leq c^n m(f\overline{B}^n(x, r))$ for $0 < r \leq r_0$. This gives $L(x, f) \leq c\overline{\mu}_f'(x)^{1/n} < \infty$. Since the partial derivatives of an ACL mapping exist a.e., 6.1 implies that f is differentiable a.e. At a point of differentiability x, $|f'(x)|^n \leq H(x, f)^{n-1}J_f(x)$. Let $D \subset\subset G$ be a domain and $c = \sup\{H(x, f) : x \in D\}$. Then $|f'(x)|^n \leq c^{n-1}J_f(x)$ a.e. in D. By I.4.11,

$$\int_D |D_if|^n \, dm \leq \int_D |f'(x)|^n \, dx \leq c^{n-1}\int_D J_f \, dm$$

$$\leq c^{n-1}N(f, D)m(fD) < \infty,$$

which shows that f is ACL^n. Hence $f|D$ is qr. Since $|f'(x)|^n \leq a^{n-1}J_f(x)$ a.e. in $G \setminus B_f$, I.4.13 implies that $K_O(f|D) \leq a^{n-1}$. We conclude that f is qr and $K_O(f) \leq a^{n-1}$. $\qquad\square$

6.5. Corollary. If $f: G \to G'$ is qc, then f^{-1} is qc, and $K_O(f^{-1}) = K_I(f)$ and $K_I(f^{-1}) = K_O(f)$.

Proof. Since $H(y, f^{-1}) = H^*(f^{-1}(y), f)$, 4.4 and 6.2 tell us that $g = f^{-1}$ is qc. By the differentiability a.e. of f and g and by the application of condition (N) to f we get $|g'(y)|^n / K_I(f) \le J_g(y) \le K_O(f)\ell(g'(y))^n$ for almost every $y \in G'$. This establishes the last part of the corollary.   □

6.6. Remark. Since we have defined also qc mappings by the so–called analytic definition and not by the geometric definition (i.e., by the quasi–invariance of the modulus of path families as is done in [V4]), we have found it appropriate to include a statement like 6.5 here. With the geometric definition 6.5 is immediate [V4, 13.2]. The equivalence of the two definitions is given in [V4, 34.6].

We shall give another characterization of quasiregularity by means of the K_O-inequality 2.4 in the following form.

6.7. Theorem. Let $f: G \to \mathbb{R}^n$ be a nonconstant mapping and let $1 \le K < \infty$. Then f is qr and $K_O(f) \le K$ if and only if the following holds:

(1) f is sense–preserving, discrete, and open.
(2) $M(\Gamma) \le K N(f, A) M(f\Gamma)$ for all path families Γ in A and all Borel sets $A \subset G$ with $N(f, A) < \infty$.

Proof. If f is qr and $K_O(f) \le K$, (1) and (2) follow from I.4.1, I.4.5, and 2.4. Suppose (1) and (2) hold. Then 4.2 yields $H(x, f) \le C$, where C depends only on n and the product $Ki(x, f)$. Theorem 6.2 implies that f is qr. If f is injective in a domain D, then $K_O(f|D) \le K$ by the argument for qc mappings in [V4, 36.1]. The condition $K_O(f) \le K$ follows then from I.4.13. The theorem is proved.   □

6.8. Theorem. If $f: G \to G'$ and $g: G' \to \mathbb{R}^n$ are qr, then $g \circ f$ is qr and $K_O(g \circ f) \le K_O(g)K_O(f)$, $K_I(g \circ f) \le K_I(g)K_I(f)$.

Proof. We may assume that f and g are nonconstant. Then $g \circ f$ is sense–preserving, discrete, and open. Let $D \subset\subset G$ be a domain. If Γ is a family of paths in a Borel set $A \subset D$, 6.7 gives

$$M(\Gamma) \le K_O(f)N(f, D)M(f\Gamma) \le K_O(f)K_O(g)N(f, D)N(g, fD)M(gf\Gamma)$$
$$\le c N(g \circ f, A)M(gf\Gamma)$$

with $c = K_O(f)K_O(g)N(f, D)N(g, fD) < \infty$, since $N(g \circ f, A) \ge 1$. By 6.7, $g \circ f|D$ is qr. Let $V \subset G \setminus B_{g \circ f} \subset G \setminus B_f$ be a domain where f is injective. Since $(f|V)^{-1}$ satisfies the condition (N) by 6.5 and I.4.14, it follows that for almost every point $x \in V$ we have f differentiable at x and g differentiable at $f(x)$. At such a point $|(g \circ f)'(x)|^n \le K_O(g)K_O(f)J_{g \circ f}(x)$

and $J_{g \circ f}(x) \leq K_I(g) K_I(f) \ell ((g \circ f)'(x))^n$. By I.4.13, $g \circ f | D$ then satisfies the required dilatation inequalities. The theorem follows.   □

6.9. Notes. The original idea of the argument of the ACL proofs in 5.3 and 6.2 is due to F.W. Gehring [G1]. The idea of using a function as defined in (5.2) is due to O. Martio [M1]. Theorem 6.2 is from [MRV1, 4.13]. In [MRV1] the ACL^n property is also obtained directly from (6.4) using an argument in [Ag] due to S. Agard. A characterization of quasiregularity by means of the inverse dilatation is given in [MRV1, 4.14].

7. Proof of Poletskiĭ's Lemma

Let $f \colon G \to \mathbb{R}^n$ be a nonconstant K-qr mapping. Recall from (5.2) the mapping g_U.

7.1. Lemma. *The mapping g_U is ACL^n.*

Proof. By 5.3 it suffices to prove that the partial derivatives $D_k g_U$ are in L^n. If $y_0 \in V' = V \smallsetminus f(B_f \cap U)$, then $f^{-1}(y_0) \cap U$ consists of m points $x_1, \ldots, x_m$. Let $U_1, \ldots, U_m$ be disjoint normal neighborhoods of the points $x_1, \ldots, x_m$. Then f is injective in each U_j. Furthermore, if W is a domain with $y_0 \in W \subset fU_1 \cap \cdots \cap fU_m$, then by 6.5 there exist K-qc inverse mappings $h_j \colon W \to f^{-1} W \cap U_j$, $j = 1, \ldots, m$. Therefore there exists a sequence of disjoint open cubes $Q_i \subset V'$, $i = 1, 2, \ldots$, whose closures cover V' such that each $f^{-1} Q_i \cap U$ has exactly m components D_{ij}, $j = 1, \ldots, m$, and f gives rise to K-qc inverse mappings $h_{ij} \colon Q_i \to D_{ij}$, $j = 1, \ldots, m$. Since for $y \in Q_i$

$$g_U(y) = \frac{1}{m} \sum_{j=1}^{m} h_{ij}(y) \, ,$$

Hölder's inequality insures that for almost every $y \in Q_i$

$$|D_k g_U(y)|^n \leq |g'_U(y)|^n \leq \left(\frac{1}{m} \sum_{j=1}^{m} |h'_{ij}(y)| \right)^n \leq \frac{1}{m} \sum_{j=1}^{m} |h'_{ij}(y)|^n$$

$$\leq \frac{K}{m} \sum_{j=1}^{m} J_{h_{ij}}(y) \, .$$

Since $m(fB_f) = 0$ by I.4.14(b), we get

$$\int_V |D_k g_U|^n dm = \sum_i \int_{Q_i} |D_k g_U|^n dm \leq \frac{K}{m} \sum_{i,j} \int_{Q_i} J_{h_{ij}} dm = \frac{K}{m} \sum_{i,j} m(h_{ij} Q_i)$$

$$= \frac{K}{m} \sum_i m(f^{-1} Q_i \cap U) \leq \frac{K}{m} m(U) < \infty \, .$$ □

We now fix a domain $D \subset\subset G$ and set $B_k = \{ x \in D : i(x,f) = k \}$, $k \geq 1$. We choose disjoint open cubes Q_j and larger open cubes Q'_j, $j \in \mathbb{N}$, such that $\overline{Q}_j \subset Q'_j \subset D \setminus B_f$, the cubes $\overline{Q}_j$ cover $D \setminus B_f$, $f|Q'_j$ is injective, and $m(Q'_j) \leq 2m(Q_j)$. The inverse mappings $h_j = (f|Q'_j)^{-1} \colon fQ'_j \to Q'_j$ are K-qc. Let A_j be the set of points $y \in fQ'_j$ where all partial derivatives of h_j – hence, the formal derivative $h'_j(y)$ – are defined. Then A_j is a Borel set and $m(fQ'_j \setminus A_j) = 0$. If we set $h'_j(y) = 0$ for $y \in fQ'_j \setminus A_j$, we get a Borel function $|h'_j| \colon fQ'_j \to \mathbb{R}^1$. With these we define the Borel function

$$\rho = \sup\{ |h'_j| \chi_{fQ'_j} : j \in \mathbb{N} \} .$$

For each point $x \in B_k$ there exists a normal neighborhood $U \subset D$ of x. We cover B_k by such normal neighborhoods U_{ki}, $i \in \mathbb{N}$, and let g_{ki} be the mapping $g_{U_{ki}}$ defined in (5.2). We fix a Borel set $F \subset \mathbb{R}^n$ of zero measure which contains all the points where at least one of the h_j is not differentiable and which also contains the set $f(C \cup B_f)$, C being the set of points where f itself is not differentiable.

7.2. Lemma. *Let Γ_0 be a family of closed paths γ in D such that either $f \circ \gamma$ is unrectifiable or $f \circ \gamma$ is rectifiable and at least one of the following conditions is not true:*

(1) $\displaystyle\int_{f \circ \gamma} \chi_F \, ds = 0 .$

(2) $\displaystyle\int_{f \circ \gamma} \rho \, ds < \infty .$

(3) *If α is a closed subpath of γ and if $|\alpha| \subset Q'_j$, h_j is absolutely continuous on $f \circ \alpha$.*

(4) *If α is a closed subpath of γ and if $|\alpha| \subset U_{ki}$, g_{ki} is absolutely continuous on $f \circ \alpha$.*

Then $\mathsf{M}(f\Gamma_0) = 0$.

Proof. Since the family of nonrectifiable paths in $\mathbb{R}^n$ is of modulus zero [V4, 6.10], we may assume that $f \circ \gamma$ is rectifiable for all $\gamma \in \Gamma_0$. Let Γ_q, $q = 1, \dots, 4$, be the family of paths $\gamma \in \Gamma_0$ for which the condition (q) is not true. It is enough to show that $\mathsf{M}(f\Gamma_q) = 0$ for all q. $\mathsf{M}(f\Gamma_1) = 0$ follows from [V4, 33.1]. We have

(7.3)
$$\int_{\mathbb{R}^n} \rho^n dm \leq \sum_j \int_{fQ'_j} |h'_j|^n dm \leq K \sum_j \int_{fQ'_j} J_{h_j} dm$$
$$\leq 2K \sum_j m(Q_j) \leq 2K m(D) < \infty .$$

Since $\rho/\nu \in \mathcal{F}(f\Gamma_2)$ and since, as a result,

$$\mathsf{M}(f\Gamma_2) \leq \nu^{-n} \int_{\mathbb{R}^n} \rho^n dm$$

holds for all $\nu \in \mathbb{N}$, $\mathrm{M}(f\Gamma_2) = 0$ follows from (7.3). Since h_j and g_{ki} are ACL^n, $\mathrm{M}(f\Gamma_3) = 0 = \mathrm{M}(f\Gamma_4)$ follows from Fuglede's theorem 2.3.   □

Proof of 5.1. Let D be as above and let $\gamma\colon I \to D$ be a closed path such that $f \circ \gamma$ is rectifiable and the conditions (1)–(4) in Lemma 7.2 are satisfied. We claim that f is absolutely precontinuous on γ.

The conditions (1)–(4) are independent of the parameter representation of γ. We can therefore assume that the f-representation γ^* of γ coincides with γ. To prove that γ is absolutely continuous, we first show that each coordinate function γ_m of γ satisfies the condition (N). Let $A \subset I$ be a set with $m_1(A) = 0$. We can cover $I \smallsetminus \gamma^{-1}B_f$ by a family $\{I_\mu : \mu \in \mathbb{N}\}$ of closed intervals with disjoint interiors in $I \smallsetminus \gamma^{-1}B_f$ such that γI_μ is contained in some Q'_{j_μ}, $\mu \in \mathbb{N}$. By (3), each $\gamma|I_\mu = h_{j_\mu} \circ f \circ \gamma|I_\mu$ is absolutely continuous. Hence $m_1\big(\gamma_m(A \cap (I \smallsetminus \gamma^{-1}B_f))\big) = 0$. The condition (1) implies that for almost every $v \in I$, $f \circ \gamma(v) \notin fB_f \subset F$, hence $m_1(\gamma^{-1}B_f) = 0$. For $u \in \gamma^{-1}(B_k \cap U_{ki})$ we have $\gamma(u) = g_{ki} \circ f \circ \gamma(u)$. Therefore by (4), $m_1\big(\gamma_m(\gamma^{-1}(B_k \cap U_{ki}))\big) = 0$ if $k \geq 2$. But

$$\gamma^{-1}B_f = \bigcup_{k \geq 2} \bigcup_i \gamma^{-1}(B_k \cap U_{ki}),$$

hence $m_1\big(\gamma_m(\gamma^{-1}B_f)\big) = 0$. It follows that $m_1(\gamma_m A) = 0$, so γ_m satisfies the condition (N).

We infer from (1) that for almost every $t \in I_\mu$, $f \circ \gamma(t) \in fQ'_{j_\mu} \smallsetminus F$ and $\gamma'(t) = h'_{j_\mu}\big(f(\gamma(t))\big)(f \circ \gamma)'(t)$. Since $\gamma^* = \gamma$, $|(f \circ \gamma)'(t)| = 1$ for almost every $t \in I$, and with (2) we get

$$\int_I |\gamma'_m|\, dm_1 \leq \int_I |\gamma'|\, dm_1 \leq \sum_\mu \int_{I_\mu} |h'_{j_\mu}\big(f(\gamma(t))\big)|\, dt \leq \int_{f \circ \gamma} \rho\, ds < \infty.$$

A function which satisfies the condition (N) and has an integrable derivative is by Bary's theorem [S, p. 285] absolutely continuous. We apply this to each γ_m and deduce that γ is absolutely continuous.

To complete the proof let $D_1 \subset D_2 \subset \ldots$ be a sequence of domains $D_i \subset\subset G$ which exhaust G. Let Γ_i be the family of closed paths α in D_i such that $f \circ \alpha$ is rectifiable and f is not absolutely precontinuous on α. Then, by the first part of the proof and by Lemma 7.2, $\mathrm{M}(f\Gamma_i) = 0$. Since $f\Gamma$ is minorized by the union of the families $f\Gamma_i$, $\mathrm{M}(f\Gamma) = 0$ and 5.1 is proved.   □

As an application of Poletskiĭ's lemma we now show that the branch set of a nonconstant qr mapping is of measure zero and that the Jacobian determinant is positive a.e.

7.4. Theorem. *If* $f: G \to \mathbb{R}^n$ *is a nonconstant qr mapping, then*

(1) $J_f > 0$ *a.e.*,

(2) $m(B_f) = 0$, *and*

(3) *for any measurable set* $E \subset G$, $m(E) = 0$ *if and only if* $m(fE) = 0$.

Proof. To prove (1) suppose $J_f(x) = 0$ for all $x \in E$, where E is a set with $m(E) > 0$. There exists by I.2.4 a closed n-interval $Q \subset G$ and a Borel set $B \subset E \cap Q$ such that $m(B) > 0$ and f is differentiable with $f'(x) = 0$ at every point x in B. Write $Q = I \times Q_0$, where $Q_0 \subset \mathbb{R}^{n-1}$ and $I \subset \mathbb{R}^1$. Let Γ be the family of paths $\gamma_z: I \to Q$, $\gamma_z(t) = (t, z)$, $z \in Q_0$, for which

$$\int_{\gamma_z} \chi_B \, ds > 0 .$$

Since $m(B) > 0$, Fubini's theorem implies that there exists a Borel set $A \subset Q_0$ with $m_{n-1}(A) > 0$ and $\gamma_z \in \Gamma$ for all $z \in A$. By 1.7(a), $\mathrm{M}(\Gamma) > 0$. The K_O-inequality 2.4 then gives $0 < \mathrm{M}(\Gamma) \le K_O(f)N(f,Q)\mathrm{M}(f\Gamma)$. Since $N(f, Q) < \infty$, Poletskiĭ's lemma 5.1 shows that there exists $\gamma \in \Gamma$ such that $f \circ \gamma$ is rectifiable and the f-representation γ^* of γ is absolutely continuous. Then

$$0 < \int_{\gamma} \chi_B \, ds = \int_0^{\ell(f \circ \gamma)} (\chi_B \circ \gamma^*)|\gamma^{*\prime}| \, dm_1 = \int_{\gamma^{*-1}B} |\gamma^{*\prime}| \, dm_1 ,$$

so $m_1(\gamma^{*-1}B) > 0$. For almost every $t \in \gamma^{*-1}B$,

$$1 = |(f \circ \gamma)^{0\prime}(t)| = |(f \circ \gamma^*)'(t)| = |f'(\gamma^*(t))\gamma^{*\prime}(t)| = 0 ,$$

a contradiction. This proves (1).

(2) follows from (1), I.2.4, and I.4.11. To prove (3), let $E \subset G$ be measurable. If $m(E) = 0$, $m(fE) = 0$ by I.4.14(a). Suppose $m(fE) = 0$. We express $G \subset B_f$ as a countable union of domains D_i where f is injective. By 6.5, $(f|D_i)^{-1}$ is qc and thus satisfies condition (N) by I.4.14(a). This and (2) then give $m(E) = 0$. $\square$

7.5. Notes. Lemma 5.1 is a slight extension of [P1, Lemma 6]. The proof presented here is a simplification of Poletskiĭ's original one in two respects. First, the path lifting result 3.4 employed in the proof of Lemma 5.3 is given a purely topological proof. Second, by the use of Bary's theorem in the final step we are able to avoid the rather cumbersome concept in [P1] of rectifiability of a path on subsets of the parameter interval. A large part of this section is taken with minor changes from an unpublished paper [Pe1] by M. Pesonen. It was also Pesonen's idea to prove 7.4 right after Poletskiĭ's lemma. The above proof is taken from [Pe3]. Originally 7.4 was proved in [MRV1, 8.2–8.4] by means of the capacity inequality 10.9.

8. Poletskiĭ's Inequality

With the help of Poletskiĭ's lemma 5.1 we are now able to prove the following important inequality by Poletskiĭ [P1], a result also known as the K_I-
inequality.

8.1. Theorem. *Let* $f\colon G \to \mathbb{R}^n$ *be a nonconstant qr mapping and let* Γ *be
a family of paths in* G. *Then*

$$(8.2) \qquad\qquad \mathsf{M}(f\Gamma) \leq K_I(f)\mathsf{M}(\Gamma) .$$

Proof. Let E be the set of points $x \in G$ at which f is differentiable with
$J_f(x) > 0$. By I.2.4, I.4.11, I.4.14, and 7.4 we have $m(G\smallsetminus E) = m(f(G\smallsetminus E)) = 0$ and $B_f \subset G \smallsetminus E$. Let $B \supset f(G \smallsetminus E)$ be a Borel set of measure zero. By
Poletskiĭ's lemma 5.1 and [V4, 33.1] we may assume that for every $\gamma \in \Gamma$

  (1) $f \circ \gamma$ is locally rectifiable,
  (2) f is locally absolutely precontinuous on γ,
  (3) $\displaystyle\int_{f\circ\gamma} \chi_B\, ds = 0$.

Let $\rho \in \mathcal{F}(\Gamma)$. We define a Borel function $\sigma\colon G \to [0,\infty]$ by

$$\sigma(x) = \begin{cases} \rho(x)/\ell(f'(x)) & \text{if } x \in G \smallsetminus f^{-1}B, \\ 0 & \text{if } x \in f^{-1}B, \end{cases}$$

and a function $\rho'\colon \mathbb{R}^n \to [0,\infty]$ by

$$\rho'(y) = \sup_{x \in f^{-1}(y)} \sigma(x)\chi_{fG}(y) .$$

We shall prove that $\rho' \in \mathcal{F}(f\Gamma)$. To show that ρ' is a Borel function,
let $G_1 \subset G_2 \subset \dots$ be a sequence of relatively compact subdomains of G
which exhaust G. Set $\rho_i = \rho\chi_{\overline{G}_i}$, $\sigma_i = \sigma\chi_{\overline{G}_i}$, $\rho_i'(y) = \sup\{\sigma_i(x) : x \in f^{-1}(y)\}\chi_{fG}(y)$. Then $\rho_i' \to \rho'$ and $\rho_i'(y) = 0$ if $y \in \mathbb{R}^n \smallsetminus f\overline{G}_i \cup f(\overline{G}_i \cap B_f)$.
Hence it suffices to show that every point $y_0 \in f\overline{G}_i \smallsetminus f(\overline{G}_i \cap B_f)$ has a
neighborhood where ρ_i' is a Borel function. To prove this let $U_1, \dots, U_k$ be
disjoint neighborhoods in $G \smallsetminus B_f$ of the points in $f^{-1}(y_0) \cap \overline{G}_i$ and let $f|\overline{U}_j$
be injective for each j. Then

$$V_0 = \left(\bigcap_{j=1}^{k} fU_j\right) \smallsetminus f\left(\overline{G}_i \smallsetminus \bigcup_{j=1}^{k} U_j\right)$$

is a neighborhood of y_0. Let $V \subset V_0$ be a connected neighborhood of y_0.
Let D be a component of $f^{-1}V$ which meets $\overline{G}_i$. Then D meets some U_i.
Since $f|\overline{U}_i$ is injective, we have $V_0 \cap f\partial U_i = \emptyset$ and hence $D \cap \partial U_i = \emptyset$.
This implies $D \subset U_i$. Therefore the components of $f^{-1}V$ which meet $\overline{G}_i$
consist of domains $D_j \subset U_j$, $j = 1, \dots, k$, and f defines homeomorphisms
$f_j\colon D_j \to V$. If $g_j = f_j^{-1}$,

$$\rho_i'(y) = \sup_{j=1,\dots,k} \sigma_i(g_j(y))$$

for $y \in V$. Since $\sigma_i \circ g_j$ is a Borel function, so is $\rho_i'|V$.

Let $\gamma \in \Gamma$, let $\alpha: I \to G$ be a closed subpath of γ, and let α^* be its f-representation. By (3),

$$\int_J \chi_B(f \circ \alpha^*(t))\, dt = 0\,,$$

where $J = s_{f \circ \alpha} I$. Hence for almost every $t \in J$, $f \circ \alpha^*(t) \in fG \setminus B$. Therefore for almost every $t \in J$,

$$1 = |(f \circ \alpha^*)'(t)| = |f'(\alpha^*(t))\alpha^{*\prime}(t)| \geq \ell\big(f'(\alpha^*(t))\big)|\alpha^{*\prime}(t)|$$

and

$$\sigma(\alpha^*(t)) = \frac{\rho(\alpha^*(t))}{\ell(f'(\alpha^*(t)))} \geq \rho(\alpha^*(t))|\alpha^{*\prime}(t)|\,.$$

Since α^* is absolutely continuous, this implies

$$\int_\alpha \rho\, ds = \int_J (\rho \circ \alpha^*)|\alpha^{*\prime}|\, dm_1 \leq \int_J \sigma \circ \alpha^*\, dm_1$$

$$\leq \int_J \rho' \circ f \circ \alpha^*\, dm_1 = \int_{f \circ \alpha} \rho'\, ds\,.$$

Taking the supremum over such α shows that $\rho' \in \mathcal{F}(f\Gamma)$.

Let $0 \leq \eta_1 \leq \eta_2 \leq \dots$ be a sequence of Borel functions converging to ρ' such that $0 < \eta_j(y) < \rho'(y)$ if $0 < \rho'(y)$. If $y \in fG$ and $j \in \mathbb{N}$, we have by the definition of ρ' that $\sigma(x) \geq \eta_j(f(x))$ for some $x \in G$. It follows that the set $P_j = \{\, x \in G : \sigma(x) \geq \eta_j \circ f(x)\,\}$ meets $f^{-1}(y)$, and thus that $N(y, f, P_j) \geq 1$. By I.4.14 and by the definition of σ we get

$$\int_{\mathbb{R}^n} \eta_j^n\, dm \leq \int_{\mathbb{R}^n} \eta_j^n(y) N(y, f, P_j)\, dy = \int_{P_j} (\eta_j \circ f)^n J_f\, dm$$

$$\leq \int_{P_j} \sigma^n J_f\, dm \leq K_I(f) \int_{\mathbb{R}^n} \rho^n dm\,.$$

The monotone convergence theorem then implies the result. □

8.3. Remarks. The proof of Poletskiĭ's inequality 8.1 that we presented here is a modified version of the one in [P1]. We have closely followed the arguments in [V5] and [Pe1]. The inequality $N(y, f, P_j) \geq 1$ in the proof is in general very crude. In the next section we shall prove Väisälä's inequality which takes this into account. For $n = 2$ and $K = 1$ inequality (8.2) appeared in [O, p. 80].

9. Väisälä's Inequality

In this section we shall prove a generalization of Poletskiĭ's inequality by J. Väisälä [V5, 3.1]. Väisälä's inequality has turned out to be an essential tool in the study of value distribution of qr mappings. We shall invoke it often in Chapters IV and V.

9.1. Theorem. *Let* $f: G \to \mathbb{R}^n$ *be a nonconstant qr mapping,* Γ *be a path family in* G *,* Γ' *be a path family in* $\mathbb{R}^n$ *, and* m *be a positive integer such that the following is true. For every path* $\beta: I \to \mathbb{R}^n$ *in* Γ' *there are paths* $\alpha_1, \ldots, \alpha_m$ *in* Γ *such that* $f \circ \alpha_j \subset \beta$ *for all* j *and such that for every* $x \in G$ *and* $t \in I$ *the equality* $\alpha_j(t) = x$ *holds for at most* $i(x, f)$ *indices* j *. Then*

$$\mathsf{M}(\Gamma') \leq \frac{K_I(f)}{m} \mathsf{M}(\Gamma) .$$

Proof. Let B be as in the proof of 8.1. Now we can assume that for every $\beta \in \Gamma'$

(a) β is locally rectifiable,
(b) if α is a path in G with $f \circ \alpha \subset \beta$, then f is locally absolutely precontinuous on α,
(c) $\displaystyle\int_\beta \chi_B \, ds = 0$.

Let $\rho \in \mathcal{F}(\Gamma)$ and let σ be as in the proof of 8.1. This time we define $\rho': \mathbb{R}^n \to [0, \infty]$ by

$$\rho'(y) = \frac{1}{m} \chi_{fG}(y) \sup_C \sum_{x \in C} \sigma(x)$$

where C runs over all subsets of $f^{-1}(y)$ such that $\operatorname{card} C \leq m$. We want to prove $\rho' \in \mathcal{F}(\Gamma')$. That ρ' is a Borel function is shown as the analogous fact was in the proof of 8.1.

Suppose that $\beta: I_0 \to \mathbb{R}^n$ is a closed path in Γ' . There exist paths $\alpha_1, \ldots, \alpha_m$ in Γ such that $f \circ \alpha_j \subset \beta$ and $\operatorname{card}\{j : \alpha_j(t) = x\} \leq i(x, f)$ for all $t \in I_0$ and $x \in G$. Let $c = \ell(\beta)$ and let $\alpha_j^*: I_j \to G$ be the f-representation of α_j with respect to β . Thus $\alpha_j(t) = \alpha_j^* \circ s_\beta(t)$ and $f \circ \alpha_j^* \subset \beta^0$.

As in the proof of 8.1 we get

$$1 \leq \int_{\alpha_j} \rho \, ds \leq \int_{I_j} \sigma \circ \alpha_j^* \, dm_1 .$$

Set $h_j(t) = \sigma(\alpha_j^*(t))\chi_{I_j}(t)$ for $t \in [0, c]$ and $J_t = \{j : t \in I_j\}$. We infer from (c) that for almost every $t \in [0, c]$ the points $\alpha_j^*(t)$, $j \in J_t$, are dictinct points in $f^{-1}(\beta^0(t))$, so

$$\rho'(\beta^0(t)) \geq m^{-1} \sum_{j=1}^{m} h_j(t) \,.$$

This gives

$$\int_{\beta}^{c} \rho' ds = \int_{0}^{c} \rho' \circ \beta^0(t)\, dt \geq \frac{1}{m} \sum_{j=1}^{m} \int_{0}^{c} h_j(t)\, dt = \frac{1}{m} \sum_{j=1}^{m} \int_{I_j} \sigma \circ \alpha_j^*\, dm_1 \geq 1 \,.$$

If β is not closed, we draw the same conclusion by taking a supremum over closed subpaths of β. We have proved that $\rho' \in \mathcal{F}(\Gamma')$.

Next we choose an exhaustion (G_i) of G and functions $\rho_i = \rho \chi_{\overline{G}_i}$, $\sigma_i = \sigma \chi_{\overline{G}_i}$, and $\rho'_i = \rho' \chi_{\overline{G}_i}$ just as we did in the proof of 8.1. Fix i. As in the aforementioned proof, let $y_0 \in f\overline{G}_i \smallsetminus f(\overline{G}_i \cap B_f)$, and let V be a connected neighborhood of y_0 such that we have k qc mappings $g_\mu \colon V \to D_\mu$ with $f \circ g_\mu = \mathrm{id}$ and $\overline{G}_i \cap f^{-1}V = \bigcup\{\overline{G}_i \cap D_\mu : 1 \leq \mu \leq k\}$. For each $y \in V$ we define a set $L_y \subset P = \{1,\ldots,k\}$ as follows. If $k \leq m$, then $L_y = P$. If $k > m$, then $\operatorname{card} L_y = m$, and for all $\mu \in L_y$, $\nu \in P \smallsetminus L_y$, either $\sigma_i(g_\mu(y)) > \sigma_i(g_\nu(y))$ or $\sigma_i(g_\mu(y)) = \sigma_i(g_\nu(y))$ and $\mu > \nu$. Then

$$\rho'_i(y) = \frac{1}{m} \sum_{\mu \in L_y} \sigma_i(g_\mu(y))$$

for $y \in V$. For $L \subset P$ the sets $V_L = \{y \in V : L_y = L\}$ are disjoint Borel sets. From Hölder's inequality, I.4.14, and the quasiconformality of the mappings $f|D_\mu$, we get

$$\int_{V_L} \rho'^n_i\, dm \leq \frac{1}{m} \sum_{\mu \in L} \int_{V_L} (\sigma_i \circ g_\mu)^n\, dm$$

$$= \frac{1}{m} \sum_{\mu \in L} \int_{g_\mu V_L} \sigma_i^n J_f\, dm \leq \frac{K_I(f)}{m} \int_{f^{-1}V_L} \rho_i^n\, dm \,.$$

Summing over all $L \subset P$ yields

$$\int_{V} \rho'^n_i\, dm \leq \frac{K_I(f)}{m} \int_{f^{-1}V} \rho_i^n\, dm \,.$$

The set $f\overline{G}_i \smallsetminus f(\overline{G}_i \cap B_f)$ can be almost covered by a countable number of disjoint sets V as described above. Hence $m(fB_f) = 0$ implies

$$\int_{\mathbb{R}^n} \rho'^n_i\, dm \leq \frac{K_I(f)}{m} \int_{\mathbb{R}^n} \rho_i^n\, dm \,.$$

Letting $i \to \infty$, we obtain the theorem. $\square$

9.2. Corollary. *Let* f *be as in 9.1,* D *be a normal domain for* f, Γ' *be a family of paths in* fD, Γ *be the family of paths* α *in* D *such that* $f \circ \alpha \in \Gamma'$, *and* $m = N(f, D)$. *Then*

$$\mathsf{M}(\Gamma') \leq \frac{K_I(f)}{m} \mathsf{M}(\Gamma) .$$

Proof. The condition in 9.1 is satisfied by 3.4.   □

9.3. Remarks. 1. Modulo slight modifications Theorem 9.1 and its proof are from [V5, 3.1]. If in 9.1 we take $\Gamma' = f\Gamma$, the conditions are satisfied for $m = 1$, hence 8.1 is a corollary of 9.1. E.A. Poletskiĭ himself gave in [P1, Theorem 2] a weaker form of Corollary 9.2, namely, he assumed the paths of Γ' to be injective.

2. All inequalities 2.4, (2.6), 8.1, and 9.1 stay valid for qr mappings between Riemannian n-manifolds. The proofs involve minor reworking of the proofs in the Euclidean case. Remarks about this are made in [MaR, Section 2].

10. Capacity Inequalities

A *condenser* in $\mathbb{R}^n$ is a pair $E = (A, C)$, where A is open in $\mathbb{R}^n$ and $C \neq \emptyset$ is a compact subset of A. If $1 \leq p < \infty$, the *p-capacity* of E is defined by

$$(10.1) \qquad \mathrm{cap}_p E = \inf_u \int_A |\nabla u|^p dm ,$$

where the infimum is taken over all nonnegative functions u in $\mathrm{ACL}^p(A)$ with compact support in A and $u|C \geq 1$. We call such functions *admissible* for E. The n-capacity of E is also called its *conformal capacity* (or, for short, its *capacity*) and is denoted by $\mathrm{cap}\, E$. A condenser (A, C) is *ringlike* if C and $\mathbb{R}^n \smallsetminus A$ are connected. The set $A \smallsetminus C$ is then also called a *ring*.

We shall make several comments on the possibility to restrict the class of admissible functions in (10.1). To this end we record some simple observations about ACL^p functions. Let u and v be ACL^p. Then $|u|$ is ACL^p and $|\nabla |u|| \leq |\nabla u|$ a.e. This is so because $y \mapsto |y|$ is 1-Lipschitz (I.1.11). Accordingly $\max(u, v) = (u + v + |u - v|)/2$ and $\min(u, v) = (u + v - |u - v|)/2$ are in ACL^p. If $u|F = 0$ and F is measurable, then a density point argument gives $\nabla u(x) = 0$ a.e. in F.

The first remark is that the infimum in (10.1) would be unchanged should we replace the condition $u|C \geq 1$ by $u|C = 1$, for if u is admissible, then $v = \min(1, u) \in \mathrm{ACL}^p(A)$ and $|\nabla v| \leq |\nabla u|$ a.e.

The second remark is that the class $\mathrm{ACL}^p(A)$ can be replaced in definition (10.1) by $C_0^\infty(A)$. To see this let $0 < \varepsilon < 1$ and let u be an admissible function for E such that $\|\nabla u\|_p^p < \mathrm{cap}_p E + \varepsilon$. By I.1.4 there exists $v \in$

$\mathcal{C}_0^\infty(A)$ such that $\|\nabla v\|_p^p < \|\nabla u\|_p^p + \varepsilon$ and $v|C \geq 1 - \varepsilon$. Then $w = v/(1 - \varepsilon)$ is admissible for E and $\|\nabla w\|_p^p \leq (1 - \varepsilon)^{-p}(\|\nabla u\|_p^p + \varepsilon)$, which proves the claim. From this argument and the first remark we see that we can as well consider functions in $\mathcal{C}_0^\infty(A)$ each of which is 1 in a neighborhood of C.

A third observation is that we can restrict to functions which are monotone in $A \setminus C$. A continuous function φ in an open set U in $\mathbb{R}^n$ is called *monotone* if its oscillation satisfies $\mathrm{osc}(\varphi, \partial D) = \mathrm{osc}(\varphi, D)$ for all domains $D \subset\subset U$. Let u be admissible for E with $u|C = 1$ and let (a_i) be a sequence consisting of all rationals. We inductively define a sequence (u_i) of functions of A as follows. Let $x \in A$. If there exists a domain $D \subset A \setminus C$ for which $x \in D$ and $u|\partial D = a_1$, we set $u_1(x) = a_1$. Otherwise let $u_1(x) = u(x)$. We define u_2 in the same way by replacing u by u_1 and a_1 by a_2 etc. By a well–known argument of H. Lebesgue [L] the sequence (u_i) converges uniformly on A to a continuous function $v \colon A \to \mathbb{R}^1$ which has compact support, which satisfies $v|C = 1$, and which is monotone in $A \setminus C$. From the construction it follows that $\mathrm{osc}(v, F) \leq \mathrm{osc}(u, F)$ for every line segment $F \subset A$. Hence v is ACL and $|\nabla v| \leq |\nabla u|$ a.e. The function v is then admissible and $\|\nabla v\|_p \leq \|\nabla u\|_p$, which proves our claim. In particular, it follows that in (10.1) we need only consider functions u with $0 \leq u \leq 1$.

The connection between the conformal capacity of a ring and extremal length was given by F.W. Gehring in [G3]. The following extension to the p-capacity of a condenser was essentially proved by W.P. Ziemer in [Z1]. Our proof differs from his only in small details.

10.2. Proposition. *Let* $E = (A, C)$ *be a condenser in* $\mathbb{R}^n$ *and let* Γ_E *be the family of all paths of the form* $\gamma \colon [a, b[\to A$ *with* $\gamma(a) \in C$ *and* $|\gamma| \cap (A \setminus F) \neq \emptyset$ *for every compact* $F \subset A$. *Then* $\mathrm{cap}_p E = \mathsf{M}_p(\Gamma_E)$.

Proof. Let $u \in \mathcal{C}_0^\infty(A)$ and $u|C \geq 1$. If we set $\rho(x) = |\nabla u(x)|$ in A and $\rho|\mathbb{R}^n \setminus A = 0$, the function ρ belongs to $\mathcal{F}(\Gamma_E)$. Hence

$$\mathsf{M}_p(\Gamma_E) \leq \int\limits_A |\nabla u|^p \, dm$$

and the inequality $\mathsf{M}_p(\Gamma_E) \leq \mathrm{cap}_p E$ follows.

To prove $\mathrm{cap}_p E \leq \mathsf{M}_p(\Gamma_E)$ we may assume $\mathsf{M}_p(\Gamma_E) < \infty$. Let $\varepsilon > 0$. For each real valued $\rho' \in \mathcal{F}(\Gamma_E)$ with $\|\rho'\|_p^p < \mathsf{M}_p(\Gamma_E) + \varepsilon$ there exists by the theorem of Vitali–Carathéodory [Ru, 2.24] a lower semicontinuous function $\rho \geq \rho'$ such that $\|\rho\|_p^p < \mathsf{M}_p(\Gamma_E) + \varepsilon$. Let $A_1 \subset A_2 \subset \dots$ be a sequence of open sets which exhaust A such that each $\overline{A}_k$ is a compact subset of A and $C \subset A_1$. Set $\rho_k = \min(\rho \chi_{A_k}, k)$ and define

$$(10.3) \qquad\qquad u_k(x) = \inf_\alpha \int\limits_\alpha \rho_k \, ds$$

for $x \in A$, where the infimum is taken over all locally rectifiable paths $\alpha \colon [a, b[\to A$ with $\alpha(a) = x$ and $|\alpha| \cap (A \setminus F) \neq \emptyset$ for all compact $F \subset A$.

The function u_k is clearly a locally Lipschitz function with Lipschitz constant k. As such, u_k is in ACL^p. By the theorem of Rademacher–Stepanov 6.1, u_k is differentiable a.e. We want to show that $|\nabla u_k| \le \rho_k$ a.e. in A. At almost every point $x \in \mathbb{R}^n$, ρ_k is approximately continuous. Fix such a point x in A. Suppose, in addition, that u_k is differentiable at x. Then there exists a dense subset P of S^{n-1} such that $\rho_k|L_w$ is approximately continuous at x for each line $L_w = \{\, tw : t \in \mathbb{R}^1 \,\}$, $w \in P$. In particular,

$$(10.4) \qquad \lim_{t \to 0} \frac{1}{t} \int_0^t \rho_k(x + tw)\, dt = \rho_k(x)$$

for $w \in P$. Let $\eta > 0$ and $v \in S^{n-1}$. Choose $w \in P$ such that $|w - v| < \eta$. Then for small $t > 0$,

$$|u_k(x + tv) - u_k(x)|$$

$$\le \int_0^t \rho_k(x + tw)\, dt + \int_0^1 \rho_k\big(x + t((1 - s)w + sv)\big) t|w - v|\, ds$$

$$\le \int_0^t \rho_k(x + tw)\, dt + kt\eta \, .$$

In conjunction with (10.4) this shows that $|\nabla u_k(x)| \le \rho_k(x)$.

Let $d_k = \min\{\, u_k(x) : x \in C \,\}$. We claim that

$$(10.5) \qquad \liminf_{k \to \infty} d_k \ge 1 \, .$$

If this is not true, there exist $d < 1$ and a sequence of locally rectifiable paths $\beta_k \colon [0, b_k[\,\to A$ parametrized by arc length and such that $x_k = \beta_k(0) \in C$, $|\beta_k| \cap (A \setminus F) \ne \emptyset$ for all compact $F \subset A$, and

$$(10.6) \qquad \int_{\beta_k} \rho_k\, ds \le d \, .$$

For each A_q we define a new sequence of paths $\beta_{qk} \colon [0, \infty[\,\to \overline{A}_q$, $k = 1, 2, \ldots$, as follows. Let $t_{qk} \in \,]0, b_k[$ be the first point t for which $\beta_k(t) \in \partial A_q$. We set $\beta_{qk}|[0, t_{qk}] = \beta|[0, t_{qk}]$ and $\beta_{qk}|[t_{qk}, \infty[\,= \text{constant}$. The family of paths β_{1k}, $k = 1, 2, \ldots$, in $\overline{A}_1$ is equicontinuous, so Ascoli's theorem gives a subsequence (β_{1k_j}) of $\beta_{11}, \beta_{12}, \ldots$, which converges to a path $\beta^1 \colon [0, \infty[\,\to \overline{A}_1$. Next we obtain a subsequence of (β_{2k_j}) which converges to a path $\beta^2 \colon [0, \infty[\,\to \overline{A}_2$. We continue this process. Let us use the same notation for the sequences and their subsequences, i.e., (β_{1k_j}) is denoted by (β_{1k}) etc. Then we can form from the sequence (β_{qq}) a restricted limit path $\beta \colon [0, b[\,\to A$ with $\beta(0) \in C$ and $|\beta| \cap (A \setminus F) \ne \emptyset$ for every compact $F \subset A$. Here $b = \liminf t_{qq} \le \liminf b_q$.

Clearly β is 1-Lipschitz, hence $|\beta'| \leq 1$ a.e. Also $\beta \in \Gamma_E$. Let $d < a < 1$ and $0 < c < b$. There exists m such that

$$(10.7) \qquad \int_\beta \rho_m \, ds \geq a \, .$$

By Fatou's lemma and the lower semicontinuity of ρ_m we get

$$\liminf_{k \to \infty} \int_{\beta_k} \rho_k \, ds \geq \liminf_{k \to \infty} \int_{\beta_k} \rho_m \, ds \geq \liminf_{k \to \infty} \int_0^c \rho_m \circ \beta_k \, dm_1$$

$$\geq \int_0^c \rho_m \circ \beta \, dm_1 \geq \int_0^c \rho_m \circ \beta |\beta'| \, dm_1 \, .$$

As $c \to b$, this and inequalities (10.6) and (10.7) imply $d \geq a$, a contradiction. We have proved (10.5). We may assume $d_k > 0$ for all k.

The function $v_k = u_k/d_k$ is admissible for E. We have $|\nabla v_k| \leq \rho_k/d_k$ a.e. in A. With (10.5) this gives

$$\mathrm{cap}_p \, E \leq \liminf_{k \to \infty} d_k^{-p} \int_{\mathbb{R}^n} \rho_k^p \, dm \leq \int_{\mathbb{R}^n} \rho^p \, dm < M_p(\Gamma_E) + \varepsilon \, ,$$

and $\mathrm{cap}_p \, E \leq M_p(\Gamma_E)$ follows. $\qquad\qquad\qquad\qquad\qquad\qquad \square$

10.8. Remarks. 1. The p-capacity of a condenser in a Riemannian n-manifold is defined in a similar way. In this setting we still have $\mathrm{cap}_p \, E = M_p(\Gamma_E)$, because the proof of 10.2 can be transferred with only formal changes.

2. If C and F are disjoint compact subsets of $\overline{\mathbb{R}}^n$, then $\mathrm{cap}(\overline{\mathbb{R}}^n \setminus F, C) = M(\Delta(C, F; \overline{\mathbb{R}}^n))$. This follows from Remark 1.15.3 because $\Delta(C, F; \overline{\mathbb{R}}^n)$ is minorized by Γ_E, where $E = (\overline{\mathbb{R}}^n \setminus F, C)$.

Let $f \colon G \to \mathbb{R}^n$ be a nonconstant qr mapping and $E = (A, C)$ a condenser in G, meaning that $A \subset G$. Since f is open, $fE = (fA, fC)$ is also a condenser. The inequalities for moduli of path families imply corresponding inequalities for condensers. The K_I-inequality is valid for arbitrary condensers (Theorem 10.10), but the K_O-inequality in 10.9 is restricted to so–called normal condensers. A condenser (A, C) is a *normal condenser* for f if A is a normal domain for f.

10.9. Theorem. *Let* $f \colon G \to \mathbb{R}^n$ *be a nonconstant qr mapping and* $E = (A, C)$ *be a normal condenser in* G *with* $N(f, A) < \infty$. *Then*

$$\mathrm{cap} \, E \leq K_O(f) N(f, A) \, \mathrm{cap} \, fE \, .$$

Proof. Since A is a normal domain, it follows that $f\Gamma_E \subset \Gamma_{fE}$, the notation being that of 10.2. By 2.4 and 10.2 we obtain

$$\text{cap } E = \mathsf{M}(\Gamma_E) \leq K_O(f)N(f,A)\mathsf{M}(f\Gamma_E)$$
$$\leq K_O(f)N(f,A)\mathsf{M}(\Gamma_{fE}) = K_O(f)N(f,A)\text{ cap } fE . \quad \square$$

10.10. Theorem. *If* $f\colon G \to \mathbb{R}^n$ *is a nonconstant qr mapping and* $E = (A,C)$ *a condenser in* G, *then*

$$\text{cap } fE \leq K_I(f)\text{ cap } E .$$

Proof. With the notation of 10.2 let Γ^* be the family of maximal $f|A$-liftings that start in C of the paths in Γ_{fE}. Then $\Gamma^* \subset \Gamma_E$. Since Γ_{fE} is minorized by $f\Gamma^*$, we get by 8.1 and 10.2

$$\text{cap } fE = \mathsf{M}(\Gamma_{fE}) \leq \mathsf{M}(f\Gamma^*) \leq K_I(f)\mathsf{M}(\Gamma^*)$$
$$\leq K_I(f)\mathsf{M}(\Gamma_E) = K_I(f)\text{ cap } E . \qquad \square$$

We can sharpen 10.10 with the aid of Väisälä's inequality 9.1 as follows. Let $f\colon G \to \mathbb{R}^n$ be a nonconstant qr mapping and let $E = (A,C)$ be a condenser in G. We call

$$M(f,C) = \inf_{y \in fC} \sum_{x \in f^{-1}(y) \cap C} i(x,f)$$

the *minimal multiplicity* of f on C.

10.11. Theorem. *If* f *and* E *are as above, then*

$$\text{cap } fE \leq \frac{K_I(f)}{M(f,C)} \text{ cap } E .$$

Proof. Set $m = M(f,C)$. Let $\beta\colon [a,b[\to fA$ be a path in Γ_{fE}. Then $C \cap f^{-1}(\beta(a))$ contains points $x_1,\dots,x_k$ such that

$$m' = \sum_{l=1}^{k} i(x_l, f) \geq m .$$

By 3.2 there exists a maximal sequence of $f|A$-liftings $\alpha_j\colon [a,c_j[\to G$ of β, $1 \leq j \leq m'$, starting at the points $x_1,\dots,x_k$. Then each path α_j belongs to Γ_E. It follows that $\Gamma = \Gamma_E$ and $\Gamma' = \Gamma_{fE}$ satisfy the condition in 9.1. The result then follows from 9.1 and 10.2. $\qquad \square$

10.12. Notes. The conformal capacity was introduced by C. Loewner [Lo]. Theorems 10.9 and 10.10 were first proved in [MRV1, 6.2 and 7.1]. The K_O-inequality 2.4 was involved in the proof of 10.10, but otherwise path families were not used because Poletskiĭ's inequality 8.1 was not known at the time. The idea to sharpen 10.10 in the manner of 10.11 is due to O. Martio. His result [M1, 5.1] is a corollary of 10.11, which is taken from [V5, 3.17]. A characterization of quasiregularity by means of the K_I-inequality 10.10 for capacities was given in [MRV1, 7.1]. Using Poletskiĭ's inequality 8.1 we can formulate an analogous result for moduli of path families. We do not need these characterizations in this monograph.

Chapter III. Applications
of Modulus Inequalities

In this chapter we shall give our first applications of the inequalities for moduli of path families proved in the preceding chapter. Further applications will be given in Chapters IV, V, and VII. We start with some global distortion results and continue by proving, among other things, that a nonconstant qr mapping of $\mathbb{R}^n$ into itself omits at most a set of zero capacity. A local form of the latter result will be used in the proof of a Picard–type theorem in Chapter IV. Next, we shall establish a generalization of the theorem of V.A. Zorich which was mentioned in the introduction. In all these results Poletskiĭ's K_I-inequality is used. The rest of this chapter is devoted to local questions. There the sharper Väisälä's inequality, or its capacity variant due to O. Martio, and the K_O-inequality play essential roles for the problems treated.

1. Global Distortion

In this section we shall apply the K_I-inequality to derive for qr mappings such properties as a counterpart of Schwarz's lemma and Hölder continuity. We need some preliminary material on modulus estimates. The connection between capacity and modulus presented in II.10.2 will be used without further notice in the following.

Let X be an open or closed set in $\mathbb{R}^n$. The *spherical symmetrization* of X with respect to the ray $\{-re_1 : r \geq 0\}$ is the set X^* defined by the following three conditions. If for $r \geq 0$ we write $X_r = S^{n-1}(r) \cap X$ and $X_r^* = S^{n-1}(r) \cap X^*$, then:

(1) $X_r^* = S^{n-1}(r)$ if and only if $S^{n-1}(r) \subset X$;
(2) $X_r^* = \emptyset$ if and only if $X_r = \emptyset$;
(3) if X is open (closed) and $\emptyset \neq X_r \neq S^{n-1}(r)$ then X_r^* is the open (closed) spherical cap in $S^{n-1}(r)$ with center at $-re_1$ and $\mathcal{H}^{n-1}(X_r^*) = \mathcal{H}^{n-1}(X_r)$.

With this definition X^* is open if X is open and closed if X is closed. Now let $E = (A, C)$ be a condenser in $\mathbb{R}^n$. Then $E^* = (A^*, C^*)$ is also a condenser, it is the *spherical symmetrization* of (A, C). We shall adopt without proof the following minimizing property of the symmetrized condenser.

1.1. Proposition. *If E is a condenser in $\mathbb{R}^n$, then $\operatorname{cap} E \geq \operatorname{cap} E^*$.*

For ringlike condensers 1.1 is a well-known result of F.W. Gehring [G2, Theorem 1]. He gave the proof for $n = 3$, but it generalizes to other dimensions in an obvious way (see [Mo, p. 87]). The proof for condensers is similar. Another proof for 1.1 was given by J. Sarvas in [Sa1]. We shall apply 1.1 only to ringlike condensers.

For $0 \leq r < 1$ the ringlike condenser $(B^n, [0, re_1])$ is called the *Grötzsch condenser* and is denoted in this book by $E_G(r)$. We write $\nu_n(r) = \operatorname{cap} E_G(r)$. If E is a condenser, we also use the notation

$$\operatorname{mod} E = \left(\frac{\omega_{n-1}}{\operatorname{cap} E} \right)^{1/(n-1)}.$$

The following lemma is due to Gehring [G4, Lemma 6], [G2, pp. 514–518] for $n = 3$. We shall follow his proof closely.

1.2. Lemma. *The function ν_n is continuous and satisfies the following conditions:*

(1) *The function $\gamma_n(r) = r \exp(\operatorname{mod} E_G(r))$ is decreasing and hence ν_n is strictly increasing.*

(2) *$\nu_n(0) = 0$ and $\nu_n(r) \to \infty$ as $r \to 1$.*

(3) *The limit*

$$\lambda_n = \lim_{r \to 0} \gamma_n(r)$$

is finite and $1 \leq \gamma_n(r) \leq \lambda_n$ holds for $0 < r < 1$.

Proof. The continuity of ν_n follows from a general convergence result [G4, Lemma 6] on capacities of rings whose boundary components converge in an appropriate sense. We shall give Gehring's arguments directly for our special case. Let $0 \leq r < s < 1$. Clearly $\nu_n(r) \leq \nu_n(s)$. Let $0 < \varepsilon < 1$ and let u be an admissible function for $E_G(r)$ which is monotone in $B^n \setminus [0, re_1]$ and for which $\|\nabla u\|_n^n \leq \nu_n(r) + \varepsilon$. There exists $\delta > 0$ such that $u|[0, se_1] \geq 1 - \varepsilon$ for $r < s < r + \delta$. Then $u/(1 - \varepsilon)$ is admissible for $E_G(s)$ and hence $\nu_n(s) \leq (1 - \varepsilon)^{-n}(\nu_n(r) + \varepsilon)$ for $r < s < r + \delta$. This confirms the continuity of ν_n from the right at r. To prove its continuity from the left at s suppose $0 < s - r < b = \min(s/2, 1 - s)$ and set $x_0 = (r + s)e_1/2$. Then every sphere $C_t = S^{n-1}(x_0, t)$ is in B^n and meets $[0, re_1]$, for $c = (s - r)/2 < t < b$. We let u be as above. By the approximation result I.1.4 and by integration from II.1.11 we get

$$\int\limits_c^b \operatorname{osc}(u, C_t)^n \frac{dt}{t} \leq A_n \|\nabla u\|_n^n,$$

so there exists $t \in \,]c, b[$ such that

(1.3) $$\operatorname{osc}(u, C_t)^n \leq A_n \left(\log \frac{b}{c} \right)^{-1} (\nu_n(s) + \varepsilon).$$

Since u is monotone in $B^n \setminus [0, re_1]$ and $[re_1, se_1] \subset B^n(x_0, t)$, it follows from (1.3) that $u|[0, se_1] \geq 1 - D$, where $D = \mathrm{osc}(u, C_t)$. Hence $u/(1 - D)$ is admissible for $E_G(s)$ and $\nu_n(s) \leq (1 - D)^{-n}(\nu_n(r) + \varepsilon)$ if $D < 1$. From the form of the bound for D^n in (1.3) we see that ν_n is continuous from the left at s.

To prove (1) we apply II.1.5(3) and II.10.2 to the condensers $E_G(s)$, $(B^n(s/r), \overline{B}^n)$, and $(B^n(s/r), [0, se_1])$ for $0 < r < s < 1$. We obtain

$$\omega_{n-1}^{1/(1-n)} \log \frac{s}{r} + \nu_n(s)^{1/(1-n)} \leq \mathrm{cap}\big(B^n(s/r), [0, se_1]\big)^{1/(1-n)} = \nu_n(r)^{1/(1-n)} .$$

This gives (1). (2) follows from II.1.7(b) and II.(1.14).

Since $[0, re_1] \subset \overline{B}^n(r)$, we have $\nu_n(r) \leq \omega_{n-1}(\log(1/r))^{1-n}$ by II.1.7(b). This implies $1 \leq \gamma_n(r)$. By (1), $\gamma_n(r)$ approaches a limit λ_n as $r \to 0$. To show that λ_n is finite we define for $a > 4$ three ringlike condensers $E = (A, J)$, $E' = (B^n(-e_1, a - 2), J)$, and $E'' = (B^n(-e_1, a + 2), J)$, where A is the ellipsoid

$$\frac{x_1^2}{a^2 + 1} + \sum_{i=2}^{n} \frac{x_i^2}{a^2} < 1$$

and J the line segment $[-e_1, e_1]$. Clearly $\mathrm{mod}\, E' \leq \mathrm{mod}\, E \leq \mathrm{mod}\, E''$, $\mathrm{mod}\, E' = \mathrm{mod}\, E_G(2/(a-2))$, and $\mathrm{mod}\, E'' = \mathrm{mod}\, E_G(2/(a+2))$. Hence

$$(1.4) \qquad \log \lambda_n = \lim_{a \to \infty} \left(\mathrm{mod}\, E - \log \frac{a}{2} \right) .$$

To estimate $\mathrm{mod}\, E$ we proceed as follows. With the complex notation $z = x_1 + i x_2$ in $\mathbb{R}^2$ we map the annulus $B^2(d) \setminus \overline{B}^2$ conformally onto $A \cap \mathbb{R}^2 \setminus J$ by the map $z \mapsto (z + 1/z)/2$, where $d = a + (a^2 + 1)^{1/2}$. Let this map in polar coordinates be $(t, \vartheta) \mapsto (t', \vartheta')$. Next we rotate about the x_1-axis and get a mapping $g \colon B^n(d) \setminus \overline{B}^n \to A \setminus J$ defined in spherical coordinates by $g(t, y, \vartheta) = (t', y, \vartheta')$, where ϑ is the angle to the positive x_1-axis and $y \in S^{n-2} \subset \{0\} \times \mathbb{R}^{n-1}$. Then g is a diffeomorphism and a simple calculation shows that for $1 < t < d$

$$D(t) = \max_{|x|=t} \big(|g'(x)|^n J_g(x)^{-1}\big)^{1/(n-1)} = \left(\frac{t^2 + 1}{t^2 - 1} \right)^{(n-2)/(n-1)} .$$

Let $v \in C_0^\infty(A)$ be such that $v|J = 1$, and let $u \colon B^n(d) \to \mathbb{R}^1$ agree with $v \circ g$ in the ring $R = B^n(d) \setminus \overline{B}^n$ and satisfy $u|\overline{B}^n = 1$. Then u is admissible for $(B^n(d), \overline{B}^n)$. For $z \in S^{n-1}$ we get by Hölder's inequality

$$1 \leq \left(\int_1^d |\nabla u(tz)|\, dt \right)^n \leq \left(\int_1^d |\nabla v(g(tz))| D(t)^{(n-1)/n} J_g(tz)^{1/n}\, dt \right)^n$$

$$\leq \left(\int_1^d |\nabla v(g(tz))|^n J_g(tz) t^{n-1}\, dt \right) \left(\int_1^d \frac{D(t)}{t}\, dt \right)^{n-1} .$$

Through integration over S^{n-1} we obtain

$$\omega_{n-1} \left(\int_1^d \frac{D(t)}{t} dt \right)^{1-n} \leq \int_R |\nabla v \circ g|^n J_g \, dm = \int_A |\nabla v|^n \, dm .$$

Hence

$$\omega_{n-1} \left(\int_1^d \frac{D(t)}{t} dt \right)^{1-n} \leq \text{cap } E$$

and

(1.5) $\text{mod } E \leq \displaystyle\int_1^d \frac{D(t)}{t} dt \leq \log d + \int_1^\infty \left(\left(\frac{t^2+1}{t^2-1} \right)^{(n-2)/(n-1)} - 1 \right) \frac{dt}{t} .$

It is easy to see that the last integral is finite. Let its value be $\log \lambda_n'$. From (1.4) and (1.5) we obtain

$$\log \lambda_n = \lim_{a \to \infty} \left(\text{mod } E - \log \frac{a}{2} \right) \leq \lim_{a \to \infty} \log \left(\frac{2\lambda_n' d}{a} \right) = \log 4\lambda_n' ,$$

which completes the proof of (3) .   □

1.6. Remark. The method in the proof of (3) above does not give the best bound for λ_n . In fact, the exact value is known only for $n = 2$, when $\lambda_2 = 4$. Rather good estimates for λ_n can be found in [An].

For $K \geq 1$ and for $0 \leq r < 1$ we set

(1.7) $\varphi(n, K, r) = \nu_n^{-1}(K\nu_n(r)) .$

Then $\varphi(n, K, r)$ is strictly increasing in r , $\varphi(n, K, 0) = 0$, and $\varphi(n, K, r) \to 1$ as $r \to 1$. Furthermore, by using the estimate $1 \leq \gamma_n(r) \leq \lambda_n$ from 1.2 we get the estimate

(1.8) $\varphi(n, K, r) \leq \lambda_n r^{K^{1/(1-n)}}$

for $0 \leq r < 1$. Here λ_n can be replaced by $\lambda_n^{1-\alpha}$, where $\alpha = K^{1/(1-n)}$ (see [Vu12, 7.47]).

1.9. Lemma. Let $E = (A, C)$ be a condenser such that $A \subset B^n(r)$, C is connected, and $0, x \in C$. Then

$$\text{cap } E \geq \nu_n(|x|/r) .$$

Proof. By performing a homothety we may assume $r = 1$. If $E^* = (A^*, C^*)$ is the symmetrized condenser, $C^* \supset [-|x|e_1, 0]$ and $A^* \subset B^n$. By the monotoneity of the capacity, we get $\text{cap } E^* \geq \nu_n(|x|)$. The result then follows from 1.1.   □

The following is an n-dimensional version of Schwarz's lemma.

1.10. Theorem. *Let* $f: B^n \to B^n$ *be qr and let* $f(0) = 0$. *Then*

$$|f(x)| \leq \varphi(n, K_I(f), |x|)$$

for all $x \in B^n$. *Here* φ *is defined in* (1.7).

Proof. We may assume that f is not constant. For $x \in B^n$ let E be the condenser $(B^n, [0, x])$. By 1.9 and II.10.10,

$$\nu_n(|f(x)|) \leq \text{cap } fE \leq K_I(f) \text{ cap } E = K_I(f)\nu_n(|x|) ,$$

which proves the theorem. □

Theorem 1.10 implies the Hölder continuity of qr mappings in the following form.

1.11. Theorem. *Let* G *be a bounded domain, let* $f: G \to \mathbb{R}^n$ *be a bounded qr mapping, and let* F *be a compact subset of* G. *Set* $\alpha = K_I(f)^{1/(1-n)}$ *and* $C = \lambda_n d(F, \partial G)^{-\alpha} d(fG)$, *where* λ_n *is as in* (1.8). *Then* f *satisfies the Hölder condition*

(1.12) $$|f(x) - f(y)| \leq C |x - y|^\alpha$$

for $x \in F$, $y \in G$. *The Hölder exponent is best possible.*

Proof. Set $r = d(F, \partial G)$. Suppose first that $|x - y| < r$. Define $g: B^n \to B^n$ by $g(z) = d(fG)^{-1}(f(x + rz) - f(x))$. Then $g(0) = 0$ and $K_I(g) \leq K_I(f)$. By (1.8) and 1.10, $|g(z)| \leq \lambda_n|z|^\alpha$. Setting $z = (y - x)/r$ we obtain (1.12). If $|x - y| \geq r$, we use $\lambda_n \geq 1$ and get

$$|f(x) - f(y)| \leq r^{-\alpha}d(fG)|x - y|^\alpha \leq C|x - y|^\alpha .$$

To show that the exponent α is best possible, define $f: B^n \to B^n$ by $f(x) = |x|^{\alpha-1}x$. Then $K_I(f) = K(f) = K$ when $\alpha = K^{1/(1-n)}$. □

1.13. Corollary. *Let* $f: \mathbb{R}^n \to \mathbb{R}^n$ *be qr and let*

$$\lim_{x \to \infty} |x|^{-\alpha}|f(x)| = 0 ,$$

where $\alpha = K_I(f)^{1/(1-n)}$. *Then* f *is constant.*

Proof. We have $|f(x)| \leq |x|^\alpha \varepsilon(|x|)$ where $\varepsilon(R) \to 0$ as $R \to \infty$. Fix $x \in \mathbb{R}^n$ and let $R > |x|$. Applying 1.11 to $G = B^n(R)$ and $F = \{0\}$, we obtain $|f(x) - f(0)| \leq C|x|^\alpha$ with $C = \lambda_n d(0, \partial G)^{-\alpha}d(fG) \leq 2\lambda_n\varepsilon(R)$. Letting $R \to \infty$ yields $f(x) = f(0)$. Hence f is constant. □

The mapping $f(x) = |x|^{\alpha-1}x$ shows that the mere boundedness of $|x|^{-\alpha}|f(x)|$ is not enough to force the conclusion in 1.13. A consequence of 1.13 is the following n-dimensional counterpart of Liouville's theorem.

1.14. Corollary. *A bounded qr mapping of* $\mathbb{R}^n$ *is constant.*

1.15. Note. The results 1.10–1.14 are from [MRV2]. For sharper versions of these see [Vu12, Ch. 11]. The Hölder continuity as well as a result in the direction of 1.13 were first proved by E.D. Callender [Ca]. Similar results appear in [Mik1].

2. Sets of Capacity Zero and Singularities

We need the following form of Poincaré's inequality. Recall that the notation $W_{p,0}^1(U)$ indicates the closure of $C_0^\infty(U)$ in the W_p^1-norm, $1 \le p < \infty$.

2.1. Lemma. *If* $u \in W_{p,0}^1(B^n(r))$ *and* $1 \le p < \infty$, *then*

$$\|u\|_p \le r\|\nabla u\|_p .$$

The proof of 2.1 can be found in [GT, p. 157].

2.2. Lemma ([Re2, Lemma 2]). *Let* F *be a compact set in* $\mathbb{R}^n$, *and let* A *and* A' *be two bounded open sets in* $\mathbb{R}^n$ *containing* F. *If* $\mathrm{cap}_p(A, F) = 0$, *then also* $\mathrm{cap}_p(A', F) = 0$.

Proof. Fix an admissible C^∞ function v for (A', F). Let u be an admissible C^∞ function for (A, F). Then $w = uv$ is also admissible for (A', F), and we have $\nabla w = v\nabla u + u\nabla v$. Let $|v|, |\nabla v| \le M < \infty$ and $A \subset B^n(r)$. With 2.1 we get then $\|\nabla w\|_p \le M(1+r)\|\nabla u\|_p$. Since M and r are independent of u, our claim follows.   □

Let F be a compact set in $\mathbb{R}^n$. We say that F is of *p-capacity zero* if $\mathrm{cap}_p(A, F) = 0$ for some (and hence by 2.2 for all) bounded open set $A \supset F$. An arbitrary set $E \subset \mathbb{R}^n$ is of *p-capacity zero* if the same is true for every compact subset of E. In this case we write $\mathrm{cap}_p E = 0$ ($\mathrm{cap}\, E = 0$ if $p = n$), otherwise $\mathrm{cap}_p E > 0$. Similarly we can define the notion of zero p-capacity in n-manifolds (see Remark II.10.8.1).

We shall prove in VII.1.15 that for a closed set E the condition $\mathrm{cap}_p E = 0$ implies that the Hausdorff dimension $\dim_{\mathcal{H}} E$ of E is at most $n - p$. In particular, $\dim_{\mathcal{H}} E = 0$ if $\mathrm{cap}\, E = 0$. At this point we prove the following weaker result, which is adequate for our purposes.

2.3. Lemma. *If* C *is a closed set in* $\mathbb{R}^n$ *of zero capacity and* P *is an orthogonal projection of* $\mathbb{R}^n$ *onto a line* L *in* $\mathbb{R}^n$, *then* $\mathcal{H}^1(PC) = 0$.

Proof. We may assume that L is the x_n-axis and $C \subset Q = \{x \in \mathbb{R}^n : |x_i| < 1, i = 1, \ldots, n\}$. Let $H = \{x \in \mathbb{R}^n : x_n > 0\}$. We can map $2Q$ onto

$(B^n(5) \smallsetminus \overline{B}^n) \cap H$ by a quasiconformal mapping g such that $g(2Q \cap T_t) = S^{n-1}(t+3) \cap H$, $-2 < t < 2$, where T_t is the $(n-1)$-plane $x_n = t$. Suppose $\mathcal{H}^1(PC) > 0$. If Γ is the family $\Delta(gC, \partial H; B^n(4) \smallsetminus \overline{B}^n(2))$, then $\mathsf{M}(\Gamma) > 0$ by II.(1.13). But then $\mathsf{M}(\Gamma_E) \geq \mathsf{M}(g^{-1}\Gamma) > 0$, where E is the condenser $(2Q, C)$. (Recall the notation Γ_E from II.10.2.) This contradicts the assumption that $\operatorname{cap} C = 0$.   □

2.4. Remark. The above proof shows also that if C is closed and of zero capacity, then for every $x \in \mathbb{R}^n$ the set $\{t > 0 : S^{n-1}(x,t) \cap C \neq \emptyset\}$ is of zero 1-measure.

2.5. Corollary. *If C is of zero capacity, then it is totally disconnected.*

In the rest of this section we are mainly interested in quasimeromorphic mappings. Recall that $q(a, b)$ is the spherical chordal distance between points $a, b \in \overline{\mathbb{R}}^n$. We need the following estimate.

2.6. Lemma. *Let E be a compact proper subset of $\overline{\mathbb{R}}^n$ and let $\operatorname{cap} E > 0$. Then for every $a > 0$ there exists $\delta > 0$ such that $\operatorname{cap}(\mathsf{C}E, C) \geq \delta$ whenever C is a continuum in $\mathsf{C}E$ with $q(C) \geq a$.*

Proof. Performing a preliminary spherical isometry we may assume $\infty \notin E$. Choose $R > 0$ such that $E \subset B^n(R)$ and such that $q(\mathsf{C}\overline{B}^n(R)) < a/2$. Then C contains a continuum $C_1 \subset B^n(R)$ with $q(C_1) \geq a/4$. Since $\operatorname{cap}(\mathsf{C}E, C) \geq \operatorname{cap}(\mathsf{C}E, C_1)$, it suffices to find a lower bound for $\operatorname{cap}(\mathsf{C}E, C_1)$.

Since $\operatorname{cap} E > 0$, $\mathsf{M}(\Gamma_1) = \delta_1 > 0$, where $\Gamma_1 = \Delta(E, S^{n-1}(2R); B^n(2R))$. Since $d(C_1) \geq q(C_1) \geq a/4$, it follows from 1.9 that $\mathsf{M}(\Gamma_2) \geq \delta_2 > 0$, where $\Gamma_2 = \Delta(C_1, S^{n-1}(2R); B^n(2R))$ and δ_2 depends only on n, R, and a. Let $\Gamma_{12} = \Delta(C_1, E; \mathbb{R}^n)$. To estimate $\operatorname{cap}(\mathsf{C}E, C_1) = \mathsf{M}(\Gamma_{12})$, we let $\rho \in \mathcal{F}(\Gamma_{12})$. If $3\rho \in \mathcal{F}(\Gamma_1)$ or $3\rho \in \mathcal{F}(\Gamma_2)$, then

$$\int_{\mathbb{R}^n} \rho^n \, dm \geq 3^{-n} \min(\delta_1, \delta_2) \,.$$

In the remaining case there exist paths $\gamma_1 \in \Gamma_1$ and $\gamma_2 \in \Gamma_2$ such that

$$\int_\gamma \rho \, ds \geq 1/3$$

for every locally rectifiable path $\gamma \in \Delta(|\gamma_1|, |\gamma_2|; B^n(2R) \smallsetminus \overline{B}^n(R)) = \Gamma$. Then $3\rho \in \mathcal{F}(\Gamma)$, and by II.(1.14),

$$\int_{\mathbb{R}^n} \rho^n \, dm \geq 3^{-n} \mathsf{M}(\Gamma) \geq 3^{-n} c(n) \log 2 \,.$$

The number $\delta = 3^{-n} \min(\delta_1, \delta_2, c(n) \log 2)$ is then a lower bound for $\operatorname{cap}(\mathsf{C}E, C)$.   □

2.7. Corollary. Let $E \subset \overline{\mathbb{R}}^n$ be a compact set of positive capacity and G be a domain in $\overline{\mathbb{R}}^n$. Then for every $K \geq 1$, the family of all K-qm mappings $f: G \to \overline{\mathbb{R}}^n \smallsetminus E$ is equicontinuous.

Proof. Let $x_0 \in G$ and $a > 0$. Let $\delta > 0$ be the number given by 2.6. Choose a connected neighborhood U of x_0 such that $\operatorname{cap}(G, \overline{U}) < \delta/K$. If $f: G \to \overline{\mathbb{R}}^n \smallsetminus E$ is K-qm and nonconstant, we obtain

$$\operatorname{cap}(\complement E, f\overline{U}) \leq \operatorname{cap}(fG, f\overline{U}) \leq K \operatorname{cap}(G, \overline{U}) < \delta$$

by II.10.10. From 2.6 it follows that $q(f\overline{U}) < a$. This proves the theorem.   □

In connection of a qm mapping $f: G \to \overline{\mathbb{R}}^n$ we always make the following agreement. For a point $x \in G \smallsetminus (\{\infty\} \cup f^{-1}(\infty))$ we mean by $|f'(x)|$, $\ell(f'(x))$, and $J_f(x)$ the values of these (if defined) with respect to the Euclidean metric. An isolated point is always removable in the following sense.

2.8. Theorem. Let $f: G \to \overline{\mathbb{R}}^n$ be K-qm, let b be an isolated point of ∂G, and let f have a continuous extension $f^*: G \cup \{b\} \to \overline{\mathbb{R}}^n$. Then f^* is K-qm.

Proof. We may assume that f is not constant and that $b = 0 = f^*(b)$. Clearly f^* is ACL, and the inequalities $|f^{*'}(x)|^n \leq K_O(f) J_{f^*}(x)$ and $J_{f^*}(x) \leq K_I(f)\ell(f^{*'}(x))^n$ hold a.e. It remains to show that $|f^{*'}|^n$ is integrable in a neighborhood of 0. Since f is discrete, we can find a ball $U = B^n(r) \subset\subset G$ such that fU is bounded and $\partial U \cap f^{*-1}(0) = \emptyset$. Let Y be the 0-component of $\complement f^* \partial U$, and let V be the 0-component of $f^{*-1}Y$. For $y \in Y \smallsetminus \{0\}$ set $A = V \cap f^{-1}(y)$. Then

$$N(y, f^*, V) = \operatorname{card} A \leq \sum_{x \in A} i(x, f) = \mu(y, f^*, V) = \mu(0, f^*, V)$$

by I.4.4. It follows that $N(y, f^*, V)$ is integrable in $\mathbb{R}^n$. Lemma I.4.11 applied to f informs us that J_{f^*} is integrable in V. The claim follows then from $|f^{*'}(x)|^n \leq K_O(f) J_{f^*}(x)$ a.e.   □

2.9. Theorem. Let G be a domain in $\overline{\mathbb{R}}^n$, let F be closed in G, and let $\operatorname{cap} F = 0$. If $f: G \smallsetminus F \to \overline{\mathbb{R}}^n$ is a qm mapping such that $\operatorname{cap} \complement f(G \smallsetminus F) > 0$, then f can be extended to a continuous mapping $f^*: G \to \overline{\mathbb{R}}^n$.

Proof. From 2.3 we conclude that $G \smallsetminus F$ is a domain. We may assume that f is not constant. To show that f has a limit at $b \in F$ we may assume $b = 0$. Let $R > 0$ be such that $B^n(R) \subset G$. If f fails to have a limit at 0, we can find two sequences (x_j) and (x_j') of points in $B^n(R) \smallsetminus F$ such that $x_j \to 0$, $x_j' \to 0$, and $q(f(x_j), f(x_j')) \geq a > 0$ for all j. Set $r_j = \max(|x_j|, |x_j'|)$. By 2.3 we can find a closed arc C_j which joins x_j and x_j' in $\overline{B}^n(r_j) \smallsetminus F$. Since the family of paths joining C_j and F is of zero modulus, the capacity of the condenser $E_j = (B^n(R) \smallsetminus F, C_j)$ equals

$\operatorname{cap}(B^n(R), C_j)$. Hence $\operatorname{cap} E_j \leq \omega_{n-1}(\log(R/r_j))^{1-n}$. Since $q(fC_j) \geq a$ and since $\operatorname{cap} \complement f(G \smallsetminus F) > 0$, it follows from 2.6 that $\operatorname{cap} fE_j \geq \delta > 0$, where δ is independent of j. But II.10.10 implies $\operatorname{cap} fE_j \leq K(f) \operatorname{cap} E_j \to 0$, a contradiction. We conclude that f has continuous extension $f^*: G \to \overline{\mathbb{R}}^n$. $\qquad\square$

2.10. Corollary. *Let $f: G \to \overline{\mathbb{R}}^n$ be K-qm, let b be an isolated point of ∂G, and let $\operatorname{cap} \complement fG > 0$. Then f can be extended to a K-qm mapping $f^*: G \cup \{b\} \to \overline{\mathbb{R}}^n$.*

Proof. The result follows by combining 2.8 and 2.9. $\qquad\square$

If $f: G \to \overline{\mathbb{R}}^n$ is qm and if b is an isolated point of ∂G, then we call b an *(isolated) essential singularity* of f if f has no limit at b.

2.11. Corollary. *Let b be an isolated essential singularity of a qm mapping $f: G \to \overline{\mathbb{R}}^n$. Then $\operatorname{cap} \complement f(U \smallsetminus \{b\}) = 0$ for every neighborhood $U \subset G \cup \{b\}$ of b. Moreover, there exists an $\mathcal{F}_\sigma$ set $E \subset \overline{\mathbb{R}}^n$ of zero capacity such that $N(y, f, U \smallsetminus \{b\}) = \infty$ for every $y \in \overline{\mathbb{R}}^n \smallsetminus E$ and all such neighborhoods U of b.*

Proof. The first assertion follows directly from 2.9. To prove the second statement we may assume that $b = 0$ and $B^n \subset G \cup \{0\}$. Let $V_k = B^n(1/k) \smallsetminus \{0\}$, $k = 1, 2, \ldots$. Set

$$E = \bigcup_{k=1}^{\infty} \complement f V_k .$$

Then $E \supset \overline{f}(U \smallsetminus \{0\})$ for every neighborhood $U \subset G \cup \{0\}$ of 0. From II.1.5(1) it follows that $\operatorname{cap} E = 0$. If $y \in \overline{\mathbb{R}}^n \smallsetminus E$, we can find a sequence (x_i) such that $x_i \in V_{k_i}$, $x_i \neq x_j$ for $i \neq j$, and $f(x_i) = y$. This proves our claim. $\qquad\square$

2.12. Corollary. *Let $f: \mathbb{R}^n \to \mathbb{R}^n$ be qr with $\operatorname{cap} \complement f \mathbb{R}^n > 0$. Then f is constant.*

Proof. By 2.10 f extends to a qm mapping $f^*: \overline{\mathbb{R}}^n \to \overline{\mathbb{R}}^n$. If f is nonconstant, $f^* \overline{\mathbb{R}}^n$ is compact and open, hence $f^* \overline{\mathbb{R}}^n = \overline{\mathbb{R}}^n$. But $\operatorname{cap} \complement f \mathbb{R}^n > 0$ implies $f^* \overline{\mathbb{R}}^n \neq \overline{\mathbb{R}}^n$, so the result follows. $\qquad\square$

2.13. Remarks. 1. The part 2.6–2.11 is a simplified version of results from [MRV2]. Corollary 2.12 was also proved by Yu.G. Reshetnyak [Re8, Theorem 2]. E.D. Callender [Ca, Theorem 3] proved 2.10 for a bounded mapping.

2. In VII.1.1 we shall generalize the removability result 2.10 to sets of capacity zero.

3. Corollary 2.12 is a step towards a Picard–type theorem on omitted values. In Chapter IV we shall prove the analogue of Picard's theorem for qr

mappings in space. The equicontinuity result 2.7 will be considerably improved in IV.3.14.

3. The Injectivity Radius of a Local Homeomorphism

In this section we prove a generalization of Zorich's theorem. We shall show that a K-qr local homeomorphism of the unit ball B^n, $n \geq 3$, is actually homeomorphic in a smaller ball $B^n(r)$, where r depends only on n and K. The basic idea in the proof goes back to [Zo1].

We need some preliminary topological lemmas concerning local homeomorphisms. A set $Q \subset \mathbb{R}^n$ is said to be *relatively locally connected* if every point in $\overline{Q}$ has arbitrarily small neighborhoods U such that $U \cap Q$ is connected. For the proof of the following result we refer to [MRV3, 2.1].

3.1. Lemma. *Let* $f: G \to \overline{\mathbb{R}}^n$ *be a local homeomorphism, let* Q *be a simply connected and locally pathwise connected set in* $\overline{\mathbb{R}}^n$, *and let* P *be a component of* $f^{-1}Q$ *such that* $\overline{P} \subset G$. *Then* f *maps* P *homeomorphically onto* Q. *If, in addition,* Q *is relatively locally connected,* f *maps* $\overline{P}$ *homeomorphically onto* $\overline{Q}$.

The proof of the following simple lemma is left to the reader.

3.2. Lemma. *Let* $f: G \to \overline{\mathbb{R}}^n$ *be a local homeomorphism and let* F *be a compact set in* G *such that* $f|F$ *is injective. Then* f *is injective in a neighborhood of* F.

3.3. Lemma. *Let* $f: G \to \overline{\mathbb{R}}^n$ *be a local homeomorphism, let* $A, B \subset G$, *and let* f *be homeomorphic in* A *and* B. *If* $A \cap B \neq \emptyset$ *and if* $fA \cap fB$ *is connected, then* f *is homeomorphic in* $A \cup B$.

Proof. Let $A' = A \cap f^{-1}C$ and $B' = B \cap f^{-1}C$, where $C = fA \cap fB$. It is sufficient to show that $A' = B'$. Clearly $A' \cap B' = A \cap B$. We claim that $f(A' \cap B') \subset C$ is both open and closed in C.

Let $x \in A' \cap B'$ and $y = f(x)$. Let U be a neighborhood of x in which f is a homeomorphism onto $V = fU$. Then V is a neighborhood of y such that $f(A' \cap B') \supset f(U \cap A' \cap B') = f(U \cap A \cap B) = V \cap fA \cap fB$, hence $f(A' \cap B')$ is open in C. To prove that $f(A' \cap B')$ is closed in C, let (x_j) be a sequence of points in $A' \cap B'$ such that $(f(x_j))$ converges to a point $c \in C$. Let a and b be the points in $f^{-1}(c) \cap A'$ and $f^{-1}(c) \cap B'$ respectively. Since f is homeomorphic on A' and B', the sequence (x_j) converges to both a and b. Then $a = b \in A' \cap B'$ and $c \in f(A' \cap B')$. Hence $f(A' \cap B')$ is closed in C.

Since C is connected and $f(A' \cap B')$ is nonempty, it follows that $f(A' \cap B') = C$ and thus $A' = B'$. □

3.4. Theorem. *If $n \geq 3$ and if $f\colon B^n \to \mathbb{R}^n$ is a K-qr local homeomorphism, then f is injective in a ball $B^n(\psi(n,K))$, where $\psi(n,K)$ is a positive number depending only on n and K.*

Proof. We may assume that $f(0) = 0$. Let $r_0 = \sup\{r : \overline{U}(0,r) \subset B^n\}$. Recall the notation from II.4. Fix $0 < r < r_0$, and set $l^* = l^*(0,r)$, $L^* = L^*(0,r)$, and $U = U(0,r)$. By 3.1, f maps $\overline{U}$ homeomorphically onto $\overline{B}^n(r)$. Thus f is injective in $B^n(l^*)$ and it suffices to find a lower bound for l^*.

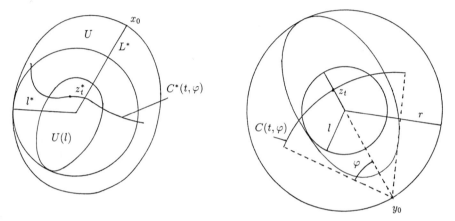

Fig. 3

Let $l = l(0,l^*)$ and suppose $l < r$. Then $E = (U, \overline{U}(0,l))$ is a ringlike condenser. Since both boundary components of $U \smallsetminus \overline{U}(0,l)$ meet $S^{n-1}(l^*)$, it follows from [V4, 11.7] that $\operatorname{cap} E \geq a_n > 0$, where a_n depends only on n. By II.10.9 and II.1.7(b),

$$a_n \leq \operatorname{cap} E \leq K \operatorname{cap} fE = K\omega_{n-1}\left(\log\frac{r}{l}\right)^{1-n}.$$

This gives an inequality

$$(3.5) \qquad\qquad \frac{r}{l} \leq \alpha(n,K),$$

which is true also for $r = l$.

Choose $x_0 \in \partial U$ such that $|x_0| = L^*$, and set $y_0 = f(x_0)$. For $r < t < r + l$ and $0 < \varphi \leq \pi$ let $C(t,\varphi)$ be the spherical cap $\{y : |y - y_0| = t, (y_0 - y) \cdot y_0 > rt \cos\varphi\}$. Then $C(t,\varphi)$ is symmetric with respect to the line segment $J = \{sy_0 : -l/r < s < 0\}$ and meets J at $z_t = (r - t)y_0/r$. Let z_t^* be the unique point in $U \cap f^{-1}(z_t)$ and let $C^*(t,\varphi)$ be the z_t^*-component of $f^{-1}C(t,\varphi)$ (Fig. 3). Let φ_t be the supremum of all $\varphi \in {]0,\pi]}$ such that f maps $C^*(t,\varphi)$ homeomorphically onto $C(t,\varphi)$. We show that $C^*(t,\varphi_t)$ meets $S^{n-1}(L^*)$ for every $t \in {]r,r+l[}$. Suppose this is not true. Then $C^*(t, \varphi_t) \subset B^n(L^*)$ for some t. By 3.1, f maps $\overline{C}^*(t, \varphi_t)$ homeomorphically onto $\overline{C}(t, \varphi_t)$. (Note that for $n = 2$ the proof breaks down here, since $C(t, \pi)$ is then not relatively locally connected.) By 3.2, f is injective in a neighborhood of $\overline{C}^*(t, \varphi_t)$. This implies $\varphi_t = \pi$, hence $\overline{C}^*(t, \varphi_t)$ is a topological $(n - 1)$-sphere. The bounded component D of $\complement\overline{C}^*(t, \varphi_t)$ is contained in $B^n(L^*)$. Since $\partial f D \subset S^{n-1}(y_0, t)$, $f D = B^n(y_0, t)$. Thus D is a component of $f^{-1} B^n(y_0, t)$. By 3.1, f maps $\overline{D}$ homeomorphically onto $\overline{B}^n(y_0, t)$. Since $z_t^* \in \overline{D} \cap \overline{U}$ and since $\overline{B}^n(y_0, t) \cap \overline{B}^n(r)$ is connected, it follows from 3.3 that f is injective in $\overline{D} \cup \overline{U}$. As $y_0 \in f D$, this implies that $x_0 \in D$, which is impossible because $D \subset B^n(L^*)$. Thus $C^*(t, \varphi_t)$ meets $S^{n-1}(L^*)$ for all $t \in {]}r, r + l[$.

For each $t \in {]}r, r + l[$ choose a point $x_t^* \in C^*(t, \varphi_t) \cap S^{n-1}(L^*)$. Let Γ_t be the family of paths joining x_t^* and z_t^* in $C^*(t, \varphi_t)$, and let Γ be the union of the families Γ_t. Since $|z_t^*| \leq l^*$,

$$(3.6) \qquad \mathsf{M}(\Gamma) \leq \omega_{n-1} \Big(\log \frac{L^*}{l^*} \Big)^{1-n} .$$

By integration of II.(1.9),

$$(3.7) \qquad \mathsf{M}(f\Gamma) \geq b(n) \log(1 + l/r) .$$

Since $\mathsf{M}(f\Gamma) \leq K\mathsf{M}(\Gamma)$ by II.8.1, we obtain from (3.5)–(3.7) an inequality $l^* \geq L^* \psi(n, K)$, where $\psi(n, K) > 0$ depends only on n and K. Since $L^* \to 1$ as $r \to r_0$, this proves the theorem. $\qquad\square$

As a corollary we get Zorich's theorem [Zo1]:

3.8. Corollary. *If $n \geq 3$ and if $f : \mathbb{R}^n \to \mathbb{R}^n$ is a qr local homeomorphism, then f is a homeomorphism.*

3.9. Corollary. *Let G be a domain in $\mathbb{R}^n$, where $n \geq 3$, and let $K \geq 1$, $\delta > 0$. If W is a family of K-qm local homeomorphisms $f : G \to \overline{\mathbb{R}}^n$ such that every $f \in W$ omits two points $a_f, b_f \in \overline{\mathbb{R}}^n$ with the spherical distance $q(a_f, b_f) \geq \delta$, then W is equicontinuous.*

Proof. For $f \in W$ let T_f be a Möbius transformation with $T_f(b_f) = \infty$. It follows from 3.4 that every point in G has a neighborhood in which every $T_f \circ f$, and hence f itself, is injective. The result follows from the corresponding result for qc mappings [V4, 19.2]. $\qquad\square$

3.10. Remarks. 1. Results 3.4, 3.8, and 3.9 break down when $n = 2$. The mappings $f_j(z) = e^{jz}$ lead to counterexamples.

2. Corollary 3.8 shows that ∞ is a removable singularity for f. A local version of this was proved by S. Agard and A. Marden in [AM] and by V.A. Zorich in [Zo2]. J. Sarvas proved in [Sa4, 6.3] that for each $n \geq 3$ there exists $K_n > 1$ such that f in 3.4 is injective in the whole ball B^n when f is K-qr.

3. Lemma 3.3 is from [Zo1]; the results 3.4 and 3.9 are from [MRV3]. The injectivity result 3.4 is given here for mappings into $\mathbb{R}^n$. It is natural to ask what other manifolds can occur as range. In [MS4] O. Martio and U. Srebro proved the corresponding result with $\overline{\mathbb{R}}^n$ in place of $\mathbb{R}^n$. Further generalizations in this direction are given in [Gro1] and [Zo3].

4. As an application of Zorich's theorem 3.8 A. Marden and S. Rickman [MarR] proved that if $f: \mathbb{R}^{2k} \to \mathbb{R}^{2k}$ is qr and holomorphic as a mapping $\mathbb{C}^k \to \mathbb{C}^k$, $k > 1$, then f is an affine homeomorphism or constant.

4. Local Distortion

One can use Väisälä's inequality II.9.1, or its capacity variant II.10.11, to improve the Hölder estimate (1.12) locally. This observation is due to O. Martio, and we shall use his idea from [M1, 6.1] in the proof. The K_O-inequality gives a corresponding estimate from below. First it is convenient to update our information about the inverse dilatation $H^*(x, f)$; namely, we shall now prove that it is bounded by a constant depending only on n and $K(f)$.

4.1. Lemma. *Let* $f: G \to \mathbb{R}^n$ *be* K-qr *and nonconstant, let* $x \in G$, *and let* $\sigma_x > 0$ *be the number given by Lemma II.4.1. With the notation of II.4 let* $r > 0$ *be such that* $L^* = L^*(x, r) < l^*(x, \sigma_x)$. *Then* $L^*/l^* \leq C^*$. *Here* $l^* = l^*(x, r)$ *and* $C^* = C^*(n, K)$ *is a constant that depends only on* n *and* K.

Proof. Set $l = l(x, l^*)$, $L = L(x, L^*)$, and $U_t = U(x, t)$ for $t > 0$. We may assume $l < r < L$. Since $L^* < l^*(x, \sigma_x)$, $L < \sigma_x$. By II.4.1, $(U_r, \overline{U}_l)$ and $(U_L, \overline{U}_r)$ are ringlike condensers. Furthermore, ∂U_l and ∂U_r meet $S^{n-1}(x, l^*)$, and ∂U_r and ∂U_L meet $S^{n-1}(x, L^*)$. Therefore $\operatorname{cap}(U_r, \overline{U}_l)$, $\operatorname{cap}(U_L, \overline{U}_r) \geq a_n > 0$, where a_n depends only on n by [V4, 11.7]. The K_O-inequality II.10.9 and II.1.7(b) imply

$$
\begin{aligned}
\operatorname{cap}(U_r, \overline{U}_l) &\leq K_O(f) i(x, f) \omega_{n-1} \left(\log \frac{r}{l} \right)^{1-n}, \\
\operatorname{cap}(U_L, \overline{U}_r) &\leq K_O(f) i(x, f) \omega_{n-1} \left(\log \frac{L}{r} \right)^{1-n}.
\end{aligned}
$$

(4.2)

Inequality II.10.11 gives

$$
\begin{aligned}
K_I(f) \operatorname{cap}(U_L, \overline{U}_l) &\geq i(x, f) \operatorname{cap}(f U_L, f \overline{U}_l) \\
&= i(x, f) \omega_{n-1} \left(\log \frac{L}{l} \right)^{1-n}.
\end{aligned}
$$

(4.3)

We also have

$$
\operatorname{cap}(U_L, \overline{U}_l) \leq \omega_{n-1} \left(\log \frac{L^*}{l^*} \right)^{1-n}.
$$

(4.4)

From (4.2) we get

$$\left(\log\frac{L}{l}\right)^{n-1} = \left(\log\frac{L}{r} + \log\frac{r}{l}\right)^{n-1} \leq 2^{n-1}K_O(f)i(x,f)\omega_{n-1}a_n^{-1}\ .$$

Inequalities (4.3) and (4.4) imply $\omega_{n-1}(\log(L^*/l^*))^{1-n} \geq a_n 2^{1-n}K^{-2}$. Hence

$$(4.5) \qquad \frac{L^*}{l^*} \leq \exp\left(2\left(\frac{\omega_{n-1}K^2}{a_n}\right)^{1/(n-1)}\right) = C^*(n,K)\ . \quad \square$$

4.6. Corollary. Let $f\colon G \to \mathbb{R}^n$ be K-qr and nonconstant. Then $H^*(x,f) \leq C^*(n,K)$, where $C^*(n,K)$ is defined in (4.5).

We formulate our local distortion result as follows.

4.7. Theorem. Let $f\colon G \to \mathbb{R}^n$ be qr and nonconstant, and let $x \in G$. Then there exist positive numbers ρ, A, and B such that for $y \in B^n(x,\rho)$

$$(4.8) \qquad A|y-x|^\nu \leq |f(y)-f(x)| \leq B|y-x|^\mu,$$

where $\nu = \left(K_O(f)i(x,f)\right)^{1/(n-1)}$ and $\mu = \left(i(x,f)/K_I(f)\right)^{1/(n-1)}$.

Proof. Let σ_x be as in 4.1 and let $r_0 > 0$ be such that $L^*(x,r_0) < l^*(x,\sigma_x) = \rho_1$. Choose $\rho = l^*(x,r_0)$ and let $0 < |x-y| < \rho$. We carry over the notation from the proof of 4.1 with $r = |f(x)-f(y)|$. Since the ring $B^n(x,\rho_1) \setminus \overline{B}^n(x,L^*)$ separates the boundary components of the ring $U_{\sigma_x} \setminus \overline{U}_r$, we obtain

$$\frac{\omega_{n-1}}{(\log(\sigma_x/r))^{n-1}} \leq \frac{K_I(f)}{i(x,f)}\,\mathrm{cap}(U_{\sigma_x},\overline{U}_r) \leq \frac{K_I(f)\omega_{n-1}}{i(x,f)(\log(\rho_1/L^*))^{n-1}}$$

by II.10.11. In conjunction with 4.1 this gives

$$r \leq \sigma_x \rho_1^{-\mu}L^{*\mu} \leq \sigma_x \rho_1^{-\mu}C^{*\mu}l^{*\mu}\ .$$

Since $l^* \leq |x-y|$, we obtain the right hand inequality in (4.8) with $B = \sigma_x\rho_1^{-\mu}C^{*\mu}$.

To prove the left hand inequality in (4.8) we observe that the ring $U_{\sigma_x}\setminus\overline{U}_r$ separates the boundary components of the ring $B^n(x,\rho_2) \setminus \overline{B}^n(x,l^*)$, where $\rho_2 = L^*(x,\sigma_x)$. By II.10.9 we then obtain

$$\frac{\omega_{n-1}}{(\log(\rho_2/l^*))^{n-1}} \leq \mathrm{cap}(U_{\sigma_x},\overline{U}_r) \leq \frac{K_O(f)i(x,f)\omega_{n-1}}{(\log(\sigma_x/r))^{n-1}}\ .$$

Hence

$$r \geq \sigma_x\rho_2^{-\nu}l^{*\nu} \geq \sigma_x\rho_2^{-\nu}C^{*-\nu}L^{*\nu} \geq A|x-y|^\nu,$$

where $A = \sigma_x\rho_2^{-\nu}C^{*-\nu}$. $\qquad\qquad \square$

4.9. Notes. Corollary 4.6 and the essence of Lemma 4.1 are from [MRV3, 5.2]. See also [Sa2, 2.3]. The right hand inequality in (4.8) is due to O. Martio [M1,

6.1] (see, too, [MRV3, 4.3]). The form of the bound B in the proof appears in [Sa2, 2.5]. The left hand inequality in (4.8) appears in [Sr2, 5.4].

5. Bounds for the Local Index

In this section we shall see that for a nonconstant K-qr mapping $f\colon G \to \mathbb{R}^n$ the local index $i(x, f)$ cannot be uniformly big in too large a set. The ideas suggesting such a result are these: on the one hand, a large value of $i(x, f)$ at a point $x \in G$ has according to 4.7 the effect of shrinking fU for neighborhoods U of x; on the other hand, the discreteness and openness of f imply that $\mathcal{H}^{n-2}(fB_f) > 0$ whenever the branch set B_f is nonempty, making excessive shrinkage impossible. Of course, we have already seen in Example I.3.2 that at individual points the local index can be arbitrarily large.

5.1. Lemma. *Let* $G \subset \mathbb{R}^n$ *be a simply connected domain, and let* A *be a set that is closed in* G *and has* $\mathcal{H}^{n-2}(A) = 0$. *Then* $G \smallsetminus A$ *is simply connected.*

Proof. First we show that, if E is a countable union of 2-dimensional planes T in $\mathbb{R}^n$, then $(E+y) \cap A = \emptyset$ for almost every $y \in \mathbb{R}^n$. Clearly it is enough to consider the case where E is itself such a plane T. If $(T+y) \cap A \neq \emptyset$, there exists $z \in T$ for which $z+y \in A$, and hence $y \in A-T$. Therefore it suffices to check that $m(A-T) = 0$. To this end, let $Q \subset T$ be a square of side length h and let $\varepsilon > 0$. There is a covering of A by balls $B_i = B^n(x_i, r_i)$, $i = 1, 2, \ldots$, such that $r_i < 1$ and $\sum_i r_i^{n-2} < \varepsilon$. Then $A - Q \subset \bigcup \{ B_i - Q : i = 1, 2, \ldots \}$ and

$$m(A - Q) \leq \sum_i m(B_i - Q) \leq (h+2)^2 2^{n-2} \sum_i r_i^{n-2} < (h+2)^2 2^{n-2} \varepsilon \, .$$

This shows that $m(A - T) = 0$.

Let now $\gamma\colon I \to G \smallsetminus A$, $I = [0, 1]$, be a path with $\gamma(0) = \gamma(1) = x_0$. By assumption there exists a homotopy $h\colon I^2 \to G$ such that $h(0, t) = \gamma(t)$, $h(1, t) = h(s, 0) = h(s, 1) = x_0$ for all $(s, t) \in I^2$. Let τ be a triangulation of I^2 and let $h_1\colon I^2 \to G$ be a simplicial approximation to h with respect to τ. Set $\gamma_1(t) = h_1(0, t)$. We assume that the triangulation is sufficiently fine so that $(s, t) \mapsto s\gamma(t) + (1 - s)\gamma_1(t)$ defines a homotopy $\gamma \simeq \gamma_1$ in $G \smallsetminus A$. By the first part of the proof there exists $y \in \mathbb{R}^n$ such that $|y| < \min\big(d(h_1 I^2, \partial G), d(|\gamma_1|, \partial(G \smallsetminus A))\big)$ and $(h_1 I^2 + y) \cap A = \emptyset$. Thus we obtain the following homotopies in $G \smallsetminus A$: $\gamma_1 \simeq \gamma_1 + y$ defined by $(s, t) \mapsto \gamma_1(t) + sy$ and $\gamma_1 + y \simeq x_0 + y$ defined by $(s, t) \mapsto h_1(s, t) + y$. This implies that $\gamma \simeq x_0 + y$ in $G \smallsetminus A$, and the lemma follows. $\qquad \square$

Let $f\colon G \to \mathbb{R}^n$ be discrete and open. Let $x \in B_f$ and let D be a normal neighborhood of x. Set $g = f|D$, $\tilde{X} = D \smallsetminus g^{-1}gB_g$, $X = g\tilde{X} = gD \smallsetminus gB_g$, and $p = g|\tilde{X}$. Then $\tilde{X}$ and X are domains, $(\tilde{X}, p)$ is a covering space of X, $\operatorname{card} p^{-1}(y) = |i(x, f)|$ for all $y \in X$, and the fundamental group $\pi_1(X, y)$ operates transitively on the right on the set $p^{-1}(y)$ for every $y \in X$. For $z \in p^{-1}(y)$ and $c \in \pi_1(X, y)$ we denote this action by $z \cdot c$, i.e., $z \cdot c \in p^{-1}(y)$ is the terminal point of the liftings of the representatives of c starting at z.

5.2. Lemma. *With the notation above, X is not simply connected.*

Proof. Let $y \in X$. Since $x \in B_f$, there exist distinct points $z, u \in p^{-1}(y)$. Let $\alpha: [0, 1] \to \tilde{X}$ be a path connecting z to u, and let $c \in \pi_1(X, y)$ be the class of $p \circ \alpha$. If c were the neutral element, we would have $u = z \cdot c = z$, a contradiction. $\qquad \square$

5.3. Proposition. *If $f: G \to \mathbb{R}^n$ is discrete open and if $B_f \neq \emptyset$, then $\mathcal{H}^{n-2}(fB_f) > 0$.*

Proof. Let $x \in B_f$ and let $D = U(x, r)$ be a normal neighborhood of x. Suppose $\mathcal{H}^{n-2}(f(D \cap B_f)) = 0$. Since the domain $fD = B^n(f(x), r)$ is simply connected and since $f(D \cap B_f)$ is closed in fD, $X = fD \smallsetminus f(D \cap B_f)$ is simply connected by 5.1. But this is in conflict with 5.2. We conclude that $0 < \mathcal{H}^{n-2}(fB_f)$. $\qquad \square$

5.4. Notes. 1. The results 5.1 and 5.3 are from [MRV3]. Lemma 5.2 is contained in [CH, Theorem 5.9].

2. In [MR2, 2.20] it is proved that, under the assumptions in 5.3, we have $\mathcal{H}^{n-2}(B_f) > 0$ for $n = 3$. This is clearly true as well for $n = 2$. The situation when $n \geq 4$ is an open question.

3. J. Sarvas [Sa3] has given in the qr setting bounds for the Hausdorff dimensions $\dim_{\mathcal{H}}$ in the other direction as follows. Let $f: G \to \mathbb{R}^n$ be a nonconstant K-qr mapping. Then $\dim_{\mathcal{H}} fB_f \leq c'(n, K) < n$ and, for each compact $F \subset G$, $\dim_{\mathcal{H}}(B_f \cap F) \leq c(n, K, k_F) < n$, where $k_F = \sup\{ i(x, f) : x \in F \}$. It is an open question whether $\dim_{\mathcal{H}} B_f$ has an upper bound less than n. In [Re10] it is proved that $\mathcal{H}^{n-1}(B_f \cap T) = 0$ for all $(n-1)$-dimensional planes $T \subset \mathbb{R}^n$, while O. Martio and S. Rickman [MR2, 3.1] have established the corresponding result for fB_f.

The following result was proved by O. Martio [M1, 6.8], whose proof we shall adopt. Proposition 5.3 is not needed at this stage.

5.5. Theorem. *Let $f: G \to \mathbb{R}^n$ be a nonconstant qr mapping and F be a compact set in B_f with $\mathcal{H}^p(fF) > 0$ for some $p > 0$. Then*

$$(5.6) \qquad \inf_{x \in F} i(x, f) < \left(\frac{n}{p}\right)^{n-1} K_I(f) .$$

Proof. By 4.6, $H^*(x, f) \leq C^*(n, K)$ for all $x \in G$. Set $c = C^*(n, K) + 1$. As in II.5.3 we conclude that there exists an integer i_0 such that the sets

$F_i = \{ x \in F : L^*(x,r)/l^*(x,r) \le c \text{ for all } r \in \,]0,1/i[\, \}$ are compact for all integers $i \ge i_0$. Recall the notation from II.4. Since F is the union of the sets F_i, we can fix $i \ge i_0$ such that $\mathcal{H}^p(fF_i) > 0$. The set F_i is the union of the Borel sets $F_{ij} = \{ x \in F_i : i(x,f) = j \}$, $j = 2,3,\ldots$. We fix j such that $\mathcal{H}^p(fF_{ij}) > 0$. Note that fF_{ij} is $\mathcal{H}^p$ measurable by [F, 2.2.13]. Pick $x_0 \in F_{ij}$ and σ with the properties (i) $0 < \sigma < 1/i$, (ii) U_σ is a normal neighborhood of x_0, and (iii) $\mathcal{H}^p(f(F_{ij} \cap U_s)) > 0$ for $0 < s \le \sigma$. Here we employ the notation $U_s = U(x_0, s)$. Choose $u > 0$ such that $\overline{B}^n(x_0, 2u) \subset U_\sigma$ and then select $s > 0$ such that $\overline{U}_s \subset B^n(x_0, u)$. The continuity of f implies that

$$\tau(r) = \sup_{x \in F} L(x,r) \to 0$$

as $r \to 0$. Let $r_0 \in \,]0, d(F, \partial G)[$ be such that $\tau(r) < d(fS^{n-1}(x_0, u), B^n(f(x_0), s))$ for $0 < r \le r_0$.

Suppose now that (5.6) is not true. Set $A = F_{ij} \cap U_s$. We show that $\mathcal{H}^p(fA) = 0$. Let $t, \varepsilon > 0$. Since $m(B_f) = 0$ by II.7.4, we can cover A by balls $B^n(x_k, r_k)$, $k = 1, 2, \ldots$, such that (1) $x_k \in A$, (2) $r_k < r_0$, (3) $\tau(r_k) < t/2$, and (4) $\sum_k r_k^n < \varepsilon$. Fix k. By (2), $E_k = (U_\sigma, V_k)$ is a condenser, where $V_k = U(x_k, L_k)$ and $L_k = L(x_k, r_k)$. Since $i(x_k, f) = i(x_0, f)$, V_k is a normal neighborhood of x_k. By the assumption and by II.10.11,

$$\operatorname{cap} fE_k \le \frac{K_I(f)}{i(x_0, f)} \operatorname{cap} E_k \le \left(\frac{p}{n}\right)^{n-1} \operatorname{cap} E_k .$$

Since $fU_\sigma \subset B^n(f(x_k), 2\sigma)$ and $\overline{B}^n(f(x_k), L_k) \subset fV_k$, we have $\operatorname{cap}(B^n(f(x_k), 2\sigma), \overline{B}^n(f(x_k), L_k)) \le \operatorname{cap} fE_k$. Similarly $\operatorname{cap} E_k \le \operatorname{cap}(B^n(x_k, u), \overline{B}^n(x_k, L_k^*))$, where $L_k^* = L^*(x_k, L_k)$. As a result,

$$\frac{\omega_{n-1}}{(\log(2\sigma/L_k))^{n-1}} \le \operatorname{cap} fE_k \le \left(\frac{p}{n}\right)^{n-1} \operatorname{cap} E_k$$

$$\le \left(\frac{p}{n}\right)^{n-1} \frac{\omega_{n-1}}{(\log(u/L_k^*))^{n-1}} .$$

From the proof of II.4.1(6) we conclude that $l_k^* = l^*(x_k, L_k) = r_k$. We thus obtain

$$(5.7) \qquad L_k \le 2\sigma u^{-n/p} L_k^{*\,n/p} \le 2\sigma u^{-n/p} c^{n/p} l_k^{*\,n/p} = br_k^{n/p},$$

where $b = 2\sigma u^{-n/p} c^{n/p}$. The set fA is covered by the sets $fB^n(x_k, r_k)$. Furthermore, by (3), $d(fB^n(x_k, r_k)) < t$. Hence (5.7) and (4) imply that (see I.1.1)

$$\mathcal{H}_t^p(fA) \le \sum_k \lambda_p d(fB^n(x_k, r_k))^p$$

$$\le 2^p \lambda_p \sum_k L_k^p \le 2^p b^p \lambda_p \sum_k r_k^n < 2^p b^p \lambda_p \varepsilon .$$

In this way we see that $\mathcal{H}^p(fA) = 0$, which contradicts (iii). The theorem is proved. $\qquad\square$

5.8. Corollary. *If* $n \geq 3$, *if* $f: G \to \mathbb{R}^n$ *is a nonconstant qr mapping, and if* $B_f \neq \emptyset$, *then*

$$\inf_{x \in B_f} i(x, f) < \left(\frac{n}{n-2}\right)^{n-1} K_I(f) \leq 9 K_I(f) .$$

Proof. The result follows from 5.3 and 5.5. □

5.9. Corollary. *Let* f *be as in 5.8. If* F *is a continuum in* B_f, *then*

$$\inf_{x \in F} i(x, f) < n^{n-1} K_I(f) .$$

Moreover, the set $E = \{ x \in B_f : i(x, f) \geq n^{n-1} K_I(f) \}$ *is totally disconnected.*

Proof. The first statement is consequence of 5.5, since the discreteness of f implies that $\mathcal{H}^1(fF) > 0$. Suppose that the closed set E contains a continuum F. Then we get a contradiction with the first statement. The final assertion of the corollary follows. □

Let $f: G \to \mathbb{R}^n$ be nonconstant and qr. Suppose there exists a nonconstant rectifiable path $\alpha: [0, 1] \to B_f$ in B_f for which the statement

$$(5.10) \qquad\qquad \inf_{x \in |\alpha|} i(x, f) \leq K_I(f)$$

is not true. Then it follows from 4.7 that $f'(x) = 0$ for *all* $x \in |\alpha|$. Integrating $f'(x)$ along α we find that f is constant on $|\alpha|$, in conflict with the discreteness of f. Hence (5.10) holds for all rectifiable paths α in B_f. The inequality (5.10) is best possible, which is seen from the example I.3.1 in the case $k = 2$. Since the left hand side of (5.10) is always at least 2, we get the inequality $K_I(f) \geq 2$ whenever such a path exists.

It is an interesting open question whether $K_I(f) \geq 2$ holds always when $B_f \neq \emptyset$. Note that the branch set need not contain any rectifiable arc. To see this let J be a quasicircle in $\mathbb{R}^2$ such that J contains no rectifiable arc [LV2, p. 109], and let $g: \mathbb{R}^2 \to \mathbb{R}^2$ be a qc mapping that takes J onto a line L. By [A4], g can be extended to a qc mapping $g^*: \mathbb{R}^3 \to \mathbb{R}^3$. If $h: \mathbb{R}^3 \to \mathbb{R}^3$ is a winding mapping with $B_h = L$ as in I.3.1, then $f = h \circ g^*$ is a qm mapping with $B_f = J$.

For a long time it was an open question whether a K-qr mapping $f: G \to \mathbb{R}^n$, where $n \geq 3$ and $K > 1$ are fixed, could have $i(x, f)$ arbitrarily large on a set with limit points in G. For $n = 3$ this question is solved in the affirmative [R12] in the following strong form: There exists $K > 1$ such that for every $c > 0$ there is a K-qm mapping $f: \mathbb{R}^3 \to \mathbb{R}^3$ with $E_c = \{ x \in \mathbb{R}^3 : i(x, f) \geq c \}$ a Cantor set. In the proof the construction from [R11] is used. On the other hand, a technique similar to the proof of 5.5 can be used to show that $i(x, f)$ cannot be large even in a finite set, provided the set is in a sense evenly distributed. For details we refer to the article [RS] by S. Rickman and U. Srebro and [R12].

5.11. Example [MRV3, 4.11]. We show with an example by using Corollary 5.9 that there are discrete open mappings in dimensions $n \geq 3$ which are not topologically equivalent to any qr mapping. We perform this for $n = 3$. Given a positive integer p, we define a map ζ_p of the cylinder $C = \{x \in \mathbb{R}^3 : x_1^2 + x_2^2 < 2\}$ onto itself in cylinder coordinates by

$$\zeta_p(r, \varphi, z) = \begin{cases} (r, (1 + 4p)(\varphi - \pi/4) + \pi/4, z) & \text{if } \pi/4 \leq \varphi \leq 3\pi/4 \\ (r, \varphi, z) & \text{elsewhere in } C. \end{cases}$$

Let α_p be the translation $\alpha_p(x) = x + 2(p - 1)e_1$ and set

$$G = \bigcup_{p=1}^{\infty} \alpha_p C .$$

We define $f: G \to G$ by $f(x) = \alpha_p(\zeta_p(\alpha_p^{-1}(x)))$ for $x \in \alpha_p C$. Then f is a sense–preserving, discrete, and open map with the property $i(x, f) = p + 1$ in the set $L_p = \{x \in \mathbb{R}^3 : x_1 = 2p, x_2 = 0\}$. Suppose that there exist homeomorphisms g_1 and g_2 such that $h = g_1 \circ f \circ g_2$ is qr. If $h_p = h|g_2^{-1}(\alpha_p C)$, we have

$$\inf_{x \in B_{h_p}} i(x, h_p) = p + 1 .$$

This is in contradiction with 5.9 with p large because $K(h_p) \leq K(h)$.

In the plane the situation is different. Given a discrete open map $f: G \to \mathbb{R}^2$ there exists by Stoïlow's theorem [Sto2, p. 120] an analytic function φ and a homeomorphism g such that $f = \varphi \circ g$. For $n \geq 3$ it is not known if a given discrete open map is locally topologically equivalent to a qr map.

Chapter IV. Mappings into the n-Sphere with Punctures

The main results of this chapter are a Picard–type theorem (Theorem 2.1) on omitted values and its variants. In dimension three it is known that this result is qualitatively best possible (Theorem 2.2). The proof of 2.2 is very technical and will not be presented here. We merely refer the reader to the article [R11]. In Section 3 we will give a quantitative growth estimate for mappings of the unit ball into the n-sphere with punctures, which delivers as a special case a counterpart to the Picard–Schottky theorem of classical function theory.

The key lemma for Sections 2 and 3, as well as for the next chapter, is a certain comparison result for averages of covering numbers over $(n-1)$-spheres. Such preparatory material is collected in Section 1.

1. Covering Averages

Let $f \colon G \to \overline{\mathbb{R}}^n$ be a nonconstant K-qm mapping. We shall be dealing with covering averages over $(n-1)$-spheres. We need some notation. In this and the next chapter we shall in general drop superscript in notation for balls and spheres. Thus the unit ball and sphere are denoted by B and S. The dilatation coefficients of f are written K_O and K_I. For $y \in \mathbb{R}^n$ and for a Borel set E such that $\overline{E}$ is a compact subset of G we set

$$n(E, y) = \sum_{x \in f^{-1}(y) \cap E} i(x, f).$$

Then $n(E, y)$ is a Borel function of y which counts the number of points in $f^{-1}(y) \cap E$ with multiplicity. If Y is an $(n-1)$-dimensional sphere in $\overline{\mathbb{R}}^n$, we let $\nu(E, Y)$ be the average of $n(E, y)$ over Y with respect to the $(n-1)$-dimensional spherical measure. We write $n(r, y)$ for $n(\overline{B}(r), y)$ and call it the *counting function*. As a function of y, $n(r, y)$ is clearly upper semicontinuous. The average $\nu(\overline{B}(r), S(s))$ will be written $\nu(r, s)$. Hence

$$\nu(r, s) = \frac{1}{\omega_{n-1}} \int_S n(r, sy) \, dy.$$

If we need to express the mapping f, we write $\nu_f(r, s)$ etc. We first give a comparison lemma for the covering averages over spheres that are concentric in the Euclidean metric.

1.1. Lemma. *If $\theta > 1$, if $r, s, t > 0$, and if $\overline{B}(\theta r) \subset G$, then*

$$(1.2) \qquad \nu(\theta r, t) \geq \nu(r, s) - \frac{K_I |\log(t/s)|^{n-1}}{(\log \theta)^{n-1}} .$$

For the proof of 1.1 we require the following.

1.3. Lemma. *Let $m: S \to \mathbb{Z}$ be a nonnegative integer valued Borel function, let $0 < s < t$, and for $y \in S$ let $\beta_y: [s, t] \to \mathbb{R}^n$ be the path $\beta_y(u) = uy$. Suppose Γ^* is a family consisting of $m(y)$ essentially separate partial f-liftings of each β_y (see II.3) when y runs over S. Then*

$$(1.4) \qquad \int_S m(y)\, dy \leq K_I \left(\log \frac{t}{s} \right)^{n-1} \mathsf{M}(\Gamma^*) .$$

Proof. Set $E_k = \{ y \in S : m(y) = k \}$, $k = 0, 1, 2, \ldots$, $\Gamma_k = \{ \beta_y : y \in E_k \}$, and let Γ_k^* be the subfamily of Γ^* consisting of lifts of $\beta_y \in \Gamma_k$. Then Väisälä's inequality II.9.1 gives

$$(1.5) \qquad k\, \mathsf{M}(\Gamma_k) \leq K_I \mathsf{M}(\Gamma_k^*) .$$

We have $\mathsf{M}(\Gamma_k) = \mathcal{H}^{n-1}(E_k) \left(\log(t/s) \right)^{1-n}$. Since the families Γ_k^* are separate, we get (1.4) by summing (1.5). □

Proof of 1.1. For $y \in S$ set $m(y) = \max\left(0, n(r, sy) - n(\theta r, ty)\right)$. We may assume $s < t$. If $m(y) > 0$, there exists by II.3.2 $m(y)$ maximal $f|B(\theta r)$-liftings of β_y starting in $\overline{B}(r)$ and ending in $\partial B(\theta r)$ which are essentially separate. If Γ^* is the family of all these liftings, Lemma 1.3 gives

$$(1.6) \qquad \int_S m(y)\, dy \leq K_I \left(\log \frac{t}{s} \right)^{n-1} \omega_{n-1} (\log \theta)^{1-n}$$

because $\mathsf{M}(\Gamma^*) \leq \omega_{n-1} (\log \theta)^{1-n}$. If $E = \{ y \in S : m(y) > 0 \}$, we get

$$\int_S n(\theta r, ty)\, dy = \int_{S \setminus E} n(\theta r, ty)\, dy + \int_E n(\theta r, ty)\, dy$$

$$\geq \int_{S \setminus E} n(r, sy)\, dy + \int_E n(r, sy)\, dy - \int_E m(y)\, dy$$

$$= \int_S n(r, sy)\, dy - \int_S m(y)\, dy ,$$

which together with (1.6) gives the claim. □

Lemma 1.1 is in a sense sharp. Let $n = 2$ and let f be the mapping given in complex notation by $z \mapsto z^k$. If $s = r^k$ and $t = (\theta + \varepsilon)^k r^k$ with $\varepsilon > 0$, the left hand side of (1.2) is 0, while the right hand side is $k - k \log(\theta + \varepsilon)/\log \theta$ which tends to 0 as $\varepsilon \to 0$. Similarly one can use Example I.3.2 to test the sharpness of the order of growth of such averages in other dimensions.

It is rather straightforward to derive from Lemma 1.1 a good covering average comparison for arbitrary spheres. To do that it is convenient to compare with averages over $\overline{\mathbb{R}}^n$. Such comparisons will be used in the next chapter. Let $A(r)$ be the average of $n(r, y)$ over $\overline{\mathbb{R}}^n$ with respect to the n-dimensional spherical measure. This means by I.4.14 (note that we use here the fact $m(fB_f) = 0$) that

$$A(r) = \frac{2^n}{\omega_n} \int\limits_{\overline{\mathbb{R}}^n} \frac{n(r, y)}{(1 + |y|^2)^n} \, dy = \frac{2^n}{\omega_n} \int\limits_{\overline{B}(r)} \frac{J_f(x)}{(1 + |f(x)|^2)^n} \, dx \; .$$

Hence $A(r)$ is continuous in r, whereas an average $\nu(r, Y)$ need not be. The number $A(r)$ is in fact the pullback by $f|\overline{B}(r)$ of the normalized volume of $\overline{\mathbb{R}}^n$ (with total volume 1). Note that $\overline{\mathbb{R}}^n$, being isometric to $S^n(1/2)$, has total spherical measure $\omega_n 2^{-n}$, i.e.,

$$\omega_n 2^{-n} = \omega_{n-1} \int\limits_0^\infty \frac{s^{n-1}}{(1 + s^2)^n} \, ds \; .$$

1.7. Lemma. *If $\theta > 1$, $r > 0$, $\overline{B}(\theta r) \subset G$, and if Y is an $(n-1)$-sphere in $\overline{\mathbb{R}}^n$ of spherical radius $u \leq \pi/4$, then*

$$(1.8) \qquad \nu(r/\theta, Y) - \frac{K_I Q |\log u|^{n-1}}{(\log \theta)^{n-1}} \leq A(r) \leq \nu(\theta r, Y) + \frac{K_I Q |\log u|^{n-1}}{(\log \theta)^{n-1}},$$

where Q is a positive constant depending only on n.

Proof. By performing a spherical isometry in $\overline{\mathbb{R}}^n$ we may assume that Y is a sphere $S(t)$. Then $t \leq 1$ and $\pi/4 \leq u/t < 1$. We multiply (1.2) by $s^{n-1}/(1 + s^2)^n$, integrate with respect to s, and get

$$\nu(\theta r, t) \omega_{n-1}^{-1} \omega_n 2^{-n} \geq \int\limits_0^\infty \frac{\nu(r, s) s^{n-1}}{(1 + s^2)^n} \, ds - \frac{K_I}{(\log \theta)^{n-1}} \int\limits_0^\infty \frac{s^{n-1} |\log(t/s)|^{n-1}}{(1 + s^2)^n} \, ds \; .$$

If $I(t)$ stands for the last integral, we obtain

$$\nu(\theta r, t) \geq A(r) - \frac{K_I \, I(t) \omega_{n-1} 2^n}{\omega_n (\log \theta)^{n-1}} \; .$$

As $|\log(t/s)|^{n-1} \leq 2^{n-1} \big(|\log t|^{n-1} + |\log s|^{n-1} \big)$, we arrive after simple estimation at

$$I(t)\omega_{n-1}\omega_n^{-1}2^n \leq 2^{n-1}\left(|\log t|^{n-1} + \omega_{n-1}\omega_n^{-1}2^n\right).$$

Since $u < t \leq 1$ and $u \leq \pi/4$, the right hand inequality follows. The left hand inequality is obtained similarly by integrating the inequality

$$\nu(r,s) \geq \nu(r/\theta,t) - \frac{K_I|\log(t/s)|^{n-1}}{(\log\theta)^{n-1}}$$

after multiplying by $s^{n-1}/(1+s^2)^n$. □

1.9. Corollary. If $\theta > 1$, $r > 0$, $\overline{B}(\theta r) \subset G$, and if Y and Z are $(n-1)$-spheres in $\overline{\mathbb{R}}^n$ of spherical radii u and v at most $\pi/4$, then

$$\nu(\theta r, Z) \geq \nu(r, Y) - \frac{2^{n-1}QK_I}{(\log\theta)^{n-1}}\left(|\log u|^{n-1} + |\log v|^{n-1}\right).$$

Proof. This follows by application of 1.7 to Y and Z with θ replaced by $\theta^{1/2}$ and r by $\theta^{1/2}r$. □

1.10. Notes. Lemmas 1.1 and 1.7 were proved in a slightly weaker form in [R4, 4.1]. The present form of Lemma 1.1 is due to M. Pesonen [Pe2] and A. Hinkkanen (independently). Corollary 1.9 appears in [R10, 2.9], except that the bound for the spherical radii should be corrected to $\pi/4$ in [R10].

2. The Analogue of Picard's Theorem

The classical result of E. Picard from 1879 [Pi] says that an analytic function $f: \mathbb{R}^2 \to \mathbb{R}^2 \setminus \{a_1, a_2\}$ must be constant if a_1 and a_2 are distinct points in $\mathbb{R}^2$. The same holds for a qr mapping f of $\mathbb{R}^2$ as well. In fact, f is then of the form $f = g \circ h$ where h is a qc mapping of $\mathbb{R}^2$ onto itself and g an analytic function of $\mathbb{R}^2$ omitting two points (see [LV2, VI.2.3]), and hence constant. In the smooth case this Picard theorem for qr mappings was proved by H. Grötzsch in 1928 [Grö].

In 1967 V.A. Zorich [Zo1] asked whether a Picard–type theorem might exist also for qr mappings in dimensions $n \geq 3$. The result III.2.12 that a nonconstant qr mapping of $\mathbb{R}^n$ into $\mathbb{R}^n$ can omit at most a set of zero capacity was proved at an early stage of the theory. The true analogue of Picard's theorem was first proved in [R5] and takes the following form.

2.1. Theorem. *For every $n \geq 3$ and $K \geq 1$ there exists a positive integer $q_0 = q_0(n, K)$ depending only on n and K such that every K-qm mapping $f: \mathbb{R}^n \to \overline{\mathbb{R}}^n \setminus \{a_1, \ldots, a_q\}$ is constant whenever $q \geq q_0$ and $a_1, \ldots, a_q$ are distinct points in $\overline{\mathbb{R}}^n$.*

After the appearance of Zorich's article [Zo1] it was long conjectured that q_0 could be taken to be three even when $n \geq 3$. However, it is now known

[R11] that, at least in dimension three, this conjecture is false and that Theorem 2.1 is in fact qualitatively best possible. More precisely, the following is true.

2.2. Theorem. *For every positive integer p there exists a nonconstant $K(p)$-qr mapping $f\colon \mathbb{R}^3 \to \mathbb{R}^3$ omitting p points.*

We will give a complete proof of Theorem 2.1. The proof of 2.2, which will be omitted, is presented in [R11].

The idea of the proof of 2.1 is to produce for a fixed sphere Y a certain growth estimate for the average $\nu\big(\overline{B}(x,r), Y\big)$ as a function of r in terms of the number of omitted values. When this average grows more rapidly than the n-measure of a ball of radius r in $\mathbb{R}^n$, we get a contradiction. The main lemma for obtaining such a growth relation is:

2.3. Lemma. *Let $M > 1$, let $f\colon G \to \overline{\mathbb{R}}^n$ be a nonconstant qm mapping of a domain G with $\overline{B}(4M/3) \subset G$, let $a_1, \ldots, a_q$, $q \geq 2$, be distinct points in $\mathbb{R}^n$, and let $2 \leq \lambda \leq q$. Set*

$$\sigma_0 = \tfrac{1}{4} \min_{j \neq k} \sigma(a_j, a_k),$$

where σ is the spherical distance, and let $0 < u < v \leq \sigma_0$. Suppose further that $F_1, \ldots, F_\lambda$ are disjoint compact sets in $\overline{B}(M)$ such that for each $j \leq \lambda$ the image fF_j is in the spherical ball $\overline{D}(a_j, u)$. Then there are positive constants b' and b'' depending only on n such that

$$(2.4) \qquad \big(\mathsf{M}(\Gamma_j) - b' K_O K_I\big)\Big(\log \frac{v}{u}\Big)^{n-1} \leq b'' K_O \nu\big(4M/3, \Sigma(a_j, v)\big)$$

for all j, where Γ_j is the family of paths in $B(M) \setminus \overline{B}$ connecting F_j to $F_j^* = \bigcup_{k \neq j} F_k$.

Proof. Let $1 \leq j \leq \lambda$. We may assume $a_j = 0$. Let s and t be the Euclidean radii of the spheres $\Sigma(0, u)$ and $\Sigma(0, v)$. Then $1 < t/v$, $s/u < 16/15$, and $v/u \leq t/s$. Note that $\sigma_0 \leq \pi/8$. We define an admissible function ρ for $f\Gamma_j$ by

$$\rho(y) = \begin{cases} \dfrac{1}{\big(\log \frac{t}{s}\big)\, |y|} & \text{if } s < |y| < t; \\[2mm] 0 & \text{elsewhere.} \end{cases}$$

Then by II.(2.6),

$$\mathsf{M}(\Gamma_j) \leq K_O \int_{\mathbb{R}^n} \rho(y)^n n(M, y)\, dy = \frac{K_O\, \omega_{n-1}}{\big(\log \frac{t}{s}\big)^n} \int_s^t \frac{\nu(M, r)}{r}\, dr .$$

By the comparison lemma 1.1,

$$\nu(M,r) \leq \nu(4M/3,t) + \frac{K_I(\log(t/s))^{n-1}}{(\log(4/3))^{n-1}} , \qquad s < r < t ,$$

hence

$$\mathsf{M}(\Gamma_j) \leq \frac{K_O\,\omega_{n-1}\nu(4M/3,t)}{(\log(t/s))^{n-1}} + \frac{K_O K_I\,\omega_{n-1}}{(\log(4/3))^{n-1}} ,$$

and the lemma follows. $\qquad\qquad\square$

In the next section we will need the succeeding variant of Lemma 2.3.

2.5. Lemma. *Let M, f, σ_0, u, and v be as in 2.3 for the special case $\lambda = 2$. Suppose F_1 and F_2 are compact disjoint sets in $\overline{B}(M) \smallsetminus B$ such that F_1 and F_2 meet every sphere $S(r)$, $1 \leq r \leq M$, and that $fF_1 \subset \overline{D}(a_1, u)$, $fF_2 \subset \overline{\mathbb{R}}^n \smallsetminus D(a_1, v)$. Then*

$$\left(c(n) \log M - b' K_O K_I \right)\left(\log \frac{v}{u} \right)^{n-1} \leq b'' K_O \nu\big(4M/3, \Sigma(a_1, v)\big)$$

where $c(n)$ is as in II.(1.14).

Proof. The proof is the same as for 2.3 except that we have estimated $\mathsf{M}(\Gamma_1)$ by II.(1.14). $\qquad\qquad\square$

2.6. Convention. In the present section and the next we will use the letter b, with subscripts or other, to represent positive constants depending only on n.

2.7. Lemma. *If $f\colon \mathbb{R}^n \to \overline{\mathbb{R}}^n$ is qr and if ∞ is an essential singularity of f, then*

$$\lim_{r \to \infty} \nu(r, Y) = \infty$$

whenever Y is an $(n-1)$-dimensional sphere in $\overline{\mathbb{R}}^n$.

Proof. By 1.9 we may assume $Y = S$. By III.2.11 there exists an $\mathcal{F}_\sigma$ set $E \subset \mathbb{R}^n$ of zero capacity such that for all $r > 0$,

$$N(y, f, \mathbb{R}^n \smallsetminus B(r)) = \infty , \qquad y \in \mathbb{R}^n \smallsetminus E .$$

Set $E_0 = S \cap E$ and $F_k(r) = \{ y \in S : n(r, y) \geq k \}$. As E_0 is an increasing limit of closed sets of capacity zero, it follows from III.2.3 that $\mathcal{H}^{n-1}(E_0) = 0$. Then

$$\lim_{r \to \infty} \nu(r, 1) \geq \lim_{j \to \infty} \frac{1}{\omega_{n-1}} \int_{F_k(j)} k\, dy = \frac{k}{\omega_{n-1}} \lim_{j \to \infty} \mathcal{H}^{n-1}(F_k(j))$$

$$\geq \frac{k}{\omega_{n-1}} \mathcal{H}^{n-1}(S \smallsetminus E_0) = k$$

for every k. The lemma follows. $\qquad\qquad\square$

2.8. Remark. If ∞ is not an essential singularity of f, then f can by definition be extended to a continuous mapping $f^*\colon \mathbb{R}^n \to \overline{\mathbb{R}}^n$. By III.2.8, f^* is qm. It follows that $\nu(r,s)$ tends to the degree $\mu(f^*)$ of f^* as $r \to \infty$.

2.9. Proof of Theorem 2.1. Let $f\colon \mathbb{R}^n \to \overline{\mathbb{R}}^n \smallsetminus \{a_1,\ldots,a_q\}$ be a nonconstant K-qm mapping and $q \geq 2$. The point ∞ is necessarily an essential singularity of f because otherwise f would extend to a qm mapping f^* of $\overline{\mathbb{R}}^n$ by III.2.8. Then $f^*\overline{\mathbb{R}}^n$ would be both a compact and a proper open subset of $\overline{\mathbb{R}}^n$, which is impossible.

Let σ_0 be as in 2.3 and set $Y_j = \Sigma(a_j,\sigma_0)$. For $0 < u \leq \sigma_0$ and $r > 0$ we obtain from 1.9 an inequality

$$(2.10) \qquad \nu\big(r,\Sigma(a_j,u)\big) \geq \nu(r/2,Y_1) - b_1 K_I\big(|\log \sigma_0|^{n-1} + |\log u|^{n-1}\big)$$

for $j = 1,\ldots,q$ with $b_1 = 2^{n-1}Q/(\log 2)^{n-1}$. By 2.7 we may choose $r > 0$ such that

$$(2.11) \qquad\qquad \nu(r/2,Y_1) > 6b_1 K_I |\log \sigma_0|^{n-1}$$

and then fix $u \in\,]0,\sigma_0^2]$ such that

$$(2.12) \qquad\qquad \nu(r/2,Y_1) = 3b_1 K_I |\log u|^{n-1}.$$

Inequality (2.10) yields

$$(2.13) \qquad\qquad \nu\big(r,\Sigma(a_j,u)\big) > \nu(r/2,Y_1)/3 > 0$$

for all j. Since a_j is omitted by f, there exists for each j an unbounded component H_j of $f^{-1}D(a_j,u)$ which meets $\overline{B}(r)$. Then H_j contains a continuum F_j in $\overline{B}(3r/2) \smallsetminus B(r)$ connecting $S(r)$ and $S(3r/2)$. We apply Lemma 2.3 with $v = \sigma_0$, with the radius 1 replaced by r, and with $M = 3/2$ to the sets F_j, $j = 1,\ldots,q$. If Γ_j is as in 2.3, we obtain from 2.3 and (2.12) the inequality

$$(2.14) \qquad \big(\mathsf{M}(\Gamma_j) - b'K_O K_I\big)\nu(r/2,Y_1) \leq b_2 K_O K_I \nu(2r,Y_j)$$

for all j. Note here that $\log(\sigma_0/u) \geq |\log u|/2$ because $u \leq \sigma_0^2$. Using 1.9 again we infer with the help of (2.10) and (2.11) that

$$(2.15) \qquad \begin{aligned} \nu(4r,Y_1) &\geq \nu(2r,Y_j) - 2b_1 K_I |\log \sigma_0|^{n-1} \\ &\geq \nu(2r,Y_j)/2. \end{aligned}$$

From (2.14) and (2.15) we extract an effective growth relation on $\rho \mapsto \nu(\rho,Y_1)$, provided $\mathsf{M}(\Gamma_j)$ is large for some j. This always happens when q is sufficiently large. To see this, let δ_j be the Euclidean distance between $F_j \cap S(5/4)$ and the union of the sets $F_k \cap S(5/4)$, $k \neq j$, and let δ_j attain its minimum for $j = i$. Then $\delta_i \leq b_3 q^{1/(1-n)}$ and it follows immediately from II.(1.14) that $\mathsf{M}(\Gamma_i) \geq \beta(n,q)$, where $\beta(n,q) \to \infty$ as $q \to \infty$. The inequalities (2.14) and (2.15) then show that the ratio $\nu(4r,Y_1)/\nu(r/2,Y_1)$

becomes arbitrarily large when q is sufficiently large. There exist positive constants γ_1 and γ_2 depending only on n such that the ball $\overline{B}(4r)$ can be covered by balls $\overline{B}(x_k, r/4)$, $k = 1, \ldots, \gamma_1$, such that $x_k \in \overline{B}(4r)$ for all k and such that each point belongs to at most γ_2 of those balls. Since $E \mapsto \nu(E, Y_1)$ is clearly a measure on Borel sets,

$$\gamma_2 \sum_{k=1}^{\gamma_1} \nu\big(\overline{B}(x_k, r/4), Y_1\big) \geq \nu\big(\overline{B}(4r), Y_1\big) \, .$$

It follows that there exists an integer q_0 such that if $q \geq q_0$, then for some $x_k = z_1$ the average $\nu\big(\overline{B}(x_k, r/4), Y_1\big)$ satisfies (2.11) and (2.12) with some $u \leq \sigma_0^2$. We can then repeat this argument with the ball $\overline{B}(r/2)$ replaced by $\overline{B}(x_k, r/4)$ and continue similarly. In this way we obtain a sequence $z_0 = 0, z_1, \ldots$ of points with $z_m \in \overline{B}(z_{m-1}, r/2^{m-3})$ such that $\nu_m = \nu\big(\overline{B}(z_{m-1}, r/2^m), Y_1\big) \geq 6b_1 K_I |\log \sigma_0|^{n-1} > 0$ for all $m \geq 1$. But the balls $\overline{B}(z_{m-1}, r/2^m)$ converge to a point, which implies that $\nu_m \to 0$, a contradiction. The theorem is proved. $\qquad \square$

From the proof of Theorem 2.1 we can obtain an estimate on the number q_0. To improve the estimate we will in 2.16 give a sharper method of bounding the modulus $\mathsf{M}(\Gamma_j)$ for some j.

2.16. Lemma. Let $F_1, \ldots, F_\lambda$, $\lambda \geq 2$, be disjoint compact sets in $\overline{B}(3/2) \setminus B$ such that $S(v) \cap F_j \neq \emptyset$ for all j and all $v \in [1, 3/2]$. Let Γ_j be the family of paths in $B(3/2) \setminus \overline{B}$ that connect F_j to the union F_j^* of the sets F_k, $k \neq j$. Then there exist $b_0 > 0$ and j, $1 \leq j \leq \lambda$, such that

$$\text{(2.17)} \qquad\qquad \mathsf{M}(\Gamma_j) \geq b_0 \lambda^{1/(n-1)} \, .$$

Proof. Let $1 < v < 3/2$ and set $E_j = S(v) \cap F_j$, $E_j^* = S(v) \cap F_j^*$, $\delta_j = d(E_j, E_j^*)$. For each j we let $\xi_j \in E_j$ and $\eta_j \in E_j^*$ be points such that $|\xi_j - \eta_j| = \delta_j$. Set $\Gamma_j^v = \Delta(E_j, E_j^*; C_j)$, where C_j is the cap $\overline{B}(\xi_j, \delta_j) \cap S(v)$ (which can be the sphere $S(v)$ if $\lambda = 2$). Then

$$\text{(2.18)} \qquad\qquad \mathsf{M}_n^v(\Gamma_j^v) \geq b_0'/\delta_j$$

where $\mathsf{M}_n^v(\Gamma_j^v)$ is the n-modulus in $S(v)$. This can be seen either from direct estimates, as in [V4, 10.2], or from II.(1.9), as follows. By II.(1.9) it is enough to consider caps C_j that are smaller than a hemisphere of $S(v)$. We can then map C_j via a smooth 2-bilipschitz map φ onto a hemisphere of $S(\delta_j)$. It follows that $\mathsf{M}_n^{\delta_j}(\varphi \Gamma_j^v) \leq 2^{2n-1} \mathsf{M}_n^v(\Gamma_j^v)$ by simple estimation. Inequality (2.18) is then implied by II.(1.9).

The smaller caps $C_j' = \overline{B}(\xi_j, \delta_j/2) \cap S(v)$ are disjoint, hence

$$\omega_{n-1} v^{n-1} \geq \sum_{j=1}^{\lambda} \mathcal{H}^{n-1}(C_j') \geq \frac{\Omega_{n-1}}{3^{n-1}} \sum_{j=1}^{\lambda} \delta_j^{n-1} \, .$$

Write $\tau_j = (\Omega_{n-1}\omega_{n-1}^{-1})^{1/(n-1)}3^{-1}v^{-1}\delta_j$. Then $1 \geq \sum_1^\lambda \tau_j^{n-1}$ and Hölder's inequality gives

$$\lambda^n = \left(\sum_{j=1}^\lambda \tau_j^{(n-1)/n}\tau_j^{-(n-1)/n}\right)^n$$

$$\leq \left(\sum_{j=1}^\lambda \tau_j^{n-1}\right)\left(\sum_{j=1}^\lambda \frac{1}{\tau_j}\right)^{n-1} \leq \left(\sum_{j=1}^\lambda \frac{1}{\tau_j}\right)^{n-1} .$$

This we write as

(2.19)
$$\frac{1}{\lambda}\sum_{j=1}^\lambda \frac{1}{\tau_j} \geq \lambda^{1/(n-1)} .$$

Inequalities (2.18) and (2.19) yield

(2.20)
$$\frac{1}{\lambda}\sum_{j=1}^\lambda \mathsf{M}_n^v(\Gamma_j^v) \geq 3b_0\lambda^{1/(n-1)} .$$

If $\rho_j \in \mathcal{F}(\Gamma_j)$, then $\rho_j|S(v) \in \mathcal{F}(\Gamma_j^v)$, so (2.20) gives

$$\frac{1}{\lambda}\sum_j \int_{\mathbb{R}^n} \rho_j^n\, dm \geq \int_1^{3/2}\left(\frac{1}{\lambda}\sum_j \int_{S(v)} \rho_j^n\, d\mathcal{H}^{n-1}\right)dv \geq b_0\lambda^{1/(n-1)} .$$

Thus

$$\frac{1}{\lambda}\sum_{j=1}^\lambda \mathsf{M}(\Gamma_j) \geq b_0\lambda^{1/(n-1)} ,$$

and the lemma follows.   □

2.21. Corollary. *There exists a constant $b_4 > 0$ such that the condition*

(2.22)
$$q \geq b_4(K_O K_I)^{n-1}$$

implies the mapping in Theorem 2.1 is constant.

Proof. We apply 2.16 to the proof of 2.1. Let Γ_j satisfy (2.17) with $\lambda = q$. Then an inequality of the form (2.22) implies via (2.14) and (2.15) the required bound for the ratio $\nu(4r, Y_1)/\nu(r/2, Y_1)$.   □

2.23. Remark. The proof of 2.1 is valid for $n = 2$, but the estimate (2.22) for q_0 has no importance, because q_0 can always be taken to be 3 . However, the method can be modified so as to give the sharp result for $n = 2$. We shall discuss this in the next chapter in connection with a defect relation V.7.1 for the case $n = 2$. If $n \geq 3$, we know that a nonconstant 1-qm mapping is a restriction of a Möbius transformation. Hence in this case the bound for q_0

says nothing either. In fact, we shall prove in VI.8.14 that for each $n \geq 3$ there exists $K_n > 1$ such that every nonconstant K_n-qr mapping is a local homeomorphism. Hence, if a K_n-qr mapping is of $\mathbb{R}^n$, $n \geq 3$, it is qc or constant by Zorich's theorem III.3.8 and the bound for q_0 is irrelevant for $1 \leq K \leq K_n$. At this point there exists no effective estimate for K_n. Also, no better bound than (2.22) is known for q_0 in the general case. Note that an explicit value for b_4 can be obtained from the proof.

With only minor modifications of the proof of 2.1 we will in 2.27 obtain a counterpart of the big Picard's theorem. For this we need the following variant for rings of the comparison lemma 1.1.

2.24. Lemma. Let $f\colon G \to \overline{\mathbb{R}}^n$ be a nonconstant qm mapping, let $0 < r_1 < r_2$, and let $\theta > 1$. Write $R = \overline{B}(r_2) \smallsetminus B(r_1)$ and $R_\theta = \overline{B}(\theta r_2) \smallsetminus B(r_1/\theta)$. If $R_\theta \subset G$, then

$$(2.25) \qquad \nu(R_\theta, t) \geq \nu(R, s) - \frac{2K_I |\log(t/s)|^{n-1}}{(\log \theta)^{n-1}}.$$

Proof. We only have to observe that in place of the family Γ^* in the proof of 1.1 we obtain a family Γ_1^* of maximal $f|R_\theta$-liftings that start in R and end in ∂R_θ. The modulus has then the estimate $M(\Gamma_1^*) \leq 2\omega_{n-1}(\log \theta)^{1-n}$, which accounts for the factor 2 in (2.25). □

2.26. Remark. From 2.24 we obtain in a straightforward manner analogues of inequalities 1.7 and 1.9 for rings. Also, Lemma 2.3 can easily be modified to the case where the balls $\overline{B}$, $\overline{B}(M)$, and $\overline{B}(4M/3)$ are replaced by rings R_γ, $\gamma = 1, M, 4M/3$, where R is some ring $R(r_1, r_2) = \overline{B}(r_2) \smallsetminus B(r_1)$.

2.27. Theorem. Let $\rho > 0$ and let $f\colon \mathbb{R}^n \smallsetminus \overline{B}(\rho) \to \overline{\mathbb{R}}^n \smallsetminus \{a_1, \ldots, a_q\}$ be a qm mapping. There exists a constant $b_4' > 0$ such that if

$$(2.28) \qquad q \geq b_4'(K_O K_I)^{n-1},$$

then f has a limit at ∞.

Proof. Suppose ∞ is an essential singularity. According to the proof of 2.1 we only have to find a ball $\overline{B}(z_0, r_0)$ for which (2.11) is satisfied in place of $\overline{B}(r/2)$ and such that $\overline{B}(x_0, 16r_0) \subset \mathbb{R}^n \smallsetminus \overline{B}(\rho)$.

For $\rho < r_1 < r_2$ we write $R(r_1, r_2) = \overline{B}(r_2) \smallsetminus B(r_1)$ as in 2.26. Set $\rho_1 = 16\rho$. By the proof of 2.7 we obtain

$$\lim_{r \to \infty} \nu(R(\rho_1, r), Y) = \infty$$

for all $(n-1)$-spheres Y in $\overline{\mathbb{R}}^n$. Let Y_j, $j = 1, \ldots, q$, be spheres as in the proof of 2.1. Fix r such that $\nu(R(\rho_1, r), Y_1)/2$, in place of $\nu(r/2, Y_1)$, satisfies (2.11) and such that $\nu(R(\rho_1/8, \rho_1), Y_1) \leq \nu(R(\rho_1, r), Y_1)/4$. By using

2.24 (instead of 1.1) together with Remark 2.26, we arrive as in the proof of 2.1 at the fact that $\nu\big(R(\rho_1/8, 8r), Y_1\big)/\nu\big(R(\rho_1, r), Y_1\big)$ becomes arbitrarily large when q is sufficiently large. Then the major part of the measure $\nu\big(R(\rho_1/8, 8r), Y_1\big)$ is concentrated in $R(r, 8r)$. This time we cover $R(r, 8r)$ by balls $\overline{B}(x_k, r/32)$, $x_k \in R(r, 8r)$, similarly to earlier. As a result we find $x_k = z_0$ such that $\nu\big(\overline{B}(x_k, r/32), Y_1\big)$ satisfies (2.11) and (2.12) for some $u \leq \sigma_0^2$. We then can continue as before to get a contradiction. Also the bound in (2.28) follows similarly.   □

2.29. Comments on existence of quasiregular mappings. Theorems 2.1 and 2.2 give a solution to a special case of the following general existence problem. Given two oriented Riemannian n-manifolds M and N does there exist a nonconstant qr map of M into N? So far very little is known about this question in general and there is much left for future research, but some problems in this area have been solved.

For $n = 3$ J. Jormakka studied in [Jo] the case $M = \mathbb{R}^3$ and N compact. K. Peltonen has in [Pel] given the existence proof for the case where M is arbitrary and $N = S^n$. Techniques similar to those in the proof of Theorem 2.1 can for example be used to show that a qr map of $\mathbb{R}^n$ into $T^n \# N'$ is constant, where T^n is the n-torus, N' is a compact Riemannian n-manifold with nontrivial cohomology in some dimension m, $1 \leq m \leq n-1$, and $\#$ is the connected sum operation (see [Pel]). An interesting open question is whether the same is true in dimension four for $N = S^2 \times S^2 \# S^2 \times S^2$. M. Gromov posed the question whether the claim in Theorem 2.1 remains true if $\overline{\mathbb{R}}^n \setminus \{a_1, \dots, a_q\}$ is provided with an arbitrary Riemannian metric. This problem was partially solved by I. Holopainen in [H1]. The final positive solution was obtained by Holopainen and Rickman [HR2]. The proof is based on a recent purely potential theoretic proof of 2.1 by A. Eremenko and J.L. Lewis [EL] and on a method from [H1] to construct extremals of variational integrals (see VI.1.10) with prescribed singularities. Extensions of Theorem 2.1 in other directions have been given by M. Vuorinen [Vu13] and by P. Järvi [J].

An interesting object in connection of the existence problem is the Heisenberg group. It is a Lie group which in dimension three has the representation

$$H_1 = \left\{ \begin{pmatrix} 1 & x & z \\ 0 & 1 & y \\ 0 & 0 & 1 \end{pmatrix} : x, y, z \in \mathbb{R}^1 \right\}$$

as a matrix group. We provide H_1 with a left invariant Riemannian metric. By [Pa2] H_1 satisfies an isoperimetric inequality

$$(2.30) \qquad\qquad \mathrm{vol}(A) \leq b\mathcal{H}^2(\partial A)^{4/3}$$

for all $A \subset\subset H_1$. It follows from (2.30) that H_1 has positive (3-)capacity at infinity (see [Pa2] or [Gro1]). Therefore, from the proof of III.2.12 we conclude that a qr map of $\mathbb{R}^3$ into H_1 is constant. On the other hand, the claims in

Theorems 2.1 and 2.2 remain true if the domain of the map is replaced by H_1 [HR3]. For Theorem 2.1 this is surprising because the capacity at infinity of H_1 is positive. The proof depends on ideas from [HR2] and on Holopainen's article [H1]. The weaker theorem that a qr map of H_1 into B^3 is constant was proved in [H2]. This latter result was earlier pointed out to the author by M. Gromov. The qc theory on the Heisenberg group with respect to the so–called Carnot–Carathéodory metric (a limiting metric of a one parameter family of Riemannian metrics) has been studied by A. Korányi and H.M. Reimann [KR1], [KR2]. A qr theory with respect to this metric has not been developed.

2.31. Notes. The original proof of 2.1 in [R5] differs from the one given here in that variational integrals are used in [R5]. The present proof is also simpler. The proof in 2.9 is a simplification of a proof in [R9], where ideas from [R7] are used. Still another proof is given in [R10] in connection with a counterpart to the Picard–Schottky theorem. We have chosen a more unified approach here. Thus we obtain the main result in [R10] in the next section by adding some preliminary steps to the proof of 2.1. Lemma 2.16 is from [R7, 6.6] and we have here simplified the proof. Theorem 2.27 without the bound (2.28) is in [R5]. The first purely potential theoretic proof of 2.1 was given by A. Eremenko and J.L. Lewis in [EL]. Very recently Lewis simplified the method of [EL] in [Le2].

3. Mappings of a Ball

We will now study qm mappings of the unit ball B into $\overline{\mathbb{R}}^n \smallsetminus \{a_1, \ldots, a_q\}$, where $a_1, \ldots, a_q$, $q \geq 2$, are distinct points in $\overline{\mathbb{R}}^n$. We will use the generic notation W_q for $\overline{\mathbb{R}}^n \smallsetminus \{a_1, \ldots, a_q\}$. If $n = 2$, if $q \geq 3$, and if $f \colon B \to W_q$ is a meromorphic function, then f is distance decreasing when B and W_q are equipped with their Poincaré metrics. This follows from the Schwarz–Pick lemma applied to a lifting $\tilde{f} \colon B \to B$ of f to the universal analytic cover B of W_q. More precisely, let $\pi \colon B \to W_q$ be an analytic covering projection. When B is given the Poincaré metric $d\rho^2$, π induces a complete metric $d\tau^2$ on W_q, called the Poincaré metric of W_q. As the lifting $\tilde{f}$ is distance decreasing, so is $f = \pi \circ \tilde{f}$. For $q = 3$ one derives from estimates on the metric $d\tau^2$ the Picard–Schottky theorem (see [A6, Theorem 1–13]).

In this section we will prove a counterpart for this distance decreasing result in dimensions $n \geq 3$. We cannot copy the method for $n = 2$, for when $n \geq 3$ W_q already is simply connected and the ball B does not appear as universal covering space of W_q. We will instead give W_q directly a metric $d\tau^2$ whose singular behavior near the points a_j mimics the Poincaré metric for $n = 2$. It is interesting to note that such a metric arises naturally as a metric induced by a qm branched covering $h \colon B \to W_q$ which is automorphic with respect to a discrete Möbius group acting on B with finite volume and which is injective in Dirichlet fundamental domains. In special cases explicit construction of such h can be given (see [R10, Section 3]).

To give one metric $d\tau^2$ of the type sought we set

$$\sigma_0 = \tfrac{1}{4} \min_{j \neq k} \sigma(a_j, a_k),$$

$U_j = D(a_j, \sigma_0) \setminus \{a_j\}$, and $U = \bigcup_j U_j$. Then we define $d\tau^2$ as a conformal metric with respect to the spherical metric $d\sigma^2$ by

$$(3.1) \qquad\qquad d\tau^2 = \gamma^2 d\sigma^2,$$

where γ is continuous in W_q, is constant in $W_q \setminus U$, and satisfies

$$(3.2) \qquad\qquad \gamma(y) = \frac{1}{\sigma(a_j, y) \log\big(1/\sigma(a_j, y)\big)} \quad \text{if } y \in U_j.$$

Up to a bounded variable factor, (3.1) is in the case $n = 2$ the Poincaré metric in W_q.

Our main result in this section is the following theorem which says that a qm mapping $B \to W_q$ is in a global sense Lipschitz when W_q is provided with the metric (3.1) and B is given the Poincaré metric $d\rho^2$ defined by

$$d\rho^2 = \frac{4|dx|^2}{(1 - |x|^2)^2} .$$

Note that locally such mappings need only be Hölder continuous.

3.3. Theorem. *For every* $n \geq 3$ *and* $K \geq 1$ *there exists a number* $\delta = \delta(n, K) > 0$ *such that the following holds. If* $f: B \to W_q = \mathbb{R}^n \setminus \{a_1, \dots, a_q\}$ *is a* K-qm *mapping and* $q \geq q_0$, *where* $q_0 = q_0(n, K)$ *is the integer in* 2.1, *then*

$$(3.4) \qquad\qquad \tau\big(f(x), f(z)\big) \leq C \max\big(\rho(x, z), \delta\big) , \quad x, z \in B .$$

Here the constant C *depends only on* n, K, *and* σ_0 *and the distance* τ *is given by the metric* (3.1) *on* W_q.

Proof. We may assume f is nonconstant. The strategy is to show that violation of (3.4) with suitable C leads to the existence of $x \in B$ and $r > 0$ such that $\overline{B}(x, 8r) \subset B$ and such that, when $\overline{B}(r/2)$ is replaced by $\overline{B}(x, r/2)$, (2.11) is satisfied. Then we can proceed as in the proof of 2.1.

We are going to apply Lemma 2.5. To this end we choose M so that $c(n) \log M = 2b' K_O K_I$. Then we set $\delta = 2^{-6} M^{-1}$. Let $x, z \in B$ be points with $\rho(x, z) = \delta$ and write $\xi = f(x)$, $\zeta = f(z)$. Because f is open, it suffices to find a satisfactory estimate for $\tau(\xi, \zeta)$. Of the points ξ, ζ let ζ be nearest some a_j in the σ distance. We let j be 1. The τ-diameter of $\mathbb{R}^n \setminus U$ is bounded by a constant $\alpha(n, \sigma_0)$ depending only on n and σ_0. We may therefore assume that $\zeta \in U_1$. For any point η in $\Sigma(a_1, \sigma_0)$ we then have $\tau(\xi, \eta) \leq \alpha(n, \sigma_0) + \tau(\eta, \zeta)$. It follows that we may assume that ξ also belongs to U_1.

Set $u = \sigma(a_1, \zeta)$, $v = \sigma(a_1, \xi)$. We write $r_0 = |x - z|$ and apply 2.5 to the ball $\overline{B}(x, r_0)$ in place of $\overline{B}$. Since a_1 and some other a_k are omitted, we find F_1 and F_2 as in 2.5. With our choice of M, 2.5 leads to an inequality

$$(3.5) \qquad \left(\log \frac{v}{u}\right)^{n-1} \leq \overline{b}\,\nu\big(\overline{B}(x, 4r_0 M/3), \Sigma(a_1, v)\big) \,.$$

By 1.9 we then obtain

$$(3.6) \qquad \left(\log \frac{v}{u}\right)^{n-1} \leq \overline{b}\left(\nu\big(\overline{B}(x, 2r_0 M), Y_1\big) + 2b_1 K_I |\log v|^{n-1}\right),$$

where b_1 is the constant in (2.10). From formula (3.2) we get the estimate

$$(3.7) \qquad \tau(\xi, \zeta) < \log \frac{|\log u|}{|\log v|} + 4\,.$$

Suppose $\tau(\xi, \zeta) > 6$. Then we can replace (3.7) by

$$\exp\!\left(\tfrac{1}{2}\tau(\xi, \zeta)\right) < \frac{1}{|\log v|}\log \frac{v}{u}\,.$$

If we now substitute this into (3.6) and use $v \leq \sigma_0$, we obtain

$$(3.8) \qquad \left(\exp\!\left(\tfrac{1}{2}\tau(\xi, \zeta)\right)\right)^{n-1} < \overline{b}\left(\frac{\nu\big(\overline{B}(x, 2r_0 M), Y_1\big)}{|\log \sigma_0|^{n-1}} + 2b_1 K_I\right).$$

Inequality (3.8) gives a constant $A(n, \sigma_0, K)$ such that $\tau(\xi, \eta) > A(n, \sigma_0, K)$ implies

$$\nu\big(\overline{B}(x, 2r_0 M), Y_1\big) > 6b_1 K_I |\log \sigma_0|^{n-1}\,,$$

which is just (2.11) with $\overline{B}(r/2)$ replaced by $\overline{B}(x, 2r_0 M)$. Write $r = 4r_0 M$. Then $8r < 1 - |x|$, as simple estimates on the hyperbolic metric show. Therefore, we can adapt the proof of 2.1 to get a contradiction. □

As a corollary we obtain a counterpart to the Picard–Schottky theorem.

3.9. Corollary. Let $f: B \to \mathbb{R}^n \setminus \{a_1, \ldots, a_{q-1}\}$, $n \geq 3$, be K-qr and let $q \geq q_0$, where q_0 is as in 2.1. Then

$$(3.10) \qquad \log|f(x)| \leq C_0\big(|\log \sigma_0| + \log^+ |f(0)|\big)\big(1 - |x|\big)^{-C_1},$$

where σ_0 is as in 3.3 and C_0 and C_1 are positive constants which depend only on n, K, and σ_0.

Proof. We set $a_q = \infty$ and apply 3.3. Since $|f(x)| < \pi/(2\sigma(f(x), \infty))$, we need only concern ourselves with the case where $f(x) \in D(\infty, \sigma_0)$. If $f(0) \in \overline{D}(\infty, \sigma_0)$, then $|f(0)| > 3/(4\sigma(f(0), \infty))$ and 3.3 yields

$$\text{(3.11)} \qquad \frac{\log|f(x)|}{\log|f(0)|} < \frac{4|\log\sigma(\infty,f(x))|}{|\log\sigma(\infty,f(0))|} \le 4\exp\tau(f(0),f(x))$$

$$\le 4\exp\big(C\rho(0,x)+C\delta\big) \le C_0\big(1-|x|\big)^{-C_1}.$$

If $f(0) \notin \overline{D}(\infty,\sigma_0)$, we choose a point $\eta \in \Sigma(\infty,\sigma_0)$ with $\tau(f(0),f(x)) > \tau(\eta,f(x))$ and obtain

$$\text{(3.12)} \qquad \frac{\log|f(x)|}{|\log\sigma_0|} < 4\exp\tau(\eta,f(x)) < 4\exp\tau(f(0),f(x))$$

$$\le C_0\big(1-|x|\big)^{-C_1}.$$

Inequalities (3.11) and (3.12) imply (3.10). $\square$

3.13. Remark. As in the classical case, Corollary 3.9 implies 2.1. It is enough to consider a K-qr mapping $f: \mathbb{R}^n \to \mathbb{R}^n \setminus \{a_1,\dots,a_{q-1}\}$ where $q \ge q_0$. Let $z \in \mathbb{R}^n$ and let h be the mapping $x \mapsto 2|z|x$ of B. Then 3.9 applied to $f \circ h$ gives

$$\log|f(z)| \le C_0(|\log\sigma_0| + \log^+|f(0)|)2^{C_1}.$$

Hence f is bounded and therefore constant by III.1.14.

By means of Theorem 3.3 we are able to sharpen the equicontinuity result III.2.7.

3.14. Corollary. *Let G be a domain in $\overline{\mathbb{R}}^n$, let q_0 be as in 2.1, and let $\mathcal{F}$ be a family of K-qm mappings $f: G \to \overline{\mathbb{R}}^n$ such that f omits distinct points $a_1^f,\dots,a_{q_0}^f$. Suppose that for some positive constant γ,*

$$\text{(3.15)} \qquad \sigma_0^f = \tfrac{1}{4}\min_{j\ne k}\sigma(a_j^f,a_k^f) \ge \gamma$$

for all $f \in \mathcal{F}$. Then $\mathcal{F}$ is equicontinuous.

Proof. To check the equicontinuity at a point $x \in G$ we may assume $x = 0$ and $\overline{B} \subset G$. Let $d\tau_f^2$ be the metric in (3.1) defined by the points $a_1^f,\dots,a_{q_0}^f$. Then by (3.15) and by 3.3 there exists $\gamma' > 0$ such that $f\overline{B}(1/2)$ can meet the ball $D(a_j^f,\gamma')$ for at most one j. Otherwise $\tau_f(f\overline{B}(1/2))$ would exceed a bound of the form $C\max(\rho(\overline{B}(1/2)),\delta)$, where C depends only on n, K, and γ. We may assume $f\overline{B}(1/2) \cap D(a_1^f,\gamma') = \emptyset$ for all $f \in \mathcal{F}$. Let T_f be a spherical isometry that takes a_1^f to ∞. Then for some $a > 0$, $T_f f\overline{B}(1/2) \subset B(a)$ for all $f \in \mathcal{F}$. The Hölder continuity result III.1.11 says that for some $D > 0$ and $\alpha = K_I^{1/(n-1)}$ we have $|T_f f(x) - T_f f(0)| \le D|x|^\alpha$ for all $x \in B(1/4)$ and $f \in \mathcal{F}$. Since $\sigma(f(x),f(0)) \le |T_f f(x) - T_f f(0)|$, the equicontinuity of $\mathcal{F}$ follows. $\square$

3.16. Notes. The results 3.3 and 3.9 are from [R10]. The proof of 3.3 is a simplified version of that in [R10]. Since the target space for the mappings

f in 3.14 is the compact space $\overline{\mathbb{R}}^n$, the family $\mathcal{F}$ is also normal according to Ascoli's theorem. Normal families of K-qm mappings have been discussed from various viewpoints by R. Miniowitz [Mi3] and M. Vuorinen [Vu8], [Vu13]. In [Vu8], [Vu13] (see too [Vu12, Section 13]) Vuorinen also considers so-called normal qm mappings, a notion introduced for meromorphic functions by O. Lehto and K.I. Virtanen [LV1]. Among other things Vuorinen shows [Vu8, 5.18] that the constant C_1 in 3.9 can be made to depend only on n and K.

Chapter V. Value Distribution

Value distribution theory is concerned with how evenly a given mapping covers points. For example, the discussion in the preceding chapter of how many points in the target space can be omitted by a mapping falls under this heading.

In this chapter we will study value distribution for qm mappings in a much sharper sense. To state precise results it is relevant to consider the covering properties with respect to some natural exhaustion of the domain of the mapping. For example, if the domain is $\mathbb{R}^n$, one natural exhaustion is by concentric balls.

The main result in this chapter is a defect relation for qm mappings which is a higher dimensional counterpart to a well–known relation discovered by L.V. Ahlfors [A2] and which implies the Picard–type result Theorem IV.2.1. Its proof will occupy the first six sections. The defect relation is known to be qualitatively sharp when $n = 3$. Because this sharpness result is even more technical than the proof in [R11] of Theorem IV.2.2, we leave it out and simply refer to [R16]. In Section 7 we will show that our method also implies the sharp result in dimension two. For this we will modify the proof given by M. Pesonen in [Pe2]. In Section 9 we study questions opposite to defect relations.

1. Defect Relation

The classical value distribution theory for meromorphic functions from $\mathbb{R}^2$ into $\overline{\mathbb{R}}^2$ developed by R. Nevanlinna [N1] gives a far reaching sharpening of Picard's theorem. One of the consequences is Nevanlinna's defect relation [N1, p. 97]. In 1935 L.V. Ahlfors [A2] established a parallel theory which has a very geometric character. According to [A2, p. 181] (see also [N2, p. 350]), if $f\colon \mathbb{R}^2 \to \overline{\mathbb{R}}^2$ is a nonconstant meromorphic function, then there exists a set $E \subset [1, \infty[$ of finite *logarithmic measure*, meaning

$$\int\limits_E \frac{dr}{r} < \infty \, ,$$

such that

$$(1.1) \qquad \limsup_{\substack{r \to \infty \\ r \notin E}} \sum_{j=1}^{q} \left(1 - \frac{n(r, a_j)}{A(r)}\right)_+ \le 2$$

whenever $a_1, \ldots, a_q$ are distinct points in $\overline{\mathbb{R}}^2$. In fact, Ahlfors allows here not just meromorphic functions, but also smooth quasiregular mappings. We recall from IV.1 the notation $n(r, y)$ for the counting function and $A(r)$ for its average with respect to the spherical measure. Here $\alpha_+ = \max(0, \alpha)$ if $\alpha \in \mathbb{R}^1$. The Picard theorem is clearly a trivial corollary of (1.1). We will call (1.1) Ahlfors' *pointwise defect relation* and refer to

$$(1.2) \qquad \delta(r, a_j) = \left(1 - \frac{n(r, a_j)}{A(r)}\right)_+$$

as the *defect of a_j in the disk* $\overline{B}(r)$ or the *defect function* of a_j. If $\delta(r, a_j)$ is positive, it means that the point a_j gets covered less by $f|\overline{B}(r)$ than does an average point of $\overline{\mathbb{R}}^2$. Inequality (1.1) tells us that asymptotically the defect functions for distinct points can contribute a sum of at most 2, provided we avoid a fixed set of radii r which is thin at ∞. This discussion is valid when we use the disks $B(r)$ to exhaust $\mathbb{R}^2$. If we replace the disks $B(r)$ by some other natural exhaustion, say disks $B(x_0, r)$ with x_0 fixed, then a similar statement is true, but the set E and the defect functions change.

The counterpart to Ahlfors' result (1.1) for qm mappings is the following.

1.3. Theorem. *Let $f \colon \mathbb{R}^n \to \overline{\mathbb{R}}^n$ be a nonconstant K-qm mapping, where $n \ge 3$. Then there exists a set $E \subset [1, \infty[$ of finite logarithmic measure and a constant $C(n, K) < \infty$ depending only on n and K such that*

$$(1.4) \qquad \limsup_{\substack{r \to \infty \\ r \notin E}} \sum_{j=1}^{q} \left(1 - \frac{n(r, a_j)}{A(r)}\right)_+ \le C(n, K)$$

whenever $a_1, \ldots, a_q$ are distinct points in $\overline{\mathbb{R}}^n$.

Theorem IV.2.1 is clearly a corollary of 1.3. In dimension three we can arbitrarily specify the limits of defect functions in accordance with Theorem 1.3 as follows.

1.5. Theorem. *Let $a_1, a_2, \ldots$ be a sequence of points in $\overline{\mathbb{R}}^3$ and let $\delta_1, \delta_2, \ldots$ be numbers such that $0 \le \delta_j \le 1$ and*

$$(1.6) \qquad \sum_j \delta_j \le P$$

for some integer P. Then there exists a set $E \subset [1, \infty[$ of finite 1-measure and a K-qm mapping $f \colon \mathbb{R}^3 \to \overline{\mathbb{R}}^3$, with K depending only on P, such that

(1.7) $$\lim_{r \to \infty} \left(1 - \frac{n(r, a_j)}{A(r)} \right) = \delta_j \, ,$$

(1.8) $$\lim_{r \to \infty} \left(1 - \frac{n(r, y)}{A(r)} \right) = 0 \, , \quad \text{if } y \in \mathbb{R}^3 \setminus \{a_1, a_2, \dots\} \, .$$

Apart from the exceptional set E, Theorem 1.5 demonstrates the qualitative sharpness of Theorem 1.3 for $n = 3$. For dimensions $n \geq 4$ the analogue of 1.5 remains an open problem. Theorems 1.3 and 1.5 (as well as Theorem 1.9) are from [R16]. A weaker defect relation was proved in [R7], but it is not sharp in the above sense. For dimension two it was a long standing problem in Nevanlinna theory whether arbitrary defects δ_j with defect sum at most two (i.e., $P = 2$ in (1.6)) could be realized by meromorphic functions. The full solution to the inverse problem of Nevanlinna theory was given by D. Drasin in [D], the realization of arbitrary defects being a part.

We will spread the proof of Theorem 1.3 over six sections. We closely follow the presentation in [R16] and also borrow a great deal from the article [R7]. Since we intend to apply the method to give a proof also for Ahlfors' result (1.1), we formulate the material in Sections 2–6 for dimensions $n \geq 2$.

The proof of Theorem 1.5 is omitted. It is presented in [R16]. The much easier case, namely, 1.5 with the bound $P = 2$, was proved in [R3] for $n = 3$. The method was later extended to all dimensions $n \geq 3$ [R6].

Our approach also allows us to establish the following defect relation for mappings of a ball subject to a growth condition on $A(r)$. (For the classical case in dimension two, see [N2, p. 352].)

1.9. Theorem. *Let* $f \colon B \to \overline{\mathbb{R}}^n$, $n \geq 3$, *be a K-qm mapping such that*

$$\limsup_{r \to 1}(1 - r)A(r)^{1/(n-1)} = \infty \, .$$

Then there exists a set $E \subset \,]0, 1[$ *with*

$$\liminf_{r \to 1} \frac{m_1(E \cap [r, 1[)}{1 - r} = 0$$

and a constant $C(n, K) < \infty$ *such that*

$$\limsup_{\substack{r \to 1 \\ r \notin E}} \sum_{j=1}^{q} \left(1 - \frac{n(r, a_j)}{A(r)} \right)_+ \leq C(n, K)$$

whenever $a_1, \dots, a_q$ *are distinct points in* $\overline{\mathbb{R}}^n$.

1.10. Remark. In classical value distribution theory – both in Nevanlinna's and Ahlfors' theory – there is a close connection of the total ramification and the spherical average, see for example [N2, p. 350]. Something similar should be true also for qr mappings in dimensions $n \geq 3$, but so far there does not exist even a good guess for a relevant theorem.

2. Coverings and Decomposition of Balls

In this section we do some preliminary work toward the proof of 1.3. The main idea of the proof is to produce, from the assumption that the left hand side of (1.4) is large, growth relations for averages over a fixed sphere as we did in the proof of IV.2.1. However, the situation is now much more delicate. We must bring into play the total contribution of the defect functions. This means that we cannot be content with the study of growth starting from one ball in the domain, as we could in the earlier proof, but we are more or less forced to consider these growth relations simultaneously in a number of different parts. The proof involves a detailed study of the behavior of liftings of paths.

Suppose now that $n \geq 2$ and that $f : \mathbb{R}^n \to \overline{\mathbb{R}}^n$ is a K-qm mapping and that ∞ is an essential singularity of f. These assumptions are retained throughout the upcoming sections, including most of Section 6. We observe from IV.1.7 and IV.2.7 that $A(r) \to \infty$ as $r \to \infty$. Remember that if ∞ is not an essential singularity, then f extends to a qm mapping f^* of $\mathbb{R}^n$ (III.2.8), in which event 1.3 becomes trivial (cf. Remark IV.2.8). We start out by giving the construction of the set E that appears in Theorem 1.3. Recall the notation $\nu(r, Y)$ from IV.1.

2.1. Lemma. *There exists a set $E \subset [1, \infty[$ of finite logarithmic measure such that the following is true: If $0 < \varepsilon_0 < 1/5$ and if for $s > 0$ we write*

$$(2.2) \qquad s' = s + \frac{s}{\varepsilon_0 A(s)^{1/(n-1)}} ,$$

then there exists an increasing function $w : [0, \infty[\to [D_{\varepsilon_0}, \infty[$ such that

$$(2.3) \qquad \left| \frac{\nu(s, Y)}{A(s')} - 1 \right| < \varepsilon_0$$

and

$$(2.4) \qquad \frac{\nu(s, Y)}{\nu(s', Y)} \geq 1 - \varepsilon_0$$

hold whenever Y is an $(n-1)$-sphere in $\overline{\mathbb{R}}^n$ with spherical radius $u \leq \pi/4$ and $s' \in [w(|\log u|), \infty[\smallsetminus E$. Here $D_{\varepsilon_0} > 0$ and $A(D_{\varepsilon_0}) > 1/\varepsilon_0^{n-1}$.

Proof. For each integer $m \geq 2$ we first construct a set H_m as follows. Set $\varphi(r) = m^{-2} A(r)^{1/(n-1)}$. We can choose $r_0'' = r_0''(m) \geq 1$ such that $\varphi(r_0'') \geq 1$. Set

$$F_m = \left\{ r > r_0'' : A\left(r + \frac{2r}{\varphi(r)}\right) > \frac{m}{m-1} A(r) \right\} .$$

Assuming first that $F_m \neq \emptyset$, we define inductively a (possible terminating) sequence $r_0'' \leq r_1 < r_1'' \leq r_2 < r_2'' \leq \ldots$ by $r_k = \inf\{r > r_{k-1}'' : r \in F_m\}$ and $r_k'' = r_k + 2r_k/\varphi(r_k)$. Then

$$F_m \subset \bigcup_{k \geq 1} [r_k, r_k''] .$$

Set $\rho_k = r_k'' + 2r_k''/\varphi(r_k)$ and define

$$E_m = \bigcup_{k \geq 1} [r_k, \rho_k] .$$

For the logarithmic measure of E_m we get the estimate

$$\int_{E_m} \frac{dr}{r} \leq \sum_{k \geq 1} \frac{\rho_k - r_k}{r_k} \leq \sum_{k \geq 1} \frac{1}{r_k} \left(r_k'' - r_k + \frac{2r_k''}{\varphi(r_k)} \right)$$

$$\leq \sum_{k \geq 1} \frac{8m^2}{A(r_k)^{1/(n-1)}} .$$

The last sum is finite because $A(r_{k+1}) \geq A(r_k'') \geq m(m-1)^{-1}A(r_k)$. If $F_m = \emptyset$, set $E_m = \emptyset$.

For any measurable $H \subset [1, \infty[$ we write

$$\eta(H) = \int_H \frac{dr}{r} .$$

We choose a sequence $d_2 < d_3 < \ldots$ of numbers such that $3r_0''(m) \leq d_m$ and $\eta(H_m) < 2^{-m}$, where $H_m = E_m \cap [d_m, \infty[$. Set $E = \bigcup_{m \geq 2} H_m$. Then $\eta(E) < \infty$.

Let $0 < \varepsilon_0 < 1/5$ and let Y be as in the lemma. We choose $m \geq 4$ to be the least integer such that $m^2/(m-1)^2 < 1 + \varepsilon_0/2$ and $2^{n-1}K_I Q |\log u|^{n-1} \leq m$, where Q is the constant in Lemma IV.1.7. Then $2/m^2 < \varepsilon_0$ also holds. Let $\rho \geq d_m$ and $\rho \notin E$. With the notation in the first part of the proof there exists $r \geq r_0''$ such that $\rho = (r + r/\varphi(r))' = s'$. Then $r \notin F_m$. For if $r \in [r_k, r_k'']$ for some k, then

$$r_k < s' = s + \frac{s}{\varepsilon_0 A(s)^{1/(n-1)}} < r + \frac{r}{\varphi(r)} + \frac{r + r/\varphi(r)}{2\varphi(r)}$$

$$\leq r + \frac{2r}{\varphi(r)} \leq \rho_k$$

and $\rho = s'$ would belong to E_m. Applying IV.1.7 with $\theta = 1 + 1/\varphi(r)$ and using $\log \theta \geq 1/(2\varphi(r))$ we then obtain

$$\nu(s, Y) \geq A(r) - \frac{K_I Q |\log u|^{n-1}}{(\log \theta)^{n-1}}$$

(2.5)
$$\geq A(r) - \frac{mA(r)}{m^{2n-2}} \geq \frac{m-1}{m} A(r)$$

$$\geq \left(\frac{m-1}{m} \right)^2 A\left(r + \frac{2r}{\varphi(r)} \right) \geq \left(\frac{m-1}{m} \right)^2 A(s') ,$$

so

$$(2.6) \qquad 1 - \frac{\varepsilon_0}{2} \le \frac{\nu(s, Y)}{A(s')} .$$

Since $F_m \subset E_m$, we also have $s' \notin F_m$. Applying IV.1.7 with $\theta = 1 + 1/\varphi(s')$ we get

$$\nu(s', Y) \le A\left(s' + \frac{s'}{\varphi(s')}\right) + \frac{1}{m} A(s')$$

$$\le \left(\frac{m}{m-1} + \frac{1}{m}\right) A(s') < \left(\frac{m}{m-1}\right)^2 A(s'),$$

whence

$$(2.7) \qquad \frac{\nu(s', Y)}{A(s')} \le 1 + \frac{\varepsilon_0}{2} .$$

From (2.6) and (2.7) we obtain (2.3) and (2.4). Since we have $2/m^2 < \varepsilon_0$ independent of Y, we can put $D_{\varepsilon_0} = d_{m_0}$, where m_0 is the least positive integer with $2/m_0^2 < \varepsilon_0$. The last statement follows since $1 \le m_0^{-2} A(d_{m_0})^{1/(n-1)} < \varepsilon_0 A(D_{\varepsilon_0})^{1/(n-1)}$. $\qquad \square$

Let E be the subset of $[1, \infty[$ constructed in 2.1. We assume that the points $a_1, \ldots, a_q$ in 1.3 lie in the ball $B(1/2)$. This assumption is made for notational convenience. In the end of Section 6 we will remove this restriction.

Averages of $n(F, y)$ over the unit sphere will henceforth play a central role. We therefore abbreviate $\nu(F) = \nu(F, 1)$ for any bounded Borel set $F \subset \mathbb{R}^n$ and write $\nu(r) = \nu(r, 1)$. To apply Lemma 2.1 we fix ε_0 such that $0 < \varepsilon_0 < \min(1/q, 1/5)$. By 2.1 there then exists $\kappa > 1$ such that

$$(2.8) \qquad \left|\frac{A(s')}{\nu(s)} - 1\right| < \frac{1}{q}$$

and

$$(2.9) \qquad \nu(s') \le \tfrac{3}{2} \nu(s)$$

whenever $s > 0$ is such that $s' \in [\kappa, \infty[\setminus E$. The number ε_0 – and with it the number κ – will undergo further adjustment in (6.2) and (6.6). Note that 2.1 gives a lower bound for $A(s')$ in terms of ε_0.

To prove 1.3 it suffices to show

$$(2.10) \qquad \sum_{j=1}^{q} \left(1 - \frac{n(s', a_j)}{A(s')}\right)_+ \le C(n, K) < \infty .$$

Set $J = \{1, \ldots, q\}$. We may assume that $q \ge 3$ and $1 - n(s', a_j)/A(s') > 0$ for all $j \in J$. Set

$$(2.11) \qquad \Delta_j = 1 - \frac{n(s', a_j)}{\nu(s)} .$$

From (2.8) we learn that

(2.12)
$$\sum_{j\in J}\left(1-\frac{n(s',a_j)}{A(s')}\right) = \sum_{j\in J}\Delta_j + \sum_{j\in J}\frac{n(s',a_j)}{A(s')}\left(\frac{A(s')}{\nu(s)}-1\right)$$

$$\leq \sum_{j\in J}\Delta_j + q\left|\frac{A(s')}{\nu(s)}-1\right| \leq \sum_{j\in J}\Delta_j + 1.$$

To prove 1.3 it is therefore enough to give for $\sum_j \Delta_j$ an upper bound depending only on n and K. We may freely assume that $\Delta_j > 0$ for all $j \in J$ and that $\sum_j \Delta_j \geq 20$.

The essence of the proof goes as follows. Since $n(s',a_j)/\nu(s) < 1$, the point a_j is covered less by $f|\overline{B}(s')$ than is an average point $y \in S$ by $f|\overline{B}(s)$. This implies that, on the average, straight paths connecting y and a_j have maximal liftings connecting $f^{-1}(y) \cap \overline{B}(s)$ and $S(s')$, and that the number of such liftings increases with Δ_j. If the total sum $\sum_j \Delta_j$ is large, it means that the average point of S generates such liftings for many j's. The hope is to arrive at a point where Lemma IV.2.3 can be applied, thereby ensuring a certain rate of growth for the averages over the unit sphere. With this in mind we decompose $\overline{B}(s)$ into sets U_i, $i = 1,\ldots,p$, by taking a Whitney type decomposition of $B(s')$ and restricting it to $\overline{B}(s)$. In Lemma IV.2.3 we choose $M = 3/2$, while the ball $\overline{B}(1/2)$ will correspond to U_i. The ball $\overline{B}(4)$ will correspond to a subset Z_i of $\overline{B}(s')$, and the sets Z_i will not overlap a great deal. Lemma IV.2.3 will deliver lower bounds for the ratios $\nu(Z_i)/\nu(U_i)$, which in turn will give rise to a lower bound for $\nu(s')$ in terms of $\nu(s)$. If the sum $\sum_j \Delta_j$ is too large, then we run into a contradiction with (2.9). In order to apply Lemma IV.2.3 we need to have detailed information about the liftings of straight paths connecting a point $y \in S$ and some a_j.

2.13. Convention. Throughout the proof of 1.3 we will use the letter c, with subscript or other, to represent a positive constant depending only on n. (We did the same with the letter b in Sections IV.2 and IV.3.)

2.14. Decomposition of $\overline{B}(s)$. We start by giving precise conditions on the decomposition of $\overline{B}(s)$. Set $d_0 = s' - s$. We decompose $\overline{B}(s)$ into disjoint Borel sets U_i, $i \in I = \{1,\ldots,p\}$, such that

(1) $0 < A_n \leq \rho_0(U_i) \leq B_n < \infty$, where $\rho_0(U_i)$ is the diameter of U_i in the hyperbolic metric $4s'^2 dx^2/(s'^2 - |x|^2)^2$ of the ball $B(s')$ and where the constants A_n and B_n depend only on n,

(2) there exist K_0-qc mappings $\varphi_i \colon \mathbb{R}^n \to \mathbb{R}^n$, K_0 depending only on n, such that $\varphi_i^{-1}B(1/2) \subset U_i \subset \varphi_i^{-1}\overline{B}(1/2)$, $Z_i = \varphi_i^{-1}\overline{B}(4) \subset \overline{B}(s')$, and each point of $\overline{B}(s')$ belongs to at most c_1 of the sets Z_i.

One easily obtains such a decomposition, for example, by first decomposing a cube in a similar fashion and then using radial stretching of the cube onto a ball. It follows that the number p of sets U_i has an upper bound of the form

(2.15) $$p \le c_2 \left(\frac{s}{d_0}\right)^{n-1} = c_2 \varepsilon_0^{n-1} A(s) \le 2 c_2 \varepsilon_0^{n-1} \nu(s) \,.$$

For each $i \in I$ we write

$$V_i = \varphi_i^{-1} B(3/4)\,, \quad W_i = \varphi_i^{-1} B\,, \quad X_i = \varphi_i^{-1} \overline{B}(3/2)\,.$$

3. Estimates on Liftings

3.1. Maximal sequences of liftings. Let

$$\delta_0 = \tfrac{1}{16} \min_{j \ne k} |a_j - a_k|\,,$$

and for $j \in J$ and $y \in S$ let $\gamma_y^j \colon [0,1] \to \overline{B}$ be the path $\gamma_y^j(t) = (1-t)y + t a_j$. Set $f_0 = f|B(s'+1)$. Fix $i \in I$ and $j \in J$. Let $y \in S$ and let $x_1, \ldots, x_k$ be the points of $f^{-1}(y) \cap U_i$. Set $m = n(U_i, y)$, and let $(\beta_1, \ldots, \beta_m)$ be a maximal sequence of essentially separate f_0-liftings of γ_y^j starting at the points $x_1, \ldots, x_k$. We recall the terminology from II.3 and extend it in an obvious way to liftings of a closed path. Let $\alpha_1, \ldots, \alpha_{\nu_y}$ be those liftings β_μ for which the locus $|\beta_\mu|$ is not contained in $\overline{B}(s')$.

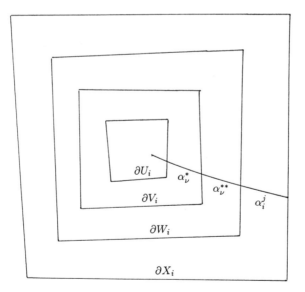

Fig. 4

For $1 \le \nu \le \nu_y$ we let $u = u_{y,\nu}$, $v = v_{y,\nu}$, $w = w_{y,\nu}$ be the smallest numbers such that $0 \le u < v < w < 1$ and $\alpha_\nu(u) \in \partial U_i$, $\alpha_\nu(v) \in \partial V_i$, $\alpha_\nu(w) \in \partial W_i$. Set (see Fig. 4)

(3.2) $$\alpha_\nu^* = \alpha_\nu | [u, v] \,, \quad \alpha_\nu^{**} = \alpha_\nu | [v, w] \,.$$

When necessary, we also write $\alpha_{y,\nu}$ etc.

We first derive an estimate for the average cardinality of the sets

(3.3) $$L_i^j(y) = \left\{ \nu \in \{1, \dots, \nu_y\} : \frac{1 - u_{y,\nu}}{1 - v_{y,\nu}} \leq \frac{1}{\delta_0} \right\}\,.$$

3.4. Lemma. *There exists a nonnegative measurable function* $l_i^j : S \to \mathbb{R}^1$ *such that*

(3.5) $$\operatorname{card} L_i^j(y) \leq l_i^j(y) \quad \text{for all } y \in S\,,$$

(3.6) $$\int_S l_i^j(y) \, dy \leq c_5 K_I \left(\log \frac{1}{\delta_0} \right)^{n-1}\,.$$

Proof. The idea is to generalize the method in Väisälä's inequality II.9.1. We want a relationship between the modulus of $\{ \alpha_{y,\nu}^* : y \in S, \ \nu \in L_i^j(y) \}$ and its image path family. Inequality II.9.1 itself is not applicable here, as it was in the proof of Lemma IV.1.1, because for fixed y there is no common path γ of which the paths $\alpha_{y,\nu}^*$ are liftings. The way to get around this is to think of the image paths as lying in different "sheets".

Let $F_0 \subset \mathbb{R}^n$ be the set of points where f is differentiable and J_f is positive. There is a Borel set F of zero n-measure such that $F \supset f(\mathbb{R}^n \setminus F_0)$. Then $\mathcal{H}^1(|\gamma_y^j| \cap F) = 0$ for almost every $y \in S$. To justify these statements we recall I.2.4 and II.7.4. We apply Poletskiĭ's lemma II.5.1 to the family Γ of paths λ such that λ is a subpath of some $\alpha_{y,\nu}$, f is not absolutely precontinuous on λ, and $a_j \notin |f\lambda|$. Then II.5.1 shows that for almost every $y \in S$, f is absolutely precontinuous on $\alpha_{y,\nu}$ for all $\nu = 1, \dots, \nu_y$. Let $T \subset S$ be a Borel set with $\mathcal{H}^{n-1}(T) = 0$ such that this precontinuity condition and $\mathcal{H}^1(|\gamma_y^j| \cap F) = 0$ are met for all $j \in J$ and $y \in S \setminus T$.

Let $\rho : \mathbb{R}^n \to \mathbb{R}^1$ be an admissible function for the condenser $E_i = (V_i, \overline{U}_i)$ (i.e., for the path family Γ_{E_i} in the notation of II.10.2) such that $\rho | (\mathbb{R}^n \setminus (V_i \setminus \overline{U}_i)) = 0$ and

$$\int_{\mathbb{R}^n} \rho^n \, dm \leq 2 \operatorname{cap}(V_i, \overline{U}_i)\,.$$

Fix $y \in S \setminus T$. For $\nu = 1, \dots, \nu_y$ we define $\rho_\nu' : |f \circ \alpha_\nu| \to \mathbb{R}^1$ by

$$\rho_\nu'(z) = \sigma(x) \quad \text{if } z = f(x) \text{ and } x \in |\alpha_\nu|\,,$$

where we have written

$$\sigma(x) = \begin{cases} \dfrac{\rho(x)}{l(f'(x))} & \text{if } x \in \mathbb{R}^n \setminus f^{-1}F, \\ 0 & \text{if } x \in f^{-1}F. \end{cases}$$

Mimicking the proof of Poletskiĭ's inequality II.8.1, we get

$$1 \leq \int\limits_{f \circ \alpha^*_\nu} \rho'_\nu \, ds \leq \tfrac{3}{2} \int\limits_u^v \rho'_\nu(\gamma^j_y(t)) \, dt \,,$$

where we have used the fact that $|\gamma^{j'}_y| \leq 3/2$. Hölder's inequality yields

$$1 \leq \left(\int\limits_{f \circ \alpha^*_\nu} \rho'_\nu \, ds \right)^n \leq \frac{3^n}{2^n} \left(\int\limits_u^v \rho'_\nu(\gamma^j_y(t))^n (1-t)^{n-1} dt \right) \left(\int\limits_u^v \frac{dt}{1-t} \right)^{n-1} ,$$

which we rewrite as

$$\left(\log \frac{1-u}{1-v} \right)^{1-n} \leq c_3 \int\limits_u^v \rho'_\nu(\gamma^j_y(t))^n (1-t)^{n-1} dt \,.$$

By summing over $L = L^j_i(y)$ we get

$$\frac{\mathrm{card}\, L}{(\log(1/\delta_0))^{n-1}} \leq \sum_{\nu \in L} \frac{1}{(\log \frac{1-u}{1-v})^{n-1}}$$

$$\leq c_3 \sum_{\nu \in L} \int\limits_u^v \rho'_\nu(\gamma^j_y(t))^n (1-t)^{n-1} dt$$

$$\leq c_3 \int\limits_0^1 \sum_{x \in f^{-1}(\gamma^j_y(t))} \sigma(x)^n (1-t)^{n-1} dt \,.$$

We define $l^j_i : S \to \mathbb{R}^1$ by

$$l^j_i(y) = \begin{cases} c_3 \left(\log \frac{1}{\delta_0} \right)^{n-1} \int\limits_0^1 \sum\limits_{x \in f^{-1}(\gamma^j_y(t))} \sigma(x)^n (1-t)^{n-1} dt & \text{if } y \in S \setminus T, \\ n(U_i, y) & \text{if } y \in T. \end{cases}$$

Then l^j_i is measurable and satisfies (3.5).

For the integral of l^j_i over S we obtain

$$\int\limits_S l^j_i(y) \, dy = c_3 \left(\log \frac{1}{\delta_0} \right)^{n-1} \int\limits_B \sum_{x \in f^{-1}(\varphi(z))} \sigma(x)^n \, dz$$

$$\leq c_4 \left(\log \frac{1}{\delta_0} \right)^{n-1} \int\limits_B \sum_{x \in f^{-1}(z)} \sigma(x)^n \, dz \,.$$

Here $\varphi \colon \overline{B} \to \overline{B}$ is the mapping defined by $\varphi((1-t)y) = \gamma^j_y(t)$ for $t \in [0,1]$, $y \in S$. As in the proof of II.8.1 we find for a given $z_0 \in B \setminus f(\overline{B}(s') \cap B_f)$ a neighborhood V of z_0 such that the components of $f^{-1}V$ that meet $\overline{B}(s')$ form a finite collection $D_1, \ldots, D_r$ and f defines qc mappings $f_h \colon D_h \to V$, $h = 1, \ldots, r$. We obtain

$$\int_V \sum_{x \in f^{-1}(z)} \sigma(x)^n \, dz = \sum_{h=1}^r \int_{D_h} \sigma(x)^n J_{f_h}(x) \, dx \leq K_I \int_{f^{-1}V} \rho^n dm \, .$$

The set $B \smallsetminus f(\overline{B}(s') \cap B_f)$ can be covered up to a set of zero n-measure by disjoint neighborhoods of this kind, hence

$$\int_S l_i^j(y) \, dy \leq c_4 \left(\log \frac{1}{\delta_0} \right)^{n-1} K_I \int_{\mathbb{R}^n} \rho^n dm \leq c_5 K_I \left(\log \frac{1}{\delta_0} \right)^{n-1} \, .$$

The lemma is proved.   □

The second estimate concerns the paths α_ν^{**}. For each $i \in I$ we let $\sigma_i \in {]}0, \delta_0]$ be a number defined by

$$(3.7) \qquad \left(\log \frac{\delta_0}{\sigma_i} \right)^{n-1} = A_i \nu(U_i),$$

where A_i is a nonnegative number to be chosen in (5.10). Set

$$(3.8) \qquad M_i^j(y) = \left\{ \nu \in \{1, \ldots, \nu_y\} : \frac{1 - v_{y,\nu}}{1 - w_{y,\nu}} \leq \frac{\delta_0}{\sigma_i} \right\} \, .$$

3.9. Lemma. *There exist measurable functions* $m_i^j \colon S \to \mathbb{R}^1$ *such that*

$$(3.10) \qquad \operatorname{card}\left(M_i^j(y) \smallsetminus L_i^j(y) \right) \leq m_i^j(y) \qquad \text{for all} \quad y \in S \, ,$$

$$(3.11) \qquad \sum_{j \in J} \int_S m_i^j(y) \, dy \leq c_6 K_I \left(\log \frac{\delta_0}{\sigma_i} \right)^{n-1} \, .$$

Proof. Let $T \subset S$ be the set determined in the proof of 3.4. We now let $\rho \colon \mathbb{R}^n \to \mathbb{R}^1$ be an admissible function for the condenser $(W_i, \overline{V}_i)$ like the one in that proof.

Fix $y \in S \smallsetminus T$. We define ρ_ν' as in the proof of 3.4. If $\nu \in M_i^j(y) \smallsetminus L_i^j(y) = M \smallsetminus L$, then $1 - v_{y,\nu} < \delta_0$. In the present situation we get

$$\operatorname{card}(M_i^j(y) \smallsetminus L_i^j(y)) \leq \frac{3^n}{2^n} \left(\log \frac{\delta_0}{\sigma_i} \right)^{n-1} \sum_{\nu \in M \smallsetminus L} \int_v^w \rho_\nu'(\gamma_u^j(t))^n (1 - t)^{n-1} dt$$

$$\leq c_3 \left(\log \frac{\delta_0}{\sigma_i} \right)^{n-1} \int_{1-\delta_0}^1 \sum_{x \in f^{-1}(\gamma_y^j(t))} \sigma(x)^n (1 - t)^{n-1} dt$$

$$= m_i^j(y) \, ,$$

where the equality defines $m_i^j(y)$ in $S \setminus T$. If for $y \in T$ we set $m_i^j(y) = n(U_i, y)$, then (3.10) is true. Integrating over S we obtain

$$\int_S m_i^j(y)\, dy \le c_4 \left(\log \frac{\delta_0}{\sigma_i}\right)^{n-1} \int_{B_j} \sum_{x \in f^{-1}(z)} \sigma(x)^n\, dz ,$$

where $B_j = B(a_j, 3\delta_0/2)$. Because the balls B_j are disjoint, we now arrive by summing over J at

$$\sum_{j \in J} \int_S m_i^j(y)\, dy \le c_4 \left(\log \frac{\delta_0}{\sigma_i}\right)^{n-1} \int_B \sum_{x \in f^{-1}(z)} \sigma(x)^n\, dz$$

$$\le c_6 K_I \left(\log \frac{\delta_0}{\sigma_i}\right)^{n-1} .$$

The lemma is proved. $\square$

3.12. Remark. Since we have not given a precise description of how the various liftings $\alpha_{y,\nu}$ are chosen, we cannot say anything about the measurability of card $L_i^j(y)$ and card $M_i^j(y)$. This is particularly true because, in general, the Hausdorff dimension of fB_f can be arbitrarily near n (cf. III.5.4). In the next section we make special choices of the liftings β_ν (see the beginning of this section) in order to ensure the measurability of certain cardinality functions on S.

4. Extremal Maximal Sequences of Liftings

As mentioned in Remark 3.12 we will now restrict the choices of f_0-liftings of γ_y^j. For this we introduce some new notation. Fix $j \in J$. The path γ_y^j is here denoted by γ_y. For $y \in S$ we consider maximal essentially separate sequences $\Lambda = (\lambda_1, \ldots, \lambda_g)$ of f_0-liftings of γ_y starting at the points of $f^{-1}(y) \cap \overline{B}(s)$. Then $g = n(s, y)$ by definition. Let the set of such sequences be Ω_y. For $k = 1, \ldots, p$, set $H_k = \{1, \ldots, k\}$. We will inductively define for each H_k a function $\psi_k(y, a_j) = \psi_k(y)$ on S, starting with $H_p = I$.

For $\Lambda \in \Omega_y$ we set

$$N(I, \Lambda) = \operatorname{card}\{\nu : |\lambda_\nu| \subset \overline{B}(s')\}$$

and define

$$\psi_p(y) = \sup_{\Lambda \in \Omega_y} N(I, \Lambda) .$$

The set of sequences Λ in Ω_y for which $N(I, \Lambda) = \psi_p(y)$ is denoted by $\Omega(I, y)$. Write

$$U^k = \bigcup_{i=1}^k \operatorname{int} U_i , \quad k = 1, \ldots, p .$$

Then $\partial U^k = \bigcup_{i=1}^k \partial U_i$.

4.1. Lemma. *If* $y_0 \in S \setminus f\partial U^p$, *then* ψ_p *is upper semicontinuous at* y_0.

Proof. Let $y_1, y_2, \ldots$ be a sequence in $S \setminus f\partial U^p$ such that $y_h \to y_0$. We want to show

$$\limsup_{h \to \infty} \psi_p(y_h) \leq \psi_p(y_0).$$

By passing to subsequences we may assume that for some integer m, $\psi_p(y_h) = m$ holds for $h \geq 1$, and that the following conditions are fulfilled:

(a) There exist normal neighborhoods $V_1, \ldots, V_k \subset U^p$ of the points $x_1, \ldots, x_k$ of $f^{-1}(y_0) \cap U^p$ such that $y_h \in fV_1 \cap \ldots \cap fV_k$ for $h \geq 1$.

(b) For each h there is a maximal sequence $\Lambda_h = (\lambda_{h,1}, \ldots, \lambda_{h,g}) \in \Omega(I, y_h)$, where $g = n(s, y_0)$, such that each $\lambda_{h,\nu}$ starts at a point in $V_\mu \cap f^{-1}(y_h)$ for some $\mu = \mu(\nu)$ that does not depend on h, and $|\lambda_{h,\nu}| \subset \overline{B}(s')$ for $1 \leq \nu \leq m$.

Let $1 \leq \nu \leq m$. We claim that the family $\{\lambda_{h,\nu} : h = 1, 2, \ldots\}$ is equicontinuous. Note that each $\lambda_{h,\nu}$ is defined on $[0,1]$ because $|\lambda_{h,\nu}| \subset \overline{B}(s')$. Let $\varepsilon > 0$. For each $t \in [0,1]$ there exists $\delta_t > 0$ such that, in the notation of II.4, $U(\xi, \rho)$ is a normal neighborhood of ξ with $d(U(\xi, \rho)) < \varepsilon$ whenever $\xi \in f^{-1}(\gamma_{y_0}(t)) \cap \overline{B}(s')$, and $\overline{B}(s') \cap f^{-1}B(\gamma_{y_0}(t), \rho)$ is contained in the union of $U(\xi, \rho)$, $\xi \in f^{-1}(\gamma_{y_0}(t)) \cap \overline{B}(s')$, whenever $0 < \rho \leq \delta_t$. We cover $|\gamma_{y_0}|$ by a finite number of balls $B(\gamma_{y_0}(t), \delta_t/2)$, say $B(\eta_u, \rho_u)$, $u = 1, \ldots, v$. We may assume that $|\gamma_{y_h}| \subset \bigcup \{B(\eta_u, \rho_u) : u = 1 \ldots, v\}$ for $h \geq 1$. Now let $0 \leq t \leq 1$. There exists u such that $\gamma_{y_h}(t') \in B(\eta_u, 2\rho_u)$ for $|t' - t| < 2\rho_u/3$ and $h \geq 1$. For each $h \geq 1$ there then exists $\xi \in f^{-1}(\eta_u) \cap \overline{B}(s')$ such that $\lambda_{h,\nu}(t') \in U(\xi, 2\rho_u)$ provided $|t' - t| < 2\rho_u/3$. This implies the desired equicontinuity at t because $d(U(\xi, 2\rho_u)) < \varepsilon$. Invoking Ascoli's theorem we may by passing to a subsequence assume that $(\lambda_{h,\nu})$ converges uniformly to a path $\lambda_\nu : [0,1] \to \overline{B}(s')$. The path λ_ν is a maximal f_0-lifting of γ_{y_0} starting in $\overline{B}(s)$.

Now consider ν for which $m + 1 \leq \nu \leq g$. If $\lambda_{h,\nu}$ is half open, it extends to a closed path in $\overline{B}(s'+1)$. Let the extended path be $\overline{\lambda}_{h,\nu} : [0, t_h] \to \overline{B}(s'+1)$. We may assume $t_h \to t_0 \in]0,1]$. If G_h maps $[0, t_0]$ affinely onto $[0, t_h]$ and $G_h(0) = 0$, arguing as above, we may assume that the paths $\overline{\lambda}_{h,\nu} \circ G_h$ converge uniformly to a path $\overline{\lambda}_\nu : [0, t_0] \to \overline{B}(s'+1)$ which is a lifting of $\gamma_{y_0} | [0, t_0]$. If $\Delta \subset [0, t_0]$ is the largest interval such that $0 \in \Delta$ and $\overline{\lambda}_\nu \Delta \subset B(s'+1)$, then $\lambda_\nu = \overline{\lambda}_\nu | \Delta$ is a maximal f_0-lifting of γ_{y_0} starting in $\overline{B}(s)$.

Next, we want to show that $\Lambda_0 = (\lambda_1, \ldots, \lambda_g) \in \Omega_{y_0}$. For this it remains to prove that the paths λ_ν are essentially separate. Let $A = \{\nu : \lambda_\nu(t) = x\} \neq \emptyset$ and let $U = U(x, \rho)$ be a normal neighborhood of x. There exists h_0 such that $|\lambda_{h,\nu}| \cap U \neq \emptyset$ for all $h \geq h_0$ and $\nu \in A$. Let $h \geq h_0$. There exists a point $\eta = \gamma_{y_h}(t')$ in $\bigcap \{f(|\lambda_{h,\nu}| \cap U) : \nu \in A\}$. Let $\xi_1, \ldots, \xi_w$ be the points in $\{\lambda_{h,\nu}(t') : \nu \in A\} \subset f^{-1}(\eta) \cap U$. Since the paths $\lambda_{h,1}, \ldots, \lambda_{h,g}$ are essentially separate, we have

$$\theta_u = \mathrm{card}\{\nu : \lambda_{h,\nu}(t') = \xi_u\} \leq i(\xi_u, f), \quad u = 1, \ldots, w.$$

Hence

$$\text{card } A = \sum_{u=1}^{w} \theta_u \leq \sum_{u=1}^{w} i(\xi_u, f) \leq i(x, f) \,,$$

which proves the claim.

Because $|\lambda_\nu| \subset \overline{B}(s')$ for $1 \leq \nu \leq m$, we have $\psi_p(y_0) \geq N(I, \Lambda_0) \geq m$. The lemma is proved. $\square$

Because each ∂U_i is a qc image of a sphere, $m(\partial U^p) = 0$, so $m(f\partial U^p) = 0$ by II.7.4. Then $\mathcal{H}^{n-1}(S(t) \cap f\partial U^p) = 0$ for almost every $t > 0$. If this is true for $t = 1$, ψ_p is measurable by Lemma 4.1. Otherwise we replace S by a nearby sphere already in Section 2. Lemma 2.1 makes this possible. Hence we may assume that $\mathcal{H}^{n-1}(S \cap f\partial U^p) = 0$ and conclude that ψ_p is measurable on S.

Next we will study how many liftings in the sequences Λ start in different U_i's. For $\Lambda \in \Omega(I, y)$ we set

$$N(H_{p-1}, \Lambda) = \text{card}\left\{ \nu : |\lambda_\nu| \subset \overline{B}(s'), \ \lambda_\nu \text{ starts in } \bigcup_{i \in H_{p-1}} U_i \right\}$$

and define

$$\psi_{p-1}(y) = \sup_{\Lambda \in \Omega(I,y)} N(H_{p-1}, \Lambda) \,.$$

We set $\Omega(H_{p-1}, y) = \{\, \Lambda \in \Omega(I, y) : N(H_{p-1}, \Lambda) = \psi_{p-1}(y) \,\}$.

4.2. Lemma. *The function ψ_{p-1} is measurable on S.*

Proof. It is enough to show that the restriction of ψ_{p-1} to each set

$$A_m = \{\, y \in S : \psi_p(y) = m \,\}, \quad m = 0, \ldots, r_0 = \max_{y \in S} n(s, y) \,,$$

is measurable. Hence it suffices to show that ψ_{p-1} is upper semicontinuous in $B_m = A_m \smallsetminus f\partial U^p$. Let $y_0 \in B_m$ and let $y_1, y_2, \ldots$ be a sequence in B_m converging to y_0. We may assume that for some integer $m_1 \leq m$, $\psi_{p-1}(y_h) = m_1$ holds for $h \geq 1$ and that, in addition to (a) in the proof of 4.1, we have:

(b$_1$) For each $h \geq 1$ a maximal sequence $\Lambda_h = (\lambda_{h,1}, \ldots, \lambda_{h,g}) \in \Omega(H_{p-1}, y_h)$, where $g = n(s, y_0)$, such that each $\lambda_{h,\nu}$ starts at a point in $V_\mu \cap f^{-1}(y_h)$ for some $\mu = \mu(\nu)$ that does not depend on h, $|\lambda_{h,\nu}| \subset \overline{B}(s')$ for $1 \leq \nu \leq m$, and $\lambda_{h,\nu}$ starts in U^{p-1} for $1 \leq \nu \leq m_1$.

As in the proof of 4.1 we get a maximal sequence $\Lambda_0 \in \Omega_{y_0}$ such that $N(I, \Lambda_0) \geq m$. Since $y_0 \in A_m$, we have $N(I, \Lambda_0) \leq \psi_p(y_0) = m$, hence $N(I, \Lambda_0) = m$ and so $\Lambda_0 \in \Omega(I, y_0)$. By the construction of Λ_0 we also have $\psi_{p-1}(y_0) \geq N(H_{p-1}, \Lambda_0) \geq m_1$. This proves the asserted semicontinuity. $\square$

In general, we inductively define subsets $\Omega(H_k, y)$ and functions $N(H_k, \Lambda)$ and $\psi_k(y)$, $1 \leq k \leq p$, as follows. Supposing that $k \leq p-1$ and that

$\Omega(H_k, y)$, $N(H_k, \Lambda)$, and $\psi_k(y)$ are defined for $y \in S$ and $\Lambda \in \Omega(H_k, y)$, we set

$$N(H_{k-1}, \Lambda) = \text{card}\left\{ \nu : |\lambda_\nu| \subset \overline{B}(s'),\ \lambda_\nu \text{ starts in } \bigcup_{i \in H_{k-1}} U_i \right\},$$

$$\psi_{k-1}(y) = \sup_{\Lambda \in \Omega(H_k, y)} N(H_{k-1}, \Lambda),$$

(4.3) $$\Omega(H_{k-1}, y) = \{ \Lambda \in \Omega(H_k, y) : N(H_{k-1}, y) = \psi_{k-1}(y) \}.$$

4.4. Lemma. *For $1 \le k \le p$ the functions ψ_k on S are measurable.*

Proof. We will use induction on k starting from p. By 4.1, ψ_p is measurable. Suppose that $\psi_k, \ldots, \psi_p$ are measurable. It suffices to prove that ψ_{k-1} is upper semicontinuous in each set $B_m = A_m \setminus f\partial U^p$, where m is now a multi–index $(m_0, m_1, \ldots, m_{p-k})$ with $m_0 \ge m_1 \ge \ldots \ge m_{p-k}$, and

$$A_m = \{ y \in S : \psi_p(y) = m_0, \ldots, \psi_k(y) = m_{p-k} \}.$$

Let $y_0 \in B_m$ and let $y_1, y_2, \ldots$ be a sequence in B_m converging to y_0. As in the proof of 4.2 we may assume that $\psi_{k-1}(y_h) = m_{p-k+1}$ for all $h \ge 1$, and as there we can by a limiting process construct a maximal sequence $\Lambda_0 \in \Omega_{y_0}$ such that

$$N(H_u, \Lambda_0) = m_{p-u}, \quad u = k, \ldots, p,$$

which then implies that $\Lambda_0 \in \Omega(H_{p-k}, y_0)$. By the construction of Λ_0 we have $\psi_{k-1}(y_0) \ge N(H_{k-1}, \Lambda_0) \ge m_{p-k+1}$, and the lemma follows.   □

In the next section we need maximal sequences Λ that belong to $\Omega(H_1, y)$.

4.5. Notes. Up to this point we have closely followed the presentation in [R7]. Apart from simplifying some notation we have made only a few changes. In Section 2 we reversed the parametrization of the path γ_y^j. The definition of σ_i in (3.7) is different from that in [R7]. This turns out to be a crucial point in obtaining the right defect relation. With the present approach we avoid using the fact that $\mathcal{H}^{n-1}(f(B_f \cap \overline{B}(s')) \cap S) = 0$ from [MR2, 3.1] (cf. III.5.4), something that is not proved in this monograph.

5. Effect of the Defect Sum on the Liftings

In this section we will establish inequalities which show that, with the notation in (2.11), a large sum of $\sum_j \Delta_j$ leads to the existence of maximal liftings of the paths γ_y^j in such a form that Lemma IV.2.3 can be applied.

Let $i \in I$ and $j \in J$. For each $y \in S$ we choose a maximal sequence $\Lambda_y = (\lambda_{y,1}, \ldots, \lambda_{y,g})$, $g = n(s, y)$, in the set $\Omega(H_1, y)$ defined in (4.3). Let $y \in S$. Those liftings $\lambda_{y,\nu}$ in Λ_y that start in U_i form a maximal sequence $(\beta_1, \ldots, \beta_m)$ of essentially separate f_0-liftings of γ_y^j as in 3.1. For this sequence we have the sequence $(\alpha_1, \ldots, \alpha_{\nu_y})$ of those liftings β_μ which satisfy $|\beta_\mu| \not\subset \overline{B}(s')$. We now write

$$(5.1) \qquad\qquad n_i^j(y) = \nu_y \ .$$

We recall the functions $\psi_k(y, a_j) = \psi_k(y)$ from the preceding section. The difference $\psi_i(y, a_j) - \psi_{i-1}(y, a_j)$ is the number of indices μ for which $|\beta_\mu| \subset \overline{B}(s')$, provided we define $\psi_0(y, a_j) = 0$. Since $m = n(U_i, y)$, we have

$$(5.2) \qquad\qquad n_i^j(y) = n(U_i, y) - \big(\psi_i(y, a_j) - \psi_{i-1}(y, a_j)\big) \ .$$

By Lemma 4.4, n_i^j is measurable on S. Summing (5.2) over $i \in I$ we obtain

$$\sum_{i \in I} n_i^j(y) = \sum_{i \in I} n(U_i, y) - \psi_p(y, a_j)$$
$$\geq n(s, y) - n(s', a_j) \ .$$

For the average over S we thus get

$$(5.3) \qquad \frac{1}{\omega_{n-1}} \int_S \sum_{i \in I} n_i^j(y)\, dy \geq \nu(s) - n(s', a_j) = \nu(s)\Delta_j \ .$$

For $i \in I$ set

$$(5.4) \qquad J_i = \left\{ j : 2\int_S n_i^j(y)\, dy > \omega_{n-1}\nu(U_i)\Delta_j \right\} .$$

Note that summing the inequalities in (5.4) over $i \in I$ gives (5.3) up to a factor 2. From (5.3) it follows that

$$\frac{\omega_{n-1}}{2}\nu(s)\sum_{j \in J}\Delta_j = \frac{\omega_{n-1}}{2}\sum_{i \in I}\sum_{j \in J}\nu(U_i)\Delta_j$$
$$\geq \frac{\omega_{n-1}}{2}\sum_{i \in I}\sum_{j \in J\smallsetminus J_i}\nu(U_i)\Delta_j \geq \sum_{i \in I}\sum_{j \in J\smallsetminus J_i}\int_S n_i^j(y)\, dy$$
$$= \sum_{i \in I}\sum_{j \in J}\int_S n_i^j(y)\, dy - \sum_{i \in I}\sum_{j \in J_i}\int_S n_i^j(y)\, dy$$
$$\geq \omega_{n-1}\nu(s)\sum_{j \in J}\Delta_j - \sum_{i \in I}\sum_{j \in J_i}\int_S n_i^j(y)\, dy \ .$$

We are thus led to the following result.

5.5. Lemma. *The functions* n_i^j *satisfy*

$$(5.6) \qquad \sum_{i \in I} \sum_{j \in J_i} \int_S n_i^j(y)\, dy \geq \frac{\omega_{n-1}}{2} \nu(s) \sum_{j \in J} \Delta_j \,.$$

Inequality (5.6) represents the first step in estimating the number of liftings in terms of the sum $\sum_j \Delta_j$. However, (5.6) says nothing about the behavior of images of different parts of the liftings, such as the paths α_ν^* and α_ν^{**} defined in (3.2). In order to refine our information on this we set

$$(5.7) \qquad J^i = \left\{ j \in J_i : 3 \int_S l_i^j \geq \int_S n_i^j \ \text{ or } \ 3 \int_S m_i^j \geq \int_S n_i^j \right\} \,.$$

By 3.4, (3.7), and 3.9,

$$\sum_{j \in J_i} \int_S n_i^j(y)\, dy \leq 3 \sum_{j \in J^i} \int_S \left(l_i^j(y) + m_i^j(y) \right) dy$$

$$\leq 3c_5 K_I q \left(\log \frac{1}{\delta_0} \right)^{n-1} + 3c_6 K_I A_i \nu(U_i) \,.$$

On the strength of Lemma 5.5 and (2.15) we then get

$$\sum_{i \in I} \sum_{j \in J_i \smallsetminus J^i} \int_S n_i^j(y)\, dy \geq \sum_{i \in I} \sum_{j \in J_i} \int_S n_i^j(y)\, dy - \sum_{i \in I} \sum_{j \in J^i} \int_S n_i^j(y)\, dy$$

$$(5.8) \qquad \geq \frac{\omega_{n-1}}{2} \nu(s) \sum_{j \in J} \Delta_j - 6c_2 c_5 K_I q \left(\log \frac{1}{\delta_0} \right)^{n-1} \varepsilon_0^{n-1} \nu(s)$$

$$- 3c_6 K_I \sum_{i \in I} A_i \nu(U_i) \,.$$

We are now going to choose the constants A_i so that inequality (5.8) yields an effective lower bound for the left hand side of (5.8). Set

$$(5.9) \qquad \lambda_i = \text{card}(J_i \smallsetminus J^i) \,.$$

As A_i increases in the range $[0, \infty[$, the number λ_i decreases from

$$\lambda_i^0 = \text{card} \left\{ j \in J_i : 3 \int_S l_i^j < \int_S n_i^j \right\}$$

to some value λ_i^∞. We may assume that at the discontinuities of the function $A_i \mapsto \lambda_i$ the jumps are 1. If this is not the case originally, we make small variations in the functions m_i^j for different j's. It then becomes clear that we can choose $A_i \geq 0$ so that

$$(5.10) \qquad \lambda_i - 1 \leq 9\omega_{n-1}^{-1} c_6 K_I A_i \leq \lambda_i \,.$$

6. Completion of the Proof of Defect Relations

With the choice of the constant A_i in (5.10) we are put in a position to make effective use of (5.8). From (5.8) and (5.10) we obtain

$$
\sum_{i \in I} \lambda_i \nu(U_i) = \sum_{i \in I} \sum_{j \in J_i \setminus J^i} \nu(U_i) \geq \sum_{i \in I} \sum_{j \in J_i \setminus J^i} \frac{1}{w_{n-1}} \int_S n_i^j(y)\, dy
$$

(6.1)
$$
\geq \frac{1}{2} \nu(s) \sum_{j \in J} \Delta_j - 6 w_{n-1}^{-1} c_2 c_5 K_I q \left(\log \frac{1}{\delta_0} \right)^{n-1} \varepsilon_0^{n-1} \nu(s)
$$

$$
- \frac{1}{3} \sum_{i \in I} \lambda_i \nu(U_i) \, .
$$

For applications of same formulas in the next section we introduce an auxiliary parameter $\varepsilon \in\ ]0, 1]$, which in this section will be 1. We are now ready to impose further limitations on our choice of ε_0. We henceforth require that

(6.2)
$$
6 w_{n-1}^{-1} c_2 c_5 K_I q \left(\log \frac{1}{\delta_0} \right)^{n-1} \varepsilon_0^{n-1} < \varepsilon \, .
$$

Then (6.1) implies that

(6.3)
$$
\sum_{i \in I} \lambda_i \nu(U_i) \geq \frac{\nu(s)}{4} \sum_{j \in J} \Delta_j \, .
$$

Set $d = \sum_j \Delta_j / 10$ and

(6.4)
$$
I_1 = \{\, i \in I : \ \lambda_i \leq d \ \text{or} \ \nu(U_i) \leq P \,\} \, ,
$$

where

(6.5)
$$
P = b_1 K_I K_O \left(|\log(\pi/4)|^{n-1} + |\log \sigma_0|^{n-1} \right)
$$

and σ_0 is as in IV.2.3. For the sum over I_1 we extract from (2.15) the estimate

$$
\sum_{i \in I_1} \lambda_i \nu(U_i) \leq \frac{1}{10} \nu(s) \sum_{j \in J} \Delta_j + 2 c_2 \varepsilon_0^{n-1} \nu(s) q P \, .
$$

We choose ε_0 such that in addition to (6.2) also

(6.6)
$$
2 c_2 \varepsilon_0^{n-1} q P < \varepsilon
$$

holds. We summarize the preceding considerations in the following proposition.

6.7. Proposition. *With the choices of A_i in* (5.10) *and ε_0 in* (6.2) *and* (6.6) *it is true that*

$$(6.8) \qquad \sum_{i \in I \setminus I_1} \lambda_i \nu(U_i) \geq \frac{\nu(s)}{8} \sum_{j \in J} \Delta_j \; .$$

We will now apply Lemma IV.2.3 to each mapping $f \circ \varphi_i^{-1}$, $i \in I \setminus I_1$. Let $i \in I \setminus I_1$. For each $j \in J_i \setminus J^i$ we have by the definitions (5.4) and (5.7) that

$$\int_S (n_i^j - l_i^j - m_i^j)(y) \, dy \geq \frac{1}{3} \int_S n_i^j(y) \, dy > \frac{\omega_{n-1}}{6} \Delta_j \nu(U_i) > 0 \; ,$$

so $n_i^j(y) - l_i^j(y) - m_i^j(y) > 0$ for some $y \in S$. By Lemmas 3.4 and 3.9 we thus have $n_i^j(y) - \mathrm{card}\big(L_i^j(y) \cup M_i^j(y)\big) > 0$. Expressed in the notation of (3.3) and (3.8) this means that there exists an index $\nu \in \{1, \dots, n_i^j(y)\}$ for which $1 - w_{y,\nu} \leq \sigma_i$. The corresponding lifting α_ν then has the following properties for some $t \leq 1$:

(1) the restriction $\alpha_i^j = \alpha_\nu | [w_{y,\nu}, t]$ is a path in X_i that connects ∂W_i and ∂X_i;

(2) $|f \circ \alpha_i^j| \subset \overline{B}(a_j, 3\sigma_i/2) \subset \overline{D}(a_j, 3\sigma_i/2)$.

Then $\varphi_i \circ \alpha_i^j$ connects S and $S(3/2)$. We now apply Lemma IV.2.3 with $M = 3/2$ to the mapping $g = f \circ \varphi_i^{-1}$ and the sets $F_j = |\varphi_i \circ \alpha_i^j|$, $j \in J_i \setminus J^i$. In addition, we use IV.2.16 to estimate $\mathrm{M}(\Gamma_j)$ in IV.2.3 for some j. As a result we find $j \in J_i \setminus J^i$ for which

$$(6.9) \qquad \big(b_0 \lambda_i^{1/(n-1)} - c_7 K_O K_I\big)\Big(\log \frac{2\sigma_0}{3\sigma_i}\Big)^{n-1} \leq c_8 K_O \nu_g\big(2, \Sigma(a_j, \sigma_0)\big) \; .$$

Note that we used also the fact that φ_i is K_O-qc for a K_O depending only on n and that $\sigma_0 > 3\sigma_i/2$, which is true because $\sigma_i \leq \delta_0 < 2\sigma_0/3$.

Next we apply IV.1.9. Since $i \in I \setminus I_1$ and P is defined by (6.5), we obtain, with the same constant b_1 that appears in IV.(2.10), the estimate

$$(6.10) \qquad \begin{aligned} \nu_g\big(2, \Sigma(a_j, \sigma_0)\big) &\leq \nu_g(4,1) + b_1 K_I K_O\big(|\log(\pi/4)|^{n-1} + |\log \sigma_0|^{n-1}\big) \\ &\leq 2\nu_g(4,1) \; . \end{aligned}$$

By (5.10) we have

$$(6.11) \qquad \frac{\omega_{n-1}\lambda_i \, \nu(U_i)}{18 c_6 K_I} \leq A_i \, \nu(U_i) = \Big(\log \frac{\delta_0}{\sigma_i}\Big)^{n-1} < \Big(\log \frac{2\sigma_0}{3\sigma_i}\Big)^{n-1} \; .$$

Inequalities (6.9)–(6.11) yield

$$(6.12) \qquad (\lambda_i^{1/(n-1)} - c_9 K_O K_I)\lambda_i \, \nu(U_i) \leq c_{10} K_O K_I \nu(Z_i) \; .$$

We now obtain the following growth relation.

6.13. Proposition. *Let c_9 and c_{10} be as in* (6.12). *Suppose that*

$$(6.14) \qquad \sum_{j \in J} \Delta_j \geq \max\left(20,\ 10(2c_9 K_O K_I)^{n-1}\right) .$$

Then, for $i \in I \smallsetminus I_1$,

$$(6.15) \qquad c_9 \lambda_i \nu(U_i) \leq c_{10} \nu(Z_i) .$$

Proof. Since $i \in I \smallsetminus I_1$, we have $\lambda_i \geq d = \sum_j \Delta_j / 10 \geq (2c_9 K_O K_I)^{n-1}$. Accordingly, $\lambda_i^{1/(n-1)} - c_9 K_O K_I \geq c_9 K_O K_I$, so (6.15) follows from (6.12). Note that $\sum_j \Delta_j \geq 20$ was needed to make certain that $\lambda_i \geq d \geq 2$ in (6.12). $\qquad\square$

6.16. Remark. Under the assumption (6.14) the stronger inequality $\lambda_i^{1+\frac{1}{n-1}} \nu(U_i) \leq 2c_{10} K_O K_I \nu(Z_i)$ is valid, but we will only need (6.15).

6.17. Proof of Theorem 1.3. Suppose that (6.14) holds. Then by (2.9), 2.14(2), 6.7, and 6.13 we obtain

$$\frac{\nu(s)}{8} \sum_{j \in J} \Delta_j \leq \sum_{i \in I \smallsetminus I_1} \lambda_i \nu(U_i) \leq c_9^{-1} c_{10} \sum_{i \in I} \nu(Z_i)$$

$$\leq c_9^{-1} c_{10} c_1 \nu(s') \leq \tfrac{3}{2} c_9^{-1} c_{10} c_1 \nu(s) .$$

The final conclusion is, therefore, that

$$(6.18) \qquad \sum_{j \in J} \Delta_j \leq \max\left(20,\ 10(2c_9 K_O K_I)^{n-1},\ 12 c_9^{-1} c_{10} c_1\right)$$

$$\leq C'(n, K) .$$

With (2.12) this proves Theorem 1.3 in the case $a_1, \ldots, a_q \in B(1/2)$.

If the points $a_1, \ldots, a_q$ do not all lie in $B(1/2)$, we replace S by another sphere $Y = \Sigma(z, u)$, $u \leq \pi/20$, such that $a_1, \ldots, a_q \in \overline{\mathbb{R}}^n \smallsetminus D(z, 5u)$. We apply Lemma 2.1 to Y. With $\nu(s)$ now replaced by $\nu(s, Y)$ we get inequality (2.12). For this we may need to increase κ. Let h be a spherical isometry of $\overline{\mathbb{R}}^n$ such that $h(z) = \infty$, and let T be a Möbius transformation of $\overline{\mathbb{R}}^n$ that keeps 0 and ∞ fixed (i.e., a homothety of $\mathbb{R}^n$) and sends hY onto the unit sphere S. Set $\psi = T \circ h \circ f$. Then by simple estimates we get $Th(\overline{\mathbb{R}}^n \smallsetminus D(z, 5u)) \subset B(1/2)$, and we have $\nu_f(s, Y) = \nu_\psi(s, 1)$. Note that the average $\nu_f(s, Y)$ is not preserved if we compose f with an arbitrary Möbius transformation. Because the dilatation coefficients of ψ are the same as those of f, we conclude from these observations that the same bound as in (6.18) is valid in the general case as well. This completes the proof of Theorem 1.3. $\qquad\square$

6.19. Remarks. 1. We see from (6.18) that the bound $C(n, K)$ in Theorem 1.3 can be taken to be of the order $(K_O K_I)^{n-1}$, as was the case in the Picard–type theorem IV.2.1 (Corollary IV.2.21).

2. From the proof of 1.3 it follows that (2.10) is true for $s' \in [\kappa, \infty[\setminus E$ where κ depends only on n, K, and σ_0. Moreover, for fixed n and K, the bound κ is increasing with σ_0^{-1}.

We now turn to the case of mappings of the unit ball and the proof of Theorem 1.9. For this we need a modification of Lemma 2.1.

6.20. Lemma. Let $f: B \to \overline{\mathbb{R}}^n$ be a K-qm mapping with the property that

$$(6.21) \qquad \limsup_{r \to \infty} (1 - r) A(r)^{1/(n-1)} = \infty .$$

Then there exists a set $E \subset [0, 1[$ such that

$$(6.22) \qquad \liminf_{s \to 1} \frac{m_1(E \cap [s, 1[)}{1 - s} = 0$$

and such that the following is true: If $0 < \varepsilon_0 < 1/5$ and if for $0 < s < 1$ we write

$$s' = s + \frac{s}{\varepsilon_0 A(s)^{1/(n-1)}} ,$$

then there exists an increasing function $w: [0, \infty[\to [D_{\varepsilon_0}, 1[$ such that for any $(n - 1)$-sphere Y in $\mathbb{R}^n$ of spherical radius $u \leq \pi/4$ and any $\rho \in [w(|\log u|), 1[\setminus E$, there is an $s \in]0, 1[$ with $s' = \rho$ for which inequalities (2.3) and (2.4) hold. Moreover $A(D_{\varepsilon_0}) > 1/\varepsilon_0^{2n-2}$.

Proof. We will inductively define an increasing sequence $t_3, t_4, t_5, \ldots$ in $]0, 1[$ tending to 1 such that $1 - t_m < (1 - t_{m-1})/m$. Simultaneously we define a sequence $E_3, E_4, \ldots$ of subsets of $]0, 1[$. Set $t_3 = 3/4$ and $E_3 = [3/4, 1[$. Let $m \geq 4$ and suppose t_{m-1} has been defined. As in 2.1 set $\varphi(r) = m^{-2} A(r)^{1/(n-1)}$. By (6.21) there exists $t = t_m$ such that $1 - t_m < (1 - t_{m-1})/m$ and

$$(6.23) \qquad (1 - t) A(t)^{1/(n-1)} > \frac{8m^3}{1 - (\frac{m-1}{m})^{1/(n-1)}} .$$

Set $t^* = t + (1 - t)/m$, $\bar{t} = 1 - (1 - t)/m$, and

$$F_m = \left\{ r \in]t, 1[\; : \; A\left(r + \frac{2r}{\varphi(r)}\right) > \frac{m}{m-1} A(r) \quad \text{or} \quad r + \frac{2r}{\varphi(r)} > 1 \right\}.$$

Supposing that $F_m \cap]t, \bar{t}] \neq \emptyset$ we define as before a sequence $t = r_0'' \leq r_1 < r_1'' \leq r_2 < \cdots \leq r_q < r_q''$ of points in $[t, 1[$ inductively by $r_k = \inf\{r \in]r_{k-1}'', 1[\; : r \in F_m\}$, $r_k'' = r_k + 2r_k/\varphi(r_k)$, where q indicates the last index k for which $F_m \cap]r_{k-1}'', 1[\neq \emptyset$ and $r_k \leq \bar{t}$. Set $\rho_k = r_k'' + 2r_k''/\varphi(r_k)$. Then $\rho_k < 1$ for $1 \leq k \leq q$ and we write

$$E_m^1 = \bigcup_{k=1}^q [r_k, \rho_k] \, .$$

If $F_m \cap]t, \bar{t}] = \emptyset$, we define $E_m^1 = \emptyset$. For the length of E_m^1 we get by (6.23) the estimate

$$m_1(E_m^1) < \sum_{k=1}^q (\rho_k - r_k) < \sum_{k=1}^q \frac{4}{\varphi(r_k)} = \sum_{k=1}^q \frac{4m^2}{A(r_k)^{1/(n-1)}}$$

$$\leq \frac{4m^2}{A(t)^{1/(n-1)} \left(1 - \left(\frac{m-1}{m}\right)^{1/(n-1)}\right)} < \frac{1}{2m}(1-t) \, .$$

Set $E_m = [t, t^*] \cup [\bar{t}, 1[\cup E_m^1$. Then $m_1(E_m) < 3(1-t)/m$. Having now defined the sequences (t_m) and (E_m) we set

$$E = \bigcup_{m \geq 3} \left(E_m \cap [t_m, t_{m+1}] \right) \, .$$

Then clearly

$$\lim_{m \to \infty} \frac{m_1(E \cap [t_m, 1[)}{1 - t_m} = 0 \, .$$

Let $0 < \varepsilon_0 < 1/5$, let Y be as in the lemma, and let $m_1 \geq 4$ be the integer m chosen in the proof of 2.1. Suppose $\rho \in [t_{m_1}, 1[\setminus E$. Then ρ belongs to some interval $[t_m, t_{m+1}]$ with $m \geq m_1$. By the definition of E, $\rho \in [t^*, \bar{t}] \setminus E_m$, where we used the notation from above with $t = t_m$. From (6.23) we get

$$\left(t + \frac{t}{\varphi(t)} \right)' = t + \frac{t}{\varphi(t)} + \frac{t + t/\varphi(t)}{\varepsilon_0 A(t)^{1/(n-1)}}$$

$$\leq t + \frac{1-t}{8m} + \frac{1 + (1-t)/8m}{16m}(1-t)$$

$$\leq t + \frac{1-t}{4m} < t^* \, .$$

It follows that there exists $r \in]t, \bar{t}[$ such that $\rho = \left(r + r/\varphi(r) \right)' = s'$. Then $r \notin F_m$. Since, in addition, $F_m \subset E_m$, inequalities (2.3) and (2.4) follow as in the proof of 2.1. If m_0 is the least positive integer with $2/m_0^2 < \varepsilon_0$, we can put $D_{\varepsilon_0} = t_{m_0}$. From (6.23) it follows that

$$A(D_{\varepsilon_0}) > \left(\frac{8m_0^3}{1 - D_{\varepsilon_0}} \right)^{n-1} > (8m_0^4)^{n-1} > 1/\varepsilon_0^{2n-2} \, .$$

The lemma is proved.   □

6.24. Proof of Theorem 1.9. Let E be the set in $]0, 1[$ given by Lemma 6.20. We again consider first the case where the points $a_1, \ldots, a_q$ lie in $B(1/2)$. We fix $0 < \varepsilon_0 < \min(1/q, 1/5)$ such that, in addition, ε_0 satisfies (6.2) and (6.6) with $\varepsilon = 1$. By Lemma 6.20 we find $\iota \in]0, 1[$ such that for every $\rho \in]\iota, 1[\setminus E$ there exists s with $s' = \rho$ and (2.8) and (2.9) hold. Fix such an s. This is all that was needed for the proof of 1.3. Therefore the proof of Theorem 1.9 goes through in the same way, except that the mapping f_0 defined in 3.1 has to be replaced by $f|B(s' + (1 - s')/2)$. If the points a_j do not all lie in $B(1/2)$, we appeal to the same remarks as in 6.17. The theorem is proved. $\square$

7. Mappings of the Plane

In this section we will show how the ideas of the proof of Theorem 1.3 give the sharp bound 2 in Ahlfors' result (1.1) for $n = 2$. We establish the following.

7.1. Theorem. *Let $f: \mathbb{R}^2 \to \mathbb{R}^2$ be a nonconstant qm mapping. Then there exists a set $E \subset [1, \infty[$ of finite logarithmic measure such that*

$$(7.2) \qquad \limsup_{\substack{r \to \infty \\ r \notin E}} \sum_{j=1}^{q} \left(1 - \frac{n(r, a_j)}{A(r)}\right)_+ \leq 2$$

holds whenever $a_1, \ldots, a_q$ are distinct points in $\mathbb{R}^2$.

A proof of Theorem 7.1 along these lines was given by M. Pesonen, whose presentation in [Pe2] we follow, except in referring to some lemmas which we have at our disposal from the general case. Since the branch set now consists of isolated points, most of the technical difficulties that in the case $n \geq 3$ demanded a lot of attention can be avoided. This comment pertains to the proofs of 3.4, 3.9, and especially the material in Section 4, which is not needed at all in the case $n = 2$.

Apart from such technical features there is one essential difference between the cases $n = 2$ and $n \geq 3$. In the case $n \geq 3$ we needed to collect different magnitudes of growth relations for the averages (Proposition 6.13). The case $n = 2$ is easier. After ruling out certain indices i we are able to concentrate on the growth behavior of averages for a single pair U_i, Z_i. The reason for this and for the fact that we get the bound 2 is that paths in the plane have a separation property. By using this separation property we obtain an arbitrary large lower bound for some modulus $\mathsf{M}(\Gamma_j)$ in Lemma IV.2.3, a bound for whose derivation only crude estimates are needed. The last remark also leads via a modification of the proof of IV.2.1 to a proof of the classical Picard theorem (which of course also follows from 7.1).

Proof of Theorem 7.1. We now also start out by applying Lemma 2.1. Let $f: \mathbb{R}^2 \to \mathbb{R}^2$ be a nonconstant qm mapping, and let $E \subset [1, \infty[$ be the set given by 2.1. We write K for the dilatation $K(f) = K_O = K_I$. Suppose that the points $a_1, \ldots, a_q$ lie in $B(1/2)$. We may also assume that $q \geq 3$. Let $0 < \varepsilon < 1/2$. To apply 2.1 we fix $\varepsilon_0 \in]0, \varepsilon/q[$ such that (6.2), (6.6), and a condition to be given later in (7.19) are in force. By 2.1 there exists $\kappa > 1$ for which

(7.3)
$$\left| \frac{A(s')}{\nu(s)} - 1 \right| < \frac{\varepsilon}{q}$$

and $\nu(s') \leq 3\nu(s)/2$ whenever $s > 0$ is such that $s' \in [\kappa, \infty[\smallsetminus E$. Fix such an s.

It suffices to show that

(7.4)
$$\sum_{j=1}^{q} \left(1 - \frac{n(s', a_j)}{A(s')} \right)_+ \leq 2 + 6\varepsilon .$$

We may assume $1 - n(s', a_j)/A(s') > 0$ for all j. In place of (2.12) we now obtain

(7.5)
$$\sum_{j \in J} \left(1 - \frac{n(s', a_j)}{A(s')} \right) < \sum_{j \in J} \Delta_j + \varepsilon .$$

Hence we may further assume that $\Delta_j > 0$ for all j and need only demonstrate that

(7.6)
$$\sum_{j \in J} \Delta_j \leq 2 + 5\varepsilon .$$

We proceed as we did in the remainder of Section 2 and in Section 3, the only difference being that the numbers A_i in (3.7) get replaced by a single constant, namely, by

(7.7)
$$A_0 = \frac{2\pi \varepsilon}{c_6 K} .$$

Hence

(7.8)
$$\log \frac{\delta_0}{\sigma_i} = A_0 \nu(U_i) .$$

Here c_6 is the constant in (3.11). As we remarked above, when $n = 2$ the proofs of 3.4 and 3.9 simplify considerably. Suppose then that (7.6) does not hold.

For $i \in I$, $j \in J$, and $y \in S$ we have, as in the beginning of Section 5, the f_0-liftings $\alpha_1, \ldots, \alpha_{\nu_y}$ of γ_y^j that start in U_i and that satisfy $|\alpha_\nu| \not\subset \overline{B}(s')$. If here $1 \leq \nu \leq \nu_y$ and $\nu \notin L_i^j(y) \cup M_i^j(y)$, we call α_ν a *lifting of the jth type*. A point $x \in \overline{B}(s) \cap f^{-1}(y)$ is y-*admissible* if it is a starting point for at least three liftings of different types.

We will rule out those indices i for which $\nu(U_i)$ is small or for which there is a large growth relation for averages, to be more precise, for which the ratio $\nu(Z_i)/\nu(U_i)$ is large. For this we set

$$I_0 = \left\{ i \in I : \ 2c_1 \frac{q}{\varepsilon} \nu(U_i) \leq \nu(Z_i) \ \text{or} \ \nu(U_i) \leq P \right\},$$

$$I^* = I \smallsetminus I_0 , \qquad U^* = \bigcup_{i \in I^*} U_i ,$$

where we recall the constant c_1 from 2.14 and P from (6.5). Then

(7.9)
$$\sum_{i \in I_0} \nu(U_i) \leq \frac{\varepsilon}{2c_1 q} \sum_{i \in I} \nu(Z_i) + 2c_2 \varepsilon_0 \, \nu(s) P$$

$$\leq \frac{\varepsilon}{2q} \nu(s') + \frac{\varepsilon}{q} \nu(s) < \frac{2\varepsilon}{q} \nu(s)$$

by (2.9), (2.15), and (6.6).

For $j \in J$ and $y \in S$ we write

(7.10)
$$n_j(y) = n(U^*, y) - n(s', a_j) - \sum_{i \in I^*} (l_i^j(y) + m_i^j(y)) .$$

We see that $n_j(y)$ gives a lower bound for the number of liftings of the jth type starting in $U^* \cap f^{-1}(y)$. In place of (5.8) we now obtain using 3.4, 3.9, and (7.9) the estimate

(7.11)
$$\sum_{j \in J} \int_S n_j(y) \, dy = \sum_{j \in J} \int_S \big(n(s, y) - n(s', a_j) \big) \, dy - 2\pi \sum_{j \in J} \sum_{i \in I_0} \nu(U_i)$$

$$- \sum_{j \in J} \sum_{i \in I^*} \int_S (l_i^j(y) + m_i^j(y)) \, dy$$

$$\geq 2\pi \sum_{j \in J} \Delta_j \nu(s) - 4\pi \varepsilon \, \nu(s)$$

$$- 2c_2 c_5 K q \Big(\log \frac{1}{\delta_0} \Big) \varepsilon_0 \, \nu(s) - c_6 K A_0 \, \nu(s) .$$

By (6.2) and the choice of A_0 in (7.7) we then obtain

$$\frac{1}{2\pi} \sum_{j \in J} \int_S n_j(y) \, dy \geq \sum_{j \in J} \Delta_j \, \nu(s) - 4\varepsilon \, \nu(s) \geq (2 + \varepsilon) \nu(s)$$

$$\geq 2 \nu(U^*) + \varepsilon \nu(s) ,$$

which we express in the form

(7.12)
$$\frac{1}{2\pi} \int_S \Big(\sum_{j \in J} n_j(y) - 2n(U^*, y) \Big) dy \geq \varepsilon \, \nu(s) .$$

Let $S_0 \subset S$ be the set of all $y \in S$ such that $|\gamma_y^j| \cap (f B_f \smallsetminus \{a_j\}) \neq \emptyset$ for some $j \in J$ or such that $|\gamma_y^j| \subset |\gamma_y^k|$ for some $j, k \in J$ with $j \neq k$. Then S_0 is plainly countable. Set $S^* = S \smallsetminus S_0$. From (7.12) it follows that there exists $y^* \in S^*$ such that

(7.13)
$$\sum_{j \in J} n_j(y^*) - 2n(U^*, y^*) \geq \varepsilon \nu(s) .$$

Let $x_1, \ldots, x_l$ be the points in $U^* \cap f^{-1}(y^*)$. Then $l = n(U^*, y^*)$ because $y^* \in S^*$. Let $x_1, \ldots, x_\mu$ be those points x_k that are y^*-admissible, and for $1 \leq k \leq l$ let $\psi(x_k)$ be the number of liftings of different types starting at x_k. Then

$$\psi(x_k) \leq 2 \quad \text{for} \quad k = \mu + 1, \ldots, l ,$$

and

$$\sum_{k=1}^{l} \psi(x_k) \geq \sum_{j \in J} n_j(y^*) .$$

From these inequalities and (7.13) we obtain

(7.14)
$$\varepsilon \nu(s) \leq \sum_{k=1}^{l} (\psi(x_k) - 2) \leq \sum_{k=1}^{\mu} (q - 2) \leq \mu q .$$

Since $\operatorname{card} I^* \leq p \leq 2c_2 \varepsilon_0 \nu(s)$, (7.14) enables us to say that there exists $i \in I^*$ for which the number μ_i of y^*-admissible points in U_i satisfies

(7.15)
$$\mu_i \geq \frac{\mu}{\operatorname{card} I^*} \geq \frac{\varepsilon}{2c_2 \varepsilon_0 q} .$$

We can make μ_i arbitrarily large by taking ε_0 at the beginning of the proof sufficiently small. We now intend to exploit this fact. Fix $i \in I^*$ such that (7.15) holds. We initially assume that $\mu_i \geq 20$.

If $x \in U_i \cap f^{-1}(y^*)$ is y^*-admissible, then there exist liftings α_{xk}, $k = 1, 2, 3$, of different types starting at x. For $1 \leq k \leq 3$ set

$$t_k = \min\{ t : \alpha_{xk}(t) \in \partial X_i \} ,$$
$$\overline{\alpha}_{xk} = \alpha_{xk} | [0, t_k] .$$

Then $X_i \setminus (\overline{\alpha}_{x1} \cup \overline{\alpha}_{x2} \cup \overline{\alpha}_{x3})$ has exactly three components Y_{xm}, $m = 1, 2, 3$. If z is another y^*-admissible point in U_i, then

$$\overline{\alpha}_{z1} \cup \overline{\alpha}_{z2} \cup \overline{\alpha}_{z3} \subset Y_{xm}$$

for some m. By a simple induction argument it follows that there are at least $\mu_i + 2$ disjoint open arcs C_r on ∂X_i such that the endpoints of each C_r belong to some pair $\overline{\alpha}_{xk}$ and $\overline{\alpha}_{zh}$, with α_{xk} and α_{zh} of different types. Here x and z may coincide. It follows that there exists r_0 such that the length of the image $\varphi_i C_{r_0}$ on the circle $S(3/2)$ satisfies $\mathcal{H}^1(\varphi_i C_{r_0}) \leq 3\pi/(\mu_i + 2) < 1/3$.

Let α_{xk} and α_{zh} correspond as above to C_{r_0}, and let ξ and ζ be the endpoints of $\varphi_i C_{r_0}$. With the earlier notation, $\alpha_{xk} = \alpha_\nu$ and $\alpha_{zh} = \alpha_{\nu'}$, where $1 \leq \nu, \nu' \leq \nu_{y^*}$. Let α_ν and $\alpha_{\nu'}$ be of types j and j', respectively. As in (1) preceding (6.9) we here write $\alpha_i^j = \alpha_\nu | [w_{y^*, \nu}, t]$ and $\alpha_i^{j'} = \alpha_{\nu'} | [w_{y^*, \nu'}, t']$ for paths in X_i that connect ∂W_i and ∂X_i. Note that t was denoted by t_k above. We apply Lemma IV.2.3 with $M = 3/2$ to the mapping $g = f \circ \varphi_i^{-1}$ and to the sets $F = |\varphi_i \circ \alpha_i^j|$, $F' = |\varphi_i \circ \alpha_i^{j'}|$. As in (6.9) we obtain

$$(7.16) \qquad \left(\mathsf{M}(\Gamma) - c_7 K^2\right)\left(\log \frac{2\sigma_0}{3\sigma_i}\right) \le c_8 K \, \nu_g\left(2, \Sigma(a_j, \sigma_0)\right),$$

where $\Gamma = \Delta(F, F'; B(3/2) \setminus \overline{B})$. Since the circular arcs $S(\xi, u) \cap B(3/2)$, $|\xi - \zeta| < u < 1/2$, meet both F and F', we get from II.(1.9) (see II.1.12) the estimate (cf., the argument used in IV.2.9)

$$(7.17) \qquad \mathsf{M}(\Gamma) \ge b(2) \log \frac{\mu_i + 2}{6\pi}.$$

Substituting (7.8) and (7.17) into (7.16), and remembering that $\delta_0/\sigma_i < 2\sigma_0/3\sigma_i$ we obtain the inequality (cf. (6.12))

$$(7.18) \qquad \left(b(2) \log \frac{\mu_i + 2}{6\pi} - c_7 K^2\right)\nu(U_i) \le c_1 c_6 c_8 K^2 \frac{q}{\varepsilon^2} \, \nu(U_i),$$

where we also used $i \in I^*$. From (7.15) we ascertain that, if ε_0 is chosen so small that, in addition to the earlier conditions (6.2) and (6.6), the inequality

$$(7.19) \qquad \exp\left(\frac{K^2}{b(2)}\left(c_1 c_6 c_8 \frac{q}{\varepsilon^2} + c_7\right)\right) < \frac{\varepsilon}{40 c_2 q \varepsilon_0}$$

is satisfied, then we get a contradiction with (7.18). This proves 7.1 in the case where the points $a_1, \ldots, a_q$ lie in $B(1/2)$. For the general case the remarks made in 6.17 are applicable. Theorem 7.1 is proved. $\square$

8. Order of Growth

Let $f: \mathbb{R}^n \to \overline{\mathbb{R}}^n$ be a nonconstant qm mapping. We define the *order* μ_f and *lower order* λ_f of f by

$$(8.1) \qquad \mu_f = \limsup_{r \to \infty} \frac{\log A(r)}{\log r},$$

$$(8.2) \qquad \lambda_f = \liminf_{r \to \infty} \frac{\log A(r)}{\log r}.$$

Recall the definition of $A(r)$ from IV.1. If $f: \mathbb{R}^n \to \mathbb{R}^n$, then μ_f and λ_f can be expressed in terms of the maximum norm function

$$M(r) = \sup_{|x| = r} |f(x)|$$

as well, namely,

$$(8.3) \qquad \mu_f = \limsup_{r \to \infty} (n-1) \frac{\log \log M(r)}{\log r} \,,$$

$$(8.4) \qquad \lambda_f = \liminf_{r \to \infty} (n-1) \frac{\log \log M(r)}{\log r} \,.$$

As in classical function theory it is interesting to study growth, not only for mappings of $\mathbb{R}^n$, but more generally for qr mappings defined in unbounded domains. In Section VII.6 we shall consider such questions and prove two types of counterparts to the Phragmén–Lindelöf theorem.

In this section we establish the statements (8.3) and (8.4) (Theorem 8.18). The proof is from [RV] by S. Rickman and M. Vuorinen. Let $f \colon \mathbb{R}^n \to \mathbb{R}^n$ be a nonconstant K-qr mapping. From 2.1 we get the following result.

8.5. Lemma. *There exists a set $E \subset [1, \infty[$ of finite logarithmic measure, i.e.,*

$$\int\limits_E \frac{dr}{r} < \infty \,,$$

such that

$$(8.6) \qquad \lim_{E \not\ni r \to \infty} \frac{\nu(r, t)}{A(r)} = 1$$

for all $t > 0$.

We want to compare the functions $\nu(r, 1)$ and $M(r)$. Let us look at the case where $f(x) \to \infty$ as $x \to \infty$. By III.2.8 f extends to a qm mapping $f^* \colon \overline{\mathbb{R}}^n \to \overline{\mathbb{R}}^n$. Then there exists $r_0 > 0$ such that $i(\infty, f) = \nu(r, 1)$ for $r \geq r_0$. From III.4.7 we deduce that

$$(8.7) \qquad \frac{1}{K} \nu(r, 1) + o(1) \leq \frac{\log M(r)^{n-1}}{(\log r)^{n-1}} \leq K \nu(r, 1) + o(1) \quad (r \to \infty) \,.$$

For the general case we obtain an upper bound for $\nu(r, 1)$ as follows. In IV.1.1 set $t > M(\theta r)$ and $s = 1$. This gives

$$0 \geq \nu(r, 1) - \frac{K |\log t|^{n-1}}{(\log \theta)^{n-1}} \,,$$

and hence

$$(8.8) \qquad |\log M(\theta r)|^{n-1} \geq \frac{(\log \theta)^{n-1}}{K} \nu(r, 1) \,.$$

We shall prove an inequality opposite to (8.8) (cf. (8.7)).

8.9. Lemma. *There exist* $\theta_0 = \theta_0(n, K) > 1$ *and* $b_1 = b_1(n)$ *such that for some* $r_0 > 1$ *we have*

(8.10)
$$\nu(\theta_0 r, 1) \geq \frac{b_1 (\log M(r))^{n-1}}{K (\log r)^{n-1}} \quad \text{if } r \geq r_0 .$$

Proof. We may assume that ∞ is an essential singularity. Let $r_1 > 0$ be such that $M(r_1) = 1$. For $r > \max(r_1, 1)$ we let Γ_r be the family of paths in $B(2r)$ which join $\overline{B}(r_1)$ and $f^{-1}(\mathbb{R}^n \smallsetminus B(M))$ where $M = M(r)$. By performing a Möbius transformation we see with the help of the symmetrization result III.1.1 that

$$\mathsf{M}(\Gamma_r) \geq \nu_n \left(\frac{s}{2s^2 + 1} \right) ,$$

where $s = r/r_1$ and where $\nu_n(t)$ is the capacity of the Grötzsch condenser $(B^n, [0, te_1])$. From III.1.2(3) we deduce the estimate

(8.11)
$$\mathsf{M}(\Gamma_r) \geq \frac{d_n}{(\log r)^{n-1}} , \quad r \geq r_0 ,$$

for some r_0, where $d_n > 0$ is a constant depending only on n. Fix $r \geq r_0$ and set

$$\rho(y) = \begin{cases} \dfrac{3}{(\log M)|y|} & \text{if } 1 < |y| < M^{1/3}, \\ 0 & \text{elsewhere.} \end{cases}$$

Then $\rho \in F(f\Gamma_r)$, and II.(2.6) gives (cf. the proof of IV.2.3)

(8.12)
$$\mathsf{M}(\Gamma_r) \leq K \int_{\mathbb{R}^n} n(2r, y) \rho(y)^n dy$$
$$= \frac{3^n K \omega_{n-1}}{(\log M)^n} \int_1^{M^{1/3}} \frac{\nu(2r, \tau)}{\tau} d\tau .$$

By (8.11) and (8.12) we get the existence of a point y, $|y| \in [1, M^{1/3}]$, such that

(8.13)
$$n(2r, y) \geq \frac{b_1 (\log M)^{n-1}}{K (\log r)^{n-1}} ,$$

where $b_1 = b_1(n)$. We separate two cases:

   Case 1. For each $x \in f^{-1}(y) \cap \overline{B}(2r)$ the x-component of $f^{-1}B(M^{2/3})$ is contained in $B(4r)$.

   Let $x \in f^{-1}(y) \cap \overline{B}(2r)$ and let D be the x-component of $f^{-1}B(M^{2/3})$. By I.4.7, D is a normal domain for f and

$$\sum_{z \in f^{-1}(\zeta) \cap D} i(z, f) = \sum_{x \in f^{-1}(y) \cap D} i(x, f)$$

for all $\zeta \in S^{n-1}$. Summing over all such components D we obtain for all $\zeta \in S^{n-1}$ that

(8.14)
$$n(4r, \zeta) \geq \sum_D \sum_{z \in f^{-1}(\zeta) \cap D} i(z, f)$$
$$= \sum_D \sum_{x \in f^{-1}(y) \cap D} i(x, f) \geq n(2r, y) .$$

Inequalities (8.13) and (8.14) give (8.10) with $\theta_0 = 4$.

Case 2. For some $x \in f^{-1}(y) \cap B(2r)$ the x-component of $f^{-1}B(M^{2/3})$ intersects $\mathbb{R}^n \setminus B(4r)$.

In this case we use an argument similar to the one in IV.2.3. Let Γ be the family of paths in $B(4r) \setminus \overline{B}(2r)$ which join $f^{-1}B(M^{2/3})$ and $f^{-1}(\mathbb{R}^n \setminus B(M))$. Then

(8.15)
$$\mathsf{M}(\Gamma) \geq c(n) \log 2 ,$$

where $c(n)$ is the constant in II.(1.14). In place of (8.12) we now obtain

(8.16)
$$\mathsf{M}(\Gamma) \leq \frac{3^n K \omega_{n-1}}{(\log M)^n} \int_{M^{2/3}}^M \frac{\nu(4r, \tau)}{\tau} d\tau .$$

By IV.1.1 we have for $\theta_0 > 4$ that

$$\nu(4r, \tau) \leq \nu(\theta_0 r, 1) + \frac{K(\log M)^{n-1}}{(\log(\theta_0/4))^{n-1}} , \quad \tau \in [1, M] .$$

Substituting this into (8.16) we get with (8.15) the inequality

(8.17)
$$\left(c(n) \log 2 - \frac{3^n \omega_{n-1} K^2}{(\log(\theta_0/4))^{n-1}} \right) (\log M)^{n-1} \leq 3^n \omega_{n-1} K \nu(\theta_0 r, 1) .$$

Finally we choose $\theta_0 > 4$ so that left hand side of (8.17) takes the value

$$\tfrac{1}{2} c(n) \log 2 (\log M)^{n-1} .$$

This completes the proof of the lemma. □

8.18. Theorem. Let $f \colon \mathbb{R}^n \to \mathbb{R}^n$ be a nonconstant qr mapping. Then (8.3) and (8.4) hold.

Proof. We shall prove (8.3). For (8.4) the proof is similar. Let the right hand side of (8.3) be α and let $r_i \to \infty$ be a sequence of positive numbers such that

$$\alpha = \lim_{i \to \infty} (n-1) \frac{\log \log M(r_i)}{\log r_i} .$$

Let $\varepsilon > 0$. By Lemma 8.9 there exists $\rho > 2$ such that

$$\alpha - \varepsilon < \frac{\log \nu(\theta_0 r_i, 1)}{\log r_i}, \qquad r_i \geq \rho.$$

Let E be the set given by 8.5. For any given $\sigma > \rho$ there exists by 8.5 $r_i \geq \sigma$ and $s \in [\theta_0 r_i, 2\theta_0 r_i] \setminus E$ such that $\nu(s, 1) \leq A(s)(1 + \varepsilon)$, and so

$$\alpha - \varepsilon < \frac{\log \nu(s, 1)}{\log r_i} \leq \frac{\log A(s) + \varepsilon}{\log s - \log(2\theta_0)}.$$

This gives $\alpha \leq \mu_f$. Similarly we get $\alpha \geq \mu_f$ by using (8.8) in place of (8.10). The theorem is proved. $\qquad \Box$

9. Further Results

In this section we study relationships opposite to the defect relation 1.3, namely we establish upper bounds for $n(r, a)$ in terms of the spherical average of the counting function.

Let $f \colon \mathbb{R}^n \to \mathbb{R}^n$ be a qm mapping. A natural question is whether there is some constant $\theta > 1$ for which

(9.1) $$\limsup_{r \to \infty} \frac{n(r, a)}{A(\theta r)}$$

is bounded. This need not be true in general.

However, at the presence of an asymptotic value the situation changes.

9.2. Theorem. *Let $f \colon \mathbb{R}^n \to \mathbb{R}^n$ be a nonconstant K-qm mapping and let f have an asymptotic value a_0, i.e., there exists a path $\alpha \colon [0, 1[\to \mathbb{R}^n$ such that $\alpha(\tau) \to \infty$ and $f(\alpha(\tau)) \to a_0$ as $\tau \to 1$. Then for each $\eta > 1$ there exists $\theta = \theta(n, K, \eta) > 1$ such that*

(9.3) $$\limsup_{r \to \infty} \sup_{a \in F} \frac{n(r, a)}{A(\theta r)} \leq \eta$$

whenever $F \subset \mathbb{R}^n \setminus \{a_0\}$ is compact.

Proof. We may assume that ∞ is an essential singularity. Let $\delta > 0$ be maximal such that the spherical ball $D(a_0, \delta)$ is contained in $\mathbb{R}^n \setminus F$ and let $\tau_1 \in]0, 1[$ be such that $f(\alpha(\tau)) \in D(a_0, \delta/2)$ for $\tau_1 \leq \tau < 1$. Let $a \in F$ and assume $a = 0$. Let $r > |\alpha(\tau_1)|$. To obtain an upper bound for $n(r, 0)/A(\theta r)$ with some $\theta > 1$ we use an argument from the proof of 8.9. We may assume $n(r, 0) > 0$. Let $t > 0$ be such that $B(t) = D(0, \delta/2)$.

Let $u > 1$ and fix $0 < s < t$ such that

(9.4) $$\left(\log \frac{t}{s} \right)^{n-1} = \frac{\omega_{n-1} K n(r, 0)}{c(n) \log u}.$$

We choose u more precisely later. We now consider two cases as in the proof of 8.9.

Case 1. For every $x \in f^{-1}(0) \cap \overline{B}(r)$ the x-component of $f^{-1}B(s)$ is contained in $B(ur)$.

As in 8.9 we conclude that

$$(9.5) \qquad n(ur, sy) \geq n(r, 0) , \quad y \in S .$$

By IV.1.1 we get from (9.4) and (9.5) for $v > u$ that

$$(9.6) \qquad \begin{aligned} \nu(vr, t) &\geq n(r, 0) - \frac{K(\log \frac{t}{s})^{n-1}}{(\log(v/u))^{n-1}} \\ &= n(r, 0)\left(1 - \frac{\omega_{n-1}K^2}{c(n)(\log u)(\log \frac{v}{u})^{n-1}}\right) . \end{aligned}$$

Note that in this case we did not use the assumption of the asymptotic value.

Case 2. There exists $x \in f^{-1}(0) \cap \overline{B}(r)$ such that the x-component of $f^{-1}B(s)$ intersects $\mathbb{R}^n \setminus B(ur)$.

By the argument in 8.9 we now obtain

$$(9.7) \qquad \left(c(n)\log u - \frac{\omega_{n-1}K^2}{(\log(v/u))^{n-1}}\right)\left(\log \frac{t}{s}\right)^{n-1} \leq \omega_{n-1}K\nu(vr, t)$$

for $v > u$. With (9.4) we can write (9.7) in the form

$$(9.8) \qquad \left(1 - \frac{\omega_{n-1}K^2}{c(n)(\log u)(\log(v/u))^{n-1}}\right) \leq \frac{\nu(vr, t)}{n(r, 0)} .$$

We now choose $v = u^2$ and u so that the left hand side in (9.8) (and the same factor in (9.6)) equals $\eta^{-1/2}$. By IV.1.7 there exists $r_0 > |\alpha(\tau_1)|$ independent of the point a chosen in F such that

$$(9.9) \qquad \eta^{1/2}A(2vr) \geq \nu(vr, t) , \quad r \geq r_0 .$$

Inequalities (9.6), (9.8), and (9.9) imply the theorem with $\theta = 2v$.   □

For the general case we obtain a result by replacing θr in (9.1) by a power of r.

9.10. Theorem. *Let* $f: \mathbb{R}^n \to \mathbb{R}^n$ *be a nonconstant K-qm mapping. Then for each $\eta > 1$ there exists $\alpha > 1$, depending only on n, K, and η such that*

$$(9.11) \qquad \limsup_{r \to \infty} \sup_{a \in \mathbb{R}^n} \frac{n(r, a)}{A(r^\alpha)} \leq \eta .$$

Proof. Let the balls $B_1 = B(s_1)$ and $B_2 = \mathbb{R}^n \setminus \overline{B}(s_2)$ have spherical radius $\pi/8$. Then for each point $a \in \mathbb{R}^n$ the spherical distance $\sigma(a, B_i)$ is at least $\pi/8$ for at least one i. There exists $\rho_0 > 2$ and balls $U_1, U_2 \subset B(\rho_0)$ such that $fU_i \subset B_i$, $i = 1, 2$. Now let $a \in \mathbb{R}^n$ and suppose $\sigma(a, B_1) \geq \pi/8$.

Let T be a spherical isometry of $\mathbb{R}^n$ such that $T(a) = 0$. Let $B(t)$ have spherical radius $\pi/8$. Then $T(fU_1) \cap B(t) = \emptyset$.

We shall apply the argument in the proof of 9.2 with the difference that in Case 2 we use an estimate of the form (8.11). As in the proof of 8.9 there exists $r_0 > \rho_0$ such that if $r \geq r_0$ and if Γ is a family of paths in $B(2r)$ joining U_1 and a continuum which connects $S(r)$ and $S(2r)$, then

$$(9.12) \qquad \mathsf{M}(\Gamma) \geq \frac{d_n}{(\log r)^{n-1}} .$$

We now apply the proof of 9.2 to $g = T \circ f$, $r \geq r_0$, $u = 2$, and $v > u$, with $c(n) \log 2$ replaced by $d_n (\log r)^{1-n}$ in (9.4). We obtain

$$(9.13) \qquad \nu_g(vr, t) \geq n_g(r, 0)\left(1 - \frac{\omega_{n-1}K^2(\log r)^{n-1}}{d_n(\log(v/2))^{n-1}}\right), \qquad r \geq r_0 .$$

There exists $r_1 > r_0$, independent of a, such that

$$(9.14) \qquad \eta^{1/2}A_f(2vr) = \eta^{1/2}A_g(2vr) \geq \nu_g(vr, t), \qquad r \geq r_1 .$$

We choose $v = v(r)$ so that the second factor in the right hand side of (9.13) is $\eta^{-1/2}$. Then $2v(r)r \leq r^\alpha$ for $r \geq r_1$ where $\alpha > 1$ is a constant depending only on n, K, and η. Then (9.13) and (9.14) give the required result $n_f(r, a) \leq \eta A_f(r^\alpha)$ for $r \geq r_1$. The proof for $\sigma(a, B_2) \geq \pi/8$ is similar. $\qquad \square$

9.15. Remark. Theorems 9.2 and 9.10 are from [R4]. It was shown by S. Toppila [To, Theorem 4] that (9.1) need not have a finite bound even for meromorphic functions in the case $n = 2$ no matter how θ is chosen. In fact, he proves in the same theorem that a power of r as in (9.11) is the best one can hope for. In [To, Theorem 5] he shows that θ cannot be chosen arbitrarily near 1 in 9.2 for $n = 2$ in the meromorphic case.

9.16. General averages of the counting function. The comparison lemma IV.1.1 can be considerably generalized. Extensions of IV.1.1 are established by P. Mattila and S. Rickman [MaR] in the setting of qr maps $f: M \to N$ where M and N are (oriented, connected) Riemannian n-manifolds, M is noncompact, and N is compact. M is assumed to be given a certain parametrized exhaustion, called admissible, and the counting function $n(r, a)$ is defined in terms of the parameter r.

In order to describe the content of [MaR] more precisely let us restrict to the case $M = \mathbb{R}^n$ with the natural exhaustion by balls $B(r)$, $0 < r < \infty$. Let μ be a measure in N such that Borel sets are μ-measurable and $0 < \mu(N) < \infty$. Given an increasing and continuous function $h: [0, \infty[\to [0, \infty[$ with $h(0) = 0$ and $h(r) > 0$ for $r > 0$, we say that μ is h-calibrated if

$$(9.17) \qquad \mu(B(x, r)) \leq h(r)$$

for all balls $B(x, r) \subset N$. Let $\nu_\mu(r)$ be the average of $n(r, y)$ with respect to μ and let $A(r)$ be the average of $n(r, y)$ with respect to the Lebesgue measure of N (if $N = \overline{\mathbb{R}}^n$, this definition of $A(r)$ thus coincides with the earlier one in IV.1). We state without proof one of the main results in [MaR] for this special case:

9.18. Theorem [MaR, 5.11(1)]. *Let N be a compact Riemannian n-manifold and let $f: \mathbb{R}^n \to N$ be a nonconstant qr mapping. Then there exists a set $E \subset [1, \infty[$ of finite logarithmic measure such that if μ is a measure as above which is h-calibrated with h satisfying*

$$(9.19) \qquad \int_0^1 \frac{h(r)^{1/pn}}{r} \, dr < \infty \quad \text{for some } p > 2,$$

then

$$(9.20) \qquad \lim_{\substack{r \to \infty \\ r \notin E}} \frac{\nu_\mu(r)}{A(r)} = 1.$$

For meromorphic functions A. Hinkkanen [Hin] showed that (9.19) can be replaced by

$$\int_0^1 \frac{h(r)}{r} \, dr < \infty.$$

Examples of meromorphic functions connected to Theorem 9.18 are presented in [MaR, 6.1, 6.5]. In [MaR, 5.11(2)] it was also shown that

$$(9.21) \qquad \liminf_{\substack{r \to \infty \\ r \notin E}} \frac{\nu_\mu(r)}{A(r)} \geq 1$$

if in 9.18 the assumption that μ be h-calibrated is replaced by

$$(9.22) \qquad \limsup_{r \to 0} \frac{\mu(B(x, r))}{h(r)} \leq 1$$

for μ almost all $x \in N$, where h satisfies (9.19). For $N = \overline{\mathbb{R}}^n$ this result can be deduced from the defect relation 1.3 and Remark 6.19.2.

The methods in [MaR] also give the following pointwise result. We refer to [MaR] for the proof.

9.23. Theorem [MaR, 5.13]. *Let f be as in 9.18. There exists a sequence $r_i \to \infty$ and a set $F \subset N$ of capacity zero such that*

$$(9.24) \qquad \lim_{i \to \infty} \frac{n(r_i, y)}{A(r_i)} = 1, \quad y \in N \setminus F.$$

9.25. Remarks. 1. For meromorphic functions stronger results than (9.24) are known. For example, J. Miles has in [Mil2, Theorem 2] shown that for a nonconstant meromorphic function $f\colon \mathbb{R}^2 \to \overline{\mathbb{R}}^2$ (9.24) can for each $\delta \in \,]0, 1/4[$ be replaced by

$$(9.26) \qquad \lim_{\substack{r \to \infty \\ r \notin E}} \frac{n(r, y)}{A(r)} = 1 , \quad y \in N \smallsetminus F ,$$

where F is of zero capacity and $E \subset [1, \infty[$ satisfies

$$(9.27) \qquad \int_E \frac{dt}{t(\log t)^{1/2+\delta}} < \infty .$$

2. For meromorphic functions J. Miles has also studied estimates for the sum in the defect relation (1.4) in the direction opposite to (1.4). In [Mil3, Theorem 1] he proves that there are absolute constants $M < \infty$ and $C \in \,]0, 1[$ such that, if $f\colon \mathbb{R}^2 \to \overline{\mathbb{R}}^2$ is a nonconstant meromorphic function, then there exists $E_1 \subset [1, \infty[$ of lower logarithmic density at least C, i.e.,

$$(9.28) \qquad \liminf_{r \to \infty} \frac{1}{\log r} \int_{E_1 \cap [1,r]} \frac{dt}{t} \geq C ,$$

such that

$$(9.29) \qquad \limsup_{\substack{r \to \infty \\ r \in E_1}} \sum_{j=1}^{q} \left(\frac{n(r, a_j)}{A(r)} - 1 \right)_+ \leq M$$

whenever $a_1, \ldots, a_q$ are distinct points in $\overline{\mathbb{R}}^2$. A relation of the type (9.29), where M is allowed to depend on n and K, is not known for qr mappings $f\colon \mathbb{R}^n \to \overline{\mathbb{R}}^n$ when $n \geq 3$. Recently S. Sastry [Sas, Theorem 6-1] has obtained a partial result in this direction. She proves a counterpart of (9.29) with a sequence of r's and with $A(r)$ replaced by $A(\theta r)$, $\theta = \theta(n, K) > 1$.

Chapter VI. Variational Integrals and Quasiregular Mappings

Extremals of certain variational integrals, like the one appearing in the definition II.(10.1) of the capacity, serve in connection with the theory of qr mappings as counterparts for harmonic functions in the plane. Nonlinearity enters in the theory for dimensions $n \geq 3$: the Euler–Lagrange equations for such variational integrals are not linear, but only quasilinear partial differential equations. For that reason methods familiar from the classical theory are for the most part not applicable to this nonlinear potential theory.

The pioneering work in the theory was done by Yu.G. Reshetnyak. His main application was the proof of the discreteness and openness of nonconstant qr mappings (Theorem I.4.1). One of the purposes of this chapter is to fill in the gap in our presentation by proving this theorem. We shall borrow some ideas of B. Bojarski and T. Iwaniec from [BI2], where they simplify the original proof of Reshetnyak. We also seize this opportunity to furnish a direct proof of the differentiability a.e. of qr mappings, which is another Reshetnyak's result quoted earlier (I.2.4) without proof. An ACL^p mapping is differentiable a.e. if $p > n$ by a result by L. Cesari and A.P. Calderón (Lemma 4.1). In [BI2] the a.e. differentiability of qr mappings is obtained as a consequence of the fact that a qr mapping is actually in ACL^p for some $p = p(n, K) > n$, but we shall avoid this result and use instead a method that has its origin in the article [GL] by F.W. Gehring and O. Lehto.

At the end of this chapter we prove a limit theorem, again credited to Reshetnyak, which states that a locally uniformly convergent sequence of K-qr mapping tends to a K-qr mapping. We then apply it to prove that for $n \geq 3$ any K-qr mapping with K sufficiently small must be a local homeomorphism.

The major recent advances in this nonlinear potential theory are due to O. Martio and his students. They have among other things developed the theory an important step further by introducing in their first fundamental article [GLM1] the notion of subextremals corresponding to subharmonic functions in the classical case. We shall use much material from [GLM1] in this chapter, but we shall postpone the treatment of such more general notions till Chapter VII, where we apply the method of variational integrals to the study of boundary behavior of qr mappings. The general reference for such potential theory is [HKM].

1. Extremals of Variational Integrals

A typical example of a variational integral which is relevant to the theory of qr mappings is the one appearing in the definition II.(10.1) of conformal capacity, namely,

$$(1.1) \qquad \int_A |\nabla u|^n \, dm \ .$$

This is the n-dimensional counterpart of the Dirichlet integral in the plane. If $f: G \to G'$ is conformal, the integral (1.1) is invariant with respect to f in the sense that

$$(1.2) \qquad \int_A |\nabla u|^n \, dm = \int_{f^{-1}A} |\nabla (u \circ f)|^n \, dm$$

for any measurable $A \subset G'$.

We extend this invariance for qc mappings by replacing the kernel $|\nabla u|^n = |h|^n$ in the integral by a kernel $F(x, h)$ which is of the order $|h|^n$ and which satisfies certain natural conditions indicated below. In addition, in the right hand side of (1.2) we insert a kernel which is a natural pullback of $F(x, h)$ under f.

1.3. Variational kernels. Let $U \subset \mathbb{R}^n$ be open and $1 < p < \infty$. We call a mapping $F: U \times \mathbb{R}^n \to \mathbb{R}^1$ a *variational kernel in U of type p* if it satisfies the following conditions:

(A) For each open $D \subset\subset U$ and $\varepsilon > 0$ there exists a compact set $C \subset D$ with $m(D \setminus C) < \varepsilon$ and $F|C \times \mathbb{R}^n$ continuous.

(B) For almost every $x \in U$ the function $h \mapsto F(x, h)$ is strictly convex and continuously differentiable.

(C) There exist positive constants α and β such that for almost every $x \in U$,

$$\alpha |h|^p \leq F(x, h) \leq \beta |h|^p \ , \quad h \in \mathbb{R}^n \ .$$

(D) For almost every $x \in U$,

$$F(x, \lambda h) = |\lambda|^p F(x, h) \ , \quad \lambda \in \mathbb{R}^1 \ , \quad h \in \mathbb{R}^n \ .$$

We shall consider only the so–called borderline case, meaning variational kernels of type n. We call such F simply *variational kernels*. However, much of the following is true for the general case too, although sometimes different proofs are required.

Let $F: U \times \mathbb{R}^n \to \mathbb{R}^1$ be a variational kernel. If $g: U \to \mathbb{R}^n$ is measurable, we conclude that the function $x \mapsto F(x, g(x))$ is measurable as follows. The condition (A) implies there exists a Borel partition of U into sets $A_0, A_1, \ldots$ with $m(A_0) = 0$ and $F|A_i \times \mathbb{R}^n$ continuous for $i \geq 1$. Since $x \mapsto (x, g(x))$ is measurable, the claim follows.

We let $\nabla_h F$ be the gradient of F with respect to the second variable. To see that $x \mapsto \nabla_h F(x, g(x))$ is measurable, we observe that each $\nabla_h F | A_i \times \mathbb{R}^n$ is a Borel mapping since $F | A_i \times \mathbb{R}^n$ is continuous.

Note that the preceding argument actually shows that both F and $\nabla_h F$ are Borel mappings.

The largest α and smallest β for which (C) holds are denoted by $\alpha(F)$ and $\beta(F)$, respectively. These are called the *structure constants* of F.

1.4. Example. Yu.G. Reshetnyak considered in [R8] kernels $F_\sigma(x, h) = (\sigma(x)h \cdot h)^{n/2}$, where $\sigma: U \to GL(n, \mathbb{R})$ is a Borel mapping with the following properties: For each $x \in U$ the bijective linear mapping $\sigma(x)$ is selfadjoint, and there exist positive constants A and B such that

$$(1.5) \qquad A|h|^2 \le \sigma(x)h \cdot h \le B|h|^2$$

for almost every $x \in U$ and all $h \in \mathbb{R}^n$. The mappings F_σ are easily seen to be variational kernels. This kind of kernel will be used in our proof of Theorem I.4.1. The special case $\sigma(x) = I = \text{id}$ for all x gives $F_I(x, h) = |h|^n$, which is the kernel in the definition of the capacity. A calculation for general σ gives

$$(1.6) \qquad \nabla_h F_\sigma(x, h) = n(\sigma(x)h \cdot h)^{n/2-1}\sigma(x)h .$$

Hence for the kernel $|h|^n$,

$$(1.7) \qquad \nabla_h F_I(x, h) = n|h|^{n-2}h .$$

1.8. Lemma. *Let* $F: U \times \mathbb{R}^n \to \mathbb{R}^1$ *be a variational kernel. Then for almost every* $x \in U$ *the following conditions hold:*

(1) $|\nabla_h F(x, h)| \le \gamma|h|^{n-1}$, $\gamma = 2^n\beta$,
(2) $\nabla_h F(x, h) \cdot k < F(x, h + k) - F(x, h)$, $k \ne 0$,
(3) $n\alpha|h|^n \le nF(x, h) = \nabla_h F(x, h) \cdot h$,
(4) $(\nabla_h F(x, k) - \nabla_h F(x, h)) \cdot (k - h) > 0$, $k \ne h$.

Proof. Let $x \in U$ be nonexceptional for the conditions (B)–(D), let $h, k \in \mathbb{R}^n$, $k \ne 0$, and set $\varphi(t) = F(x, h + tk)$. Then φ is a strictly convex, nonnegative C^1 function and $\varphi'(0) = \nabla_h F(x, h) \cdot k$. If $t \ne 0$, $\varphi(t) \ge \varphi(t) - \varphi(0) > \varphi'(0)t$. Setting $t = 1$ we get (2). Let $|k| = 1$. From (C) we infer that $\varphi(t) \le \beta|h + tk|^n \le 2^{n-1}\beta(|h|^n + |t|^n)$. If $\varphi'(0) \ne 0$, set $t = |h| \operatorname{sgn} \varphi'(0)$. Then $|\varphi'(0)| \le 2^n\beta|h|^{n-1}$. Inequality (1) follows. To prove (3) we use (C) and (D) to get

$$\nabla_h F(x, h) \cdot h = \lim_{t \to 0} \frac{1}{t}\big(F(x, h + th) - F(x, h)\big)$$

$$= F(x, h) \lim_{t \to 0} \frac{1}{t}\big((1 + t)^n - 1\big) = nF(x, h) \ge n\alpha|h|^n .$$

To prove (4), set $\psi(t) = F(x, h + t(k - h))$. Then $\psi'(0) = \nabla_h F(x, h) \cdot (k - h)$ and $\psi'(1) = \nabla_h F(x, k) \cdot (k - h)$. The strict convexity of ψ implies $\psi'(1) > \psi'(0)$, and (4) follows. $\square$

Let $F: U \times \mathbb{R}^n \to \mathbb{R}^1$ be a variational kernel. If $u: U \to \mathbb{R}^1$ belongs to some $W^1_{p,\mathrm{loc}}(U)$, $1 \le p < \infty$, we can form the variational integral

$$(1.9) \qquad I_F(u, A) = I(u, A) = \int_A F(x, \nabla u(x))\, dx$$

for any measurable $A \subset U$. If $u \in W^1_{n,\mathrm{loc}}(U)$ and $\overline{A} \subset U$ is compact, then $I(u, A) < \infty$ by (C). We sometimes write $F(x, \nabla u)$ for $F(x, \nabla u(x))$.

1.10. Extremals. Let $v \in W^1_n(U)$. We say that $u \in W^1_n(U)$ is F-*extremal* (or simply *extremal*) *for the boundary values v* if

  (1) $u - v \in W^1_{n,0}(U)$ and
  (2) $I(u, U) \le I(w, U)$ for all $w \in W^1_n(U)$ with $w - v \in W^1_{n,0}(U)$.

A function $u \in W^1_{n,\mathrm{loc}}(U)$ is a *(free) F-extremal* if $u|V$ is F-extremal for the boundary values $u|V$ whenever $V \subset\subset U$. From the definitions it follows that an F-extremal with boundary values is a free F-extremal. We also observe that $u \in W^1_{n,\mathrm{loc}}(U)$ is a free F-extremal if and only if $u|V$ is an F-extremal with boundary values $u|V$ for all open $V \subset U$ for which $u|V \in W^1_n(V)$.

An extremal always has a continuous representative in $W^1_{n,\mathrm{loc}}(U)$. However, we shall postpone the proof of this fact to VII.3.6. For continuous functions, extremals are characterized as follows.

1.11. Lemma. *Let $u \in \mathrm{ACL}^n(U)$. Then u is an extremal if and only if $I(u, V) \le I(w, V)$ whenever $V \subset\subset U$ and $w \in \mathcal{C}(\overline{V}) \cap W^1_n(V)$ satisfies $w|\partial V = u|\partial V$.*

Proof. Suppose $V \subset\subset U$, $w \in \mathcal{C}(\overline{V}) \cap W^1_n(V)$, and $w|\partial V = u|\partial V$. We claim that $w - u \in W^1_{n,0}(V)$. Let $\varphi = (w - u)^+$ and $a > 0$. The function $v = (\varphi - a)^+$ has compact support in V. By I.1.4 there exists a sequence of functions $v_i \in \mathcal{C}_0^\infty(V)$ such that $v_i \to v$ in $W^1_n(V)$. On the other hand, $\|v - \varphi\|_{1,n,V} \le a\, m(V)^{1/n} + \|\varphi\|_{1,n,V_a} \to 0$ as $a \to 0$, where $V_a = \{x \in V : \varphi(x) < a\}$. It follows that there exists a sequence (φ_j) in $\mathcal{C}_0^\infty(V)$ such that $\varphi_j \to \varphi$ in $W^1_n(V)$, i.e. $\varphi \in W^1_{n,0}(V)$. Similarly $\psi = (w - u)^- \in W^1_{n,0}(V)$, and our claim follows. Hence, if u is an extremal, it satisfies the condition.

Suppose that the stated condition holds and that $V \subset\subset U$, $w \in W^1_n(V)$, $w - u \in W^1_{n,0}(V)$. Let φ_i be a sequence in $\mathcal{C}_0^\infty(V)$ such that $\varphi_i \to w - u$ in $W^1_n(V)$. Then $\varphi_i + u \in \mathcal{C}(\overline{V}) \cap W^1_n(V)$ and $(\varphi_i + u)|\partial V = u|\partial V$. By assumption, $I(u, V) \le I(u + \varphi_i, V)$. Hence it suffices to show $I(u + \varphi_i, V) \to I(w, V)$. If $x \in U$ is nonexceptional for (B)–(D), $F(x, h + k) - F(x, h) = k \cdot \nabla_h F(x, h + tk)$ for some $t \in\,]0, 1[$. With 1.8 we get $|F(x, h + k) - F(x, h)| \le$

$|k|\gamma \cdot 2^{n-2}(|h|^{n-1} + |k|^{n-1})$ a.e. in U and for all $h, k \in \mathbb{R}^n$. We apply this to $\nabla w(x) = h$ and $\nabla(u + \varphi_i) = h + k$. Hölder's inequality then gives

$$|I(u + \varphi_i, V) - I(w, V)|$$

$$\leq C \int_V |\nabla(u + \varphi_i - w)| \left(|\nabla w|^{n-1} + |\nabla(u + \varphi_i - w)|^{n-1} \right) dm$$

$$\leq C \|\nabla(u + \varphi_i - w)\|_{n,V} \left(\|\nabla w\|_{n,V}^{n-1} + \|\nabla(u + \varphi_i - w)\|_{n,V}^{n-1} \right) \to 0$$

as $i \to \infty$. The lemma follows. $\qquad\qquad\qquad\qquad\qquad\qquad\qquad\qquad \square$

The connection between F-extremals and partial differential equations is expressed in the following result.

1.12. Proposition. *Let* $u \in W^1_{n,\text{loc}}(U)$. *Then* u *is an* F-extremal *if and only if* u *satisfies the Euler–Lagrange equation*

$$(1.13) \qquad\qquad\qquad \text{div} \, \nabla_h F(x, \nabla u) = 0$$

in the weak sense, i.e., if and only if

$$(1.14) \qquad \int_U \nabla_h F(\,\cdot\,, \nabla u) \cdot \nabla \varphi \, dm = 0 \quad \text{for all } \varphi \in C_0^\infty(U).$$

1.15. Remark. The partial differential equation (1.13) is of quasilinear degenerate elliptic type. If $u \in W_n^1(U)$, it is equivalent to require (1.14) for all test functions in $W^1_{n,0}(U)$. To see this let $w \in W^1_{n,0}(U)$ and $\varphi \in C_0^\infty(U)$. Then by 1.8 and Hölder's inequality,

$$\left| \int_U \left(\nabla_h F(\,\cdot\,, \nabla u) \cdot \nabla \varphi - \nabla_h F(\,\cdot\,, \nabla u) \cdot \nabla w \right) dm \right|$$

$$\leq \gamma \int_U |\nabla u|^{n-1} |\nabla \varphi - \nabla w| \, dm \leq \gamma \|\nabla u\|_n^{n-1} \|\nabla \varphi - \nabla w\|_n \to 0$$

as $\varphi \to w$ in $W_n^1(U)$.

Proof of 1.12. Let u be an extremal. If $\varphi \in C_0^\infty(U)$, let $V \subset\subset U$ be such that $\text{spt}\,\varphi \subset V$ and set $\psi(t) = I(u + t\varphi, V)$, $t \in \mathbb{R}^1$. Since u is extremal, $\psi(t) \geq \psi(0)$ for all t. We claim that ψ is differentiable at 0. Set $g_t = F(\,\cdot\,, \nabla u + t\nabla \varphi) - F(\,\cdot\,, \nabla u)$. As in the proof of 1.11, $|g_t|/|t| \leq 2^{n-2}\gamma|\nabla\varphi|(|\nabla u|^{n-1} + |\nabla \varphi|^{n-1})$ a.e. for $0 < |t| < 1$. The left hand side of this inequality is integrable in V and

$$\lim_{t \to 0} \frac{g_t(x)}{t} = \nabla_h F(x, \nabla u(x)) \cdot \nabla \varphi(x) \text{ a.e.}$$

Then Lebesgue's dominated convergence theorem implies

$$\psi'(0) = \lim_{t \to 0} \int_V \frac{g_t}{t} \, dm = \int_V \nabla_h F(\,\cdot\,, \nabla u) \cdot \nabla \varphi \, dm \, .$$

Since ψ has a minimum at 0, $\psi'(0) = 0$ and (1.13) follows.

Suppose (1.13) is true. Let $V \subset\subset U$ and $\varphi \in W^1_{n,0}(V)$. By 1.8, $F(\,\cdot\,, \nabla u + \nabla \varphi) - F(\,\cdot\,, \nabla u) \geq \nabla_h F(\,\cdot\,, \nabla u) \cdot \nabla \varphi$ a.e. Hence with 1.15 we get

(1.16)
$$\int_V F(\,\cdot\,, \nabla u + \nabla \varphi) \, dm - \int_V F(\,\cdot\,, \nabla u) \, dm$$
$$\geq \int_V \nabla_h F(\,\cdot\,, \nabla u) \cdot \nabla \varphi \, dm = 0 \, .$$

If $w \in W^1_n(V)$ is such that $w - u \in W^1_{n,0}(V)$, then (1.16) applied to $\varphi = w - u$ gives $I(w, V) \geq I(u, V)$, so u is extremal. $\qquad\square$

1.17. Corollary. *Let* $u \in W^1_{n,\mathrm{loc}}(U)$. *Then* u *is an* F-*extremal if and only if every* $x \in U$ *has a neighborhood* $V \subset U$ *such that* $u|V$ *is* F-*extremal.*

Proof. Let u be an extremal locally, as indicated, and let $\varphi \in C^\infty_0(U)$. We cover $\mathrm{spt}\,\varphi$ by neighborhoods $V_1, \ldots, V_k$ in each of which u is extremal. Let $\sum_{i=1}^k \psi_i = 1$ be a C^∞ partition of unity subordinate to $\{V_1, \ldots, V_k\}$. Then 1.12 yields

$$\int_U \nabla_h F(\,\cdot\,, \nabla u) \cdot \nabla \varphi \, dm = \sum_{i=1}^k \int_{V_i} \nabla_h F(\,\cdot\,, \nabla u) \cdot \nabla(\psi_i \varphi) \, dm = 0 \, ,$$

and u is extremal. On the other hand, if u is extremal, it is locally extremal. $\qquad\square$

1.18. Example. If $F = F_\sigma$ is the kernel in 1.4, (1.13) and (1.14) take the forms

(1.19)
$$\operatorname{div}\big[(\sigma \nabla u \cdot \nabla u)^{n/2-1} \sigma \nabla u\big] = 0$$

and

(1.20)
$$\int_U (\sigma \nabla u \cdot \nabla u)^{n/2-1} (\sigma \nabla u \cdot \nabla \varphi) \, dm \, , \quad \varphi \in C^\infty_0(U) \, .$$

Here we have used the notation $(\sigma \nabla u)(x) = \sigma(x) \nabla u(x)$. If $F = F_I$, (1.19) is called the *n-harmonic equation* and its solutions are *n-harmonic functions*.

1.21. Note. In this section we have followed quite closely the presentation in [GLM1].

2. Extremals and Quasiregular Mappings

In this section we shall prove some results on the extremality of functions with respect to kernels that are pullbacks of the special kernel $|h|^n$ under qr mappings. Corresponding results are true for general kernels too, but we must postpone their treatment until we have proved I.2.4, I.4.1, and I.4.5.

We recall from Remark I.1.10 that, if $f: U \to \mathbb{R}^n$ is ACL^{n-1}, then each column of the adjunct Jacobian $\mathrm{ad}\, f'$ is divergence free in the sense of I.(1.9). We apply this fact to get the following result.

2.1. Lemma. *Let U and V be open in $\mathbb{R}^n$, let $f: U \to V$ be ACL^n, and let $g: V \to \mathbb{R}^n$ be a divergence free C^1 mapping. Then $\eta = (\mathrm{ad}\, f')(g \circ f)$ is divergence free. Here $(\mathrm{ad}\, f')(g \circ f)(x) = (\mathrm{ad}\, f'(x))g(f(x))$.*

Proof. The mapping $\mathrm{ad}\, f'$ is in $L_{\mathrm{loc}}^{n/(n-1)}$ and hence η is also. Let $\varphi \in C_0^\infty(U)$. We have to show that

$$\int_U \eta \cdot \nabla\varphi \, dm = 0 .$$

By I.1.11, $g \circ f$ is ACL^n, hence each $(g_j \circ f)\varphi$ is an ACL^n function with compact support. Since each $\mathrm{ad}\, f' e_j$ is divergence free, we obtain

$$\int_U \sum_i (\mathrm{ad}\, f')_{ij} D_i((g_j \circ f)\varphi) \, dm = 0$$

by approximating $D_i((g_j \circ f)\varphi)$ in L^n by functions in $C_0^\infty(U)$ as in Remark 1.15. The ith coordinate function of η is $\eta_i = \sum_j (\mathrm{ad}\, f')_{ij} g_j \circ f$, hence by I.(1.6),

$$-\int_U \eta \cdot \nabla\varphi \, dm = \int_U \sum_{i,j} (\mathrm{ad}\, f')_{ij} D_i(g_j \circ f)\varphi \, dm$$

$$= \int_U \sum_{i,j} (\mathrm{ad}\, f')_{ij} \sum_k (D_k g_j \circ f)(D_i f_k)\varphi \, dm$$

$$= \int_U J_f \sum_{j,k} \delta_{kj} (D_k g_j \circ f)\varphi \, dm$$

$$= \int_U J_f \left(\sum_k D_k g_k \circ f \right)\varphi \, dm = 0 .$$

The last equality follows from $\sum_k (D_k g_k \circ f) = 0$, which is true because g is divergence free. The lemma is proved. $\qquad\square$

Now let G and G' be domains in $\mathbb{R}^n$ and let $f: G \to G'$ be qr. We define the pullback $f^\sharp F_I$ of the special kernel $F_I(y, k) = |k|^n$ as the mapping $f^\sharp F_I: G \times \mathbb{R}^n \to \mathbb{R}^1$ defined by

$$(2.2) \qquad f^{\sharp} F_I(x, h) = \begin{cases} F_I\big(f(x), J_f(x)^{1/n} f'(x)^{-1*} h\big) & \text{if } x \in A, \\ |h|^n & \text{if } x \in G \setminus A. \end{cases}$$

Here A is the Borel set of points $x \in G$ where all the partial derivatives of f exist and $J_f(x) > 0$, and T^* is the adjoint of a linear mapping T. We remind the reader that we do not have the fact that f is differentiable with $J_f > 0$ a.e. available here.

We notice that

$$(2.3) \qquad f^{\sharp} F_I(x, h) = \big(\sigma(x)h \cdot h\big)^{n/2}$$

where

$$(2.4) \qquad \sigma(x) = \begin{cases} J_f^{2/n}(x) f'(x)^{-1} f'(x)^{-1*} & \text{if } x \in A, \\ I & \text{if } x \in G \setminus A. \end{cases}$$

The mapping $f^{\sharp} F_I$ is therefore a variational kernel F_σ of the type in Example 1.4. Its structure constants are easily seen to be

$$(2.5) \qquad \alpha(F_\sigma) = K_O(f)^{-1}, \quad \beta(F_\sigma) = K_I(f).$$

2.6. Proposition. Let G and G' be domains in $\mathbb{R}^n$, let $f: G \to G'$ be qr, and let v be an F_I-extremal in G' of class C^2 with $\nabla v(y) \neq 0$ for all $y \in G'$. Then $u = v \circ f$ is $f^{\sharp} F_I$-extremal.

Proof. By I.1.11, u is ACL^n. Furthermore, $\nabla u(x) = f'(x)^* \nabla v(f(x))$ a.e., which we write $\nabla u = f'^*(\nabla v \circ f)$. We shall apply 1.12 and 2.1. By (1.6), $\nabla_h F_\sigma(\,\cdot\,, \nabla u) = n(\sigma \nabla u \cdot \nabla u)^{n/2-1} \sigma \nabla u$, where σ is given in (2.4). In the set A we have

$$\sigma \nabla u \cdot \nabla u = J_f^{2/n} f'^{-1} f'^{-1*} f'^* (\nabla v \circ f) \cdot f'^* (\nabla v \circ f) = J_f^{2/n} |\nabla v \circ f|^2,$$
$$\sigma \nabla u = J_f^{2/n} f'^{-1} f'^{-1*} f'^* (\nabla v \circ f) = J_f^{2/n} f'^{-1} (\nabla v \circ f),$$

and hence by I.(1.6),

$$(2.7) \qquad \begin{aligned} (\sigma \nabla u \cdot \nabla u)^{n/2-1} \sigma \nabla u &= J_f f'^{-1} |\nabla v \circ f|^{n-2} \nabla v \circ f \\ &= \operatorname{ad} f' |\nabla v \circ f|^{n-2} \nabla v \circ f. \end{aligned}$$

In the set where $J_f(x) = 0$ we have $f'(x) = 0$ and thus $\operatorname{ad} f'(x) = 0$ at almost every point x. At such a point all terms in (2.7) are zero. It follows that (2.7) holds a.e. in G. Since v is an F_I-extremal, $g = |\nabla v|^{n-2} \nabla v$ is divergence free by (1.7) and 1.12. Moreover, g is in C^1. Then 1.12 and 2.1 give the result. $\qquad \square$

2.8. Corollary. *Let* $f: G \to \mathbb{R}^n$ *be qr and set* $G_0 = G \setminus f^{-1}(0)$. *Then*

(1) $\log|f|$ *is* $f^{\sharp}F_I$*-extremal in* G_0,
(2) *each* f_j *is* $f^{\sharp}F_I$*-extremal.*

Proof. It is easy to show by direct calculation that $\mathrm{div}(|\nabla v|^{n-2}\nabla v) = 0$ if v is $y \mapsto \log|y|$ in $\mathbb{R}^n \setminus \{0\}$ or $y \mapsto y_j$ in $\mathbb{R}^n$. Hence in both cases v satisfies the conditions in 2.6, and the corollary follows from 2.6. $\square$

The fact that $\log|f|$ is an $f^{\sharp}F_I$-extremal in $G \setminus f^{-1}(0)$ for a qr f will play a crucial role in our proof of I.4.1, which will be completed in 5.7.

2.9. Notes. 1. The results 2.1 and 2.6, together with their proofs, are from [BI2]. The result 2.8(1) was first proved by Yu.G. Reshetnyak in [R4, Lemma 2] with the help of a somewhat different procedure.

2. In Section 6 we shall consider pullbacks of general variational kernels F, and establish the analogue of 2.6 for general F-extremals v.

3. Growth Estimates for Extremals

In this section we let U be an open set in $\mathbb{R}^n$ and $F: U \times \mathbb{R}^n \to \mathbb{R}^1$ a variational kernel with structure constants α and β. Let u be a continuous F-extremal. If $\lambda \in \mathbb{R}^1$, clearly $u + \lambda$ is an F-extremal. It follows from (D) in the definition 1.3 of a variational kernel that also λu is an F-extremal.

Inequality (1) in the following lemma is known as the *standard estimate*. Recall the notation $\mathrm{osc}(u, A)$ for the oscillation of a function u in a set A.

3.1. Lemma. *Suppose* u *is a continuous F-extremal, that* (A, C) *is a condenser in* U, *and that* u *is bounded in* A. *Then*

(1) $\displaystyle\int_C |\nabla u|^n \, dm \leq C_n \beta \alpha^{-1} \, \mathrm{osc}(u, A)^n \, \mathrm{cap}(A, C)$;

(2) *if* $u > 0$ *and if* $v = \log u$, *we have*

$$\int_C |\nabla v|^n \, dm \leq C_n \beta \alpha^{-1} \, \mathrm{cap}(A, C).$$

Here C_n *depends only on* n.

Proof. We may assume that $A \subset\subset U$. We shall first prove (2). For this we may assume $u \geq (n-1)^{1/n}$ in A. If not, we replace u by the extremal λu with λ sufficiently large. Let $\varphi \in C_0^{\infty}(A)$ be such that $0 \leq \varphi \leq 1$ and $\varphi|C = 1$. Set $w = u + \varphi^n u^{1-n}$. Then $w = u$ in ∂A and $\nabla w = (1 - (n-1)\varphi^n u^{-n})\nabla u + n\varphi^{n-1}u^{1-n}\nabla\varphi$. We have $0 \leq (n-1)\varphi^n u^{-n} \leq 1$. Therefore (B) and (C) in 1.3 yield

$$F(\,\cdot\,,\nabla w) \le (1 - (n-1)\varphi^n u^{-n}) F(\,\cdot\,,\nabla u) + \beta n^n (n-1)^{1-n} |\nabla \varphi|^n$$

a.e. in A. Since u is extremal, we get

$$I_F(u, A) \le I_F(w, A) \le I_F(u, A) - (n-1) \int_A \varphi^n u^{-n} F(\,\cdot\,,\nabla u)\, dm$$

$$+ \beta n^n (n-1)^{1-n} \int_A |\nabla \varphi|^n\, dm\ .$$

As $\varphi|C = 1$, condition (C) in 1.3 gives

$$(n-1)\alpha \int_C \frac{|\nabla u|^n}{u^n}\, dm \le \beta n^n (n-1)^{1-n} \int_A |\nabla \varphi|^n\, dm\ .$$

Since $\nabla v = \nabla u / u$, we obtain (2) with $C_n = (n/(n-1))^n$.

To prove (1) let $\varepsilon > 0$ and set $\eta = u - \inf_A u + \varepsilon$. Then $\eta > 0$ in A and $\operatorname{osc}(u, A) \ge \sup_A \eta - \varepsilon$. We apply (2) to $\eta|A$ in place of u and get

$$(3.2) \qquad \frac{1}{(\sup_A \eta)^n} \int_C |\nabla u|^n\, dm \le \int_C \frac{|\nabla \eta|^n}{\eta^n}\, dm \le C_n \beta \alpha^{-1} \operatorname{cap}(A, C)\ .$$

Letting $\varepsilon \to 0$ in (3.2) we get (1).   □

3.3. Lemma. *A continuous F-extremal u is monotone.*

Proof. Suppose there exists a domain $D \subset\subset U$ and $x_0 \in D$ such that

$$u(x_0) > \max_{x \in \partial D} u(x) = a\ .$$

Let $v = \min(u, a)$. Then $\nabla v(x) = 0$ in the open set $V = \{x \in D : u(x) > a\}$. Since u is ACL,

$$\int_V |\nabla u|^n\, dm > 0\ .$$

Owing to (C) in 1.3 we then get $I_F(v, D) = I_F(v, D \smallsetminus V) = I_F(u, D \smallsetminus V) < I_F(u, D)$, which contradicts the extremality of u. Hence $\sup_D u = \sup_{\partial D} u$. Similarly $\inf_D u = \inf_{\partial D} u$, revealing that u is monotone.   □

We need the following integrated form of the oscillation lemma II.1.11.

3.4. Lemma. Let $u \in \mathrm{ACL}^n(U)$, $0 < r_1 < r_2$, and for every $r \in \,]r_1, r_2[$ let C_r be a spherical cap in $U \cap S^{n-1}(r)$. If $r \mapsto \mathrm{osc}(u, C_r)$ is measurable in $]r_1, r_2[$, then

$$(3.5) \qquad \int_{r_1}^{r_2} \mathrm{osc}(u, C_r)^n r^{-1}\, dr \leq A_n \int_U |\nabla u|^n\, dm .$$

Proof. We may assume $\|\nabla u\|_n < \infty$. By I.1.11 and the spherical Fubini's theorem we conclude that for almost every $r \in \,]r_1, r_2[$ u is ACL^n in $S^{n-1}(r) \cap U$ and

$$\int_{S(r) \cap U} |\nabla u|^n\, d\mathcal{H}^{n-1} < \infty .$$

For such r let $\nabla' u$ be the gradient in $S^{n-1}(r)$. Approximating u in $S^{n-1}(r) \cap U$ by C^1 functions in the sense of I.1.4 we obtain by II.1.11 the estimate

$$\mathrm{osc}(u, C_r)^n / r \leq A_n \int_{S(r) \cap U} |\nabla' u|^n\, dm \leq A_n \int_{S(r) \cap U} |\nabla u|^n\, dm .$$

The lemma then follows by integration. □

3.6. Remark. With the argument in the proof of Lemma IV.2.16 we get a variant of (3.5) in the form

$$(3.7) \qquad \int_{r_1}^{r_2} \mathrm{osc}(u, C_r)^n d(C_r)^{-1} dr \leq A'_n \int_U |\nabla u|^n\, dm$$

where we now assume that $r \mapsto \mathrm{osc}(u, C_r)^n d(C_r)^{-1}$ is measurable in $]r_1, r_2[$.

Next we prove a result that guarantees integrability locally up to the boundary.

3.8. Lemma. Let u be a continuous F-extremal in U and let D be open in $\mathbb{R}^n$ such that $E = D \cap \partial U \neq \emptyset$. Suppose $v \in C(\overline{U}_1) \cap W_n^1(U_1)$ where $U_1 = U \cap D$, and

$$\lim_{x \to y} u(x) = v(y) \qquad \text{if } y \in E .$$

Then each $y \in E$ has a neighborhood V such that $u|V \cap U \in W_n^1(V \cap U)$.

Proof. For $y \in E$ we choose $V = B^n(y, r)$ such that $\overline{B}^n(y, 2r) \subset D$. Let $\varphi \in C_0^\infty(B^n(y, 2r))$ be such that $0 \leq \varphi \leq 1$ and $\varphi|V = 1$. Set $W = B^n(y, 2r) \cap U$ and $W_\varepsilon = \{x \in W : u(x) > \varepsilon + v(x)\}$, $0 \leq \varepsilon < 1$. Let $0 < \varepsilon < 1$. Then $W_\varepsilon \subset\subset U$ and the function $w = \varphi^n(u - \varepsilon - v)$ in W_ε belongs to $W_{n,0}^1(W_\varepsilon)$. This follows from the proof of 1.11. Then

$$(3.9) \qquad \int_{W_\varepsilon} \nabla_h F(\,\cdot\,, \nabla u) \cdot \nabla w \, dm = 0$$

by 1.15. Substituting $\nabla w = (u - \varepsilon - v) n \varphi^{n-1} \nabla \varphi + \varphi^n \nabla(u - \varepsilon - v)$ into (3.9) we obtain by 1.8 and Hölder's inequality the estimate

$$n\alpha \|\varphi|\nabla u|\|_{n,W_\varepsilon}^n \leq \int_{W_\varepsilon} \varphi^n \nabla_h F(\,\cdot\,, \nabla u) \cdot \nabla u \, dm$$

$$\leq n \int_{W_\varepsilon} \varphi^{n-1} |\nabla_h F(\,\cdot\,, \nabla u)| |\nabla \varphi| |u - \varepsilon - v| \, dm$$

$$(3.10) \qquad + \int_{W_\varepsilon} \varphi^n |\nabla_h F(\,\cdot\,, \nabla u)| |\nabla v| \, dm$$

$$\leq n M_1 M_2 \gamma \int_{W_\varepsilon} \varphi^{n-1} |\nabla u|^{n-1} + \gamma \int_{W_\varepsilon} \varphi^{n-1} |\nabla u|^{n-1} |\nabla v| \, dm$$

$$\leq C \|\varphi|\nabla u|\|_{n,W_\varepsilon}^{n-1} (1 + \|\nabla v\|_{n,W}) \,,$$

where $M_1 = \sup_W (|u - v| + 1)$, $M_2 = \sup |\nabla \varphi|$, and the constant C does not depend on ε. Inequality (3.10) gives

$$\int_{W_\varepsilon \cap V} |\nabla u|^n \, dm \leq \int_{W_\varepsilon} \varphi^n |\nabla u|^n \, dm \leq \frac{1}{n^n \alpha^n} C^n (1 + \|\nabla v\|_{n,W})^n \,.$$

Letting $\varepsilon \to 0$ we get

$$\int_{W_0 \cap V} |\nabla u|^n \, dm < \infty \,.$$

Similarly the integral of $|\nabla u|^n$ over $\{x \in W : u(x) < v(x)\} \cap V$ is finite. Since $\nabla u(x) = \nabla v(x)$ a.e. in $\{x \in W : u(x) = v(x)\}$, we conclude $u|V \cap U \in W_n^1(V \cap U)$. $\qquad \square$

Near the boundary we shall employ the following estimate when U is a domain.

3.11. Lemma. Let U be a domain, let $u \in \mathcal{C}(\overline{U})$ be nonnegative and F-extremal in U, let $x_0 \in \mathbb{R}^n$, and let $0 < t < R$. Set $U_s = U \cap B^n(x_0, s)$ for $s > 0$. If $u|\partial U \cap B^n(x_0, R) = 0$ and if $S^{n-1}(x_0, s) \cap \partial U \neq \emptyset$ for $t < s < R$, then

$$\sup_{U_t} u \leq d_n \left(\frac{\beta}{\alpha}\right)^{1/n} \left(\log \frac{R}{t}\right)^{-1} \sup_U u \,,$$

where the constant d_n depends only on n.

Proof. We may asume $x_0 = 0$. Set $r = \sqrt{tR}$ and let $\varphi \in \mathcal{C}_0^\infty(B^n(R))$ be such that $0 \leq \varphi \leq 1$ and $\varphi|B^n(r) = 1$. Following closely the argument in the proof of 3.1, we define $v = u - \varphi^n u$ and observe that $v \in C(\overline{U}_R)$ and $v|\partial U_R = u|\partial U_R$. First we get

$$F(\,\cdot\,,\nabla v) \leq (1 - \varphi^n) F(\,\cdot\,,\nabla u) + n^n u^n \beta |\nabla\varphi|^n$$

and then

$$I_F(u, U_R) \leq I_F(v, U_R)$$
$$\leq I_F(u, U_R) - \int_{U_R} \varphi^n F(\,\cdot\,,\nabla u)\, dm + n^n \beta \int_{U_R} u^n |\nabla\varphi|^n\, dm\,,$$

where all terms are finite because of Lemma 3.8. Hence,

$$(3.12) \qquad \int_{U_R} \varphi^n F(\,\cdot\,,\nabla u)\, dm \leq n^n \beta \int_{U_R} u^n |\nabla\varphi|^n\, dm\,.$$

Since $\varphi|B^n(r) = 1$, we obtain

$$(3.13) \qquad I_F(u, U_r) \leq \beta n^n \left(\sup_U u\right)^n \omega_{n-1} \left(\log \frac{R}{r}\right)^{1-n}.$$

Since U is a domain, $S^{n-1}(s) \cap U \neq \emptyset$ for every $s \in \,]t, r[$. By assumption we can choose for every $s \in \,]t, r[$ a spherical cap C_s in $S^{n-1}(s) \cap U$, centered at a point x_s for which $u(x_s) = \max\{u(x) : x \in S^{n-1}(s) \cap U\}$, such that $\overline{C}_s$ meets ∂U. Since u is monotone, it follows that $u(x_s) \geq u(x_{s'})$ for $s \geq s'$. Since $\mathrm{osc}(u, C_s) = u(x_s)$ for all s, 3.4 together with (3.13) and (C) yields

$$\left(\sup_{U_t} u\right)^n = u(x_t)^n \leq \left(\log \frac{r}{t}\right)^{-1} \int_t^r \frac{u(x_s)^n}{s}\, ds \leq A_n \left(\log \frac{r}{t}\right)^{-1} \int_{U_r} |\nabla u|^n\, dm$$

$$\leq A_n (\beta/\alpha) n^n \omega_{n-1} \left(\sup_U u\right)^n \left(\log \frac{r}{t}\right)^{-1} \left(\log \frac{R}{r}\right)^{1-n}.$$

The lemma follows by substituting $r = \sqrt{tR}$. $\square$

3.14. Remark. For future reference we observe that the proof of 3.11 also gives the following. Let U be open and let u be a continuous F-extremal in U. Then (cf. (3.12))

$$(3.15) \qquad \int_U \varphi^n |\nabla u|^n\, dm \leq \frac{n^n \beta}{\alpha} \int_U |u|^n |\nabla\varphi|^n$$

for all nonnegative $\varphi \in C_0^\infty(U)$. Inequality (3.15) is sometimes called a *Caccioppoli type estimate*.

By a process of iteration we now easily obtain a Hölder estimate for extremals in the following form.

3.16. Theorem. Let u be a continuous F-extremal. If $0 < r \leq R$ and $B^n(x_0, R) \subset U$, then

$$\operatorname{osc}(u, B^n(x_0, r)) \leq 3\left(\frac{r}{R}\right)^\kappa \operatorname{osc}(u, B^n(x_0, R)),$$

where $\kappa = d'_n(\alpha/\beta)^{1/n}$ and $d'_n > 0$ depends only on n.

Proof. We may assume $x_0 = 0$. Let $0 < r_1 < r_2 \leq R$. Applying 3.4 to the whole spheres $S^{n-1}(r)$, instead of to caps, we obtain via an argument similar to the proof of 3.11 the inequality

$$(3.17) \qquad \operatorname{osc}(u, B^n(r_1)) \leq d_n \left(\frac{\beta}{\alpha}\right)^{1/n} \left(\log \frac{r_2}{r_1}\right)^{-1} \operatorname{osc}(u, B^n(r_2)).$$

Set $a = d_n(\beta/\alpha)^{1/n}$ and $\lambda = e^{ea}$. If $R > \lambda r$, we let $m \geq 1$ be the integer such that $\lambda^m r < R \leq \lambda^{m+1} r$. We iterate (3.17) using the radii $r, \lambda r, \lambda^2 r, \ldots, \lambda^m r$ and get

$$\operatorname{osc}(u, B^n(r)) \leq \left(\frac{a}{\log \lambda}\right)^m \operatorname{osc}(u, B^n(\lambda^m r)).$$

Since $(a/\log \lambda)^m = (r/\lambda^m r)^{1/\log \lambda} \leq \lambda^{\frac{1}{ea}} (r/R)^{\frac{1}{ea}} < 3(r/R)^{\frac{1}{ea}}$, we obtain the claim with $\kappa = e^{-1} d_n^{-1} (\alpha/\beta)^{1/n}$. If $r < R \leq \lambda r$, $3(r/R)^{\frac{1}{ea}} > e(1/\lambda)^{\frac{1}{ea}} = 1$, and the estimate follows trivially. The theorem is proved. □

3.18. Notes. Lemma 3.4 is essentially [G4, Lemma 2], Lemma 3.8 is from [M6], and Lemma 3.11 is a slightly modified version of [GLM2, 4.4]. The other material of this section is from [GLM1]. For 3.1 see also [BI2, p. 292].

4. Differentiability of Quasiregular Mappings

In this section we shall prove that a qr mapping f is differentiable a.e. This result by Yu.G. Reshetnyak [Re2] was stated without proof as Theorem I.2.4. Our proof differs from the one in [Re2], being based on the fact that each coordinate function of f is monotone. A method by J. Väisälä from [V1] then applies. Väisälä's approach is itself an n-dimensional version of a technique used by F.W. Gehring and O. Lehto [GL].

We start with the following well-known result by L. Cesari [Ce] and A.P. Calderón [Cald].

4.1. Lemma. If U is open in $\mathbb{R}^n$ and if $u: U \to \mathbb{R}^1$ is in ACL^p for some $p > n$, then u is differentiable a.e.

Proof. Let $V \subset\subset U$ be a ball and $x, y \in V$. If $u \in \mathcal{C}^1$, then

$$u(x) - u(y) = \int_0^1 (x - y) \cdot \nabla u(tx + (1-t)y) \, dt.$$

Integrating with respect to x over V and invoking Hölder's inequality we get

$$|u_V - u(y)| = \left| \fint_V \left(\int_0^1 (x-y) \cdot \nabla u(tx + (1-t)y)\, dt \right) dx \right|$$

$$\leq d(V) \int_0^1 \left(\fint_V |\nabla u(tx + (1-t)y)|^p dx \right)^{1/p} dt ,$$

where u_V is the average of u over V and

$$\fint_V v\, dm = \frac{1}{m(V)} \int_V v\, dm .$$

We introduce a new variable $\xi = tx + (1-t)y$ and get

$$\int_V |\nabla u(tx + (1-t)y)|^p dx \leq \frac{1}{t^n} \int_V |\nabla u(\xi)|^p d\xi .$$

Hence

$$(4.2) \qquad |u_V - u(y)| \leq d(V) \frac{p}{p-n} \left(\fint_V |\nabla u|^p \right)^{1/p} .$$

By I.1.4 this also holds under the weaker assumption that $u \in \mathrm{ACL}^p$.

Since $u \in \mathrm{ACL}^p$, we can quote Lebesgue's theorem to justify the claim that

$$(4.3) \qquad \lim_{r \to 0} \fint_{B^n(x_0, r)} |\nabla u(x) - \nabla u(x_0)|^p dx = 0$$

for almost every $x_0 \in U$. Let $x_0 \in U$ be a point where (4.3) holds. Set $v(x) = u(x) - u(x_0) - \nabla u(x_0) \cdot (x - x_0)$. Then $v \in \mathrm{ACL}^p$, $v(x_0) = 0$, and $\nabla v = \nabla u - \nabla u(x_0)$. Inequality (4.2) applied to v and $V = B^n(x_0, r)$ together with (4.3) gives

$$|v(x)| = |v(x) - v(x_0)| \leq |v(x) - v_V| + |v(x_0) - v_V|$$

$$\leq \frac{4rp}{p-n} \left(\fint_V |\nabla v|^p \right)^{1/p} = o(r) .$$

This proves that u is differentiable at x_0, and the lemma follows. $\qquad \square$

4.4. Lemma. *Let* $U \subset \mathbb{R}^n$ *be open, let* $p > n - 1$, *and let* $g\colon U \to \mathbb{R}^1$ *be a monotone* ACL^p *function. Then* g *is differentiable a.e.*

Proof. We shall follow the presentation in [V1, Lemma 4]. For $i = 1, \ldots, n$ set $\mathbb{R}_i^{n-1} = \{ x \in \mathbb{R}^n : x_i = 0 \}$. We say that f has a partial derivative with respect to $\mathbb{R}_i^{n-1}$ at $x \in U$ if $g|((x + \mathbb{R}_i^{n-1}) \cap U)$ is differentiable at x. The corresponding derivative is then denoted by $D_i' g(x)$.

We may assume that U is bounded. Let $C \subset U$ be the set where $D_i' g(x)$ exists. We claim that C is a Borel set. Let B be the set of points x where $D_j g(x)$ exists for all $j \neq i$. Then B is a Borel set and the partial derivatives $D_j g$ are Borel functions in B [V4, 25.2]. For each $x \in B$ we define the linear mapping $A(x)\colon \mathbb{R}_i^{n-1} \to \mathbb{R}^n$ by $A(x)e_j = D_j g(x)$ and set

$$\psi_k(x) = \sup \frac{|g(x + h) - g(x) - A(x)h|}{|h|} , \quad k = 1, 2, \ldots,$$

where the supremum is taken over all $h \in \mathbb{R}_i^{n-1}$ such that $x + h \in U$ and $0 < |h| < 1/k$. Since g is continuous, each ψ_k is a Borel function in B and so is $\psi = \inf \psi_k = \lim \psi_k$. The set C is $\{ x \in B : \psi(x) = 0 \}$ and therefore a Borel set. Clearly $A(x) = D_i' g(x)$ for $x \in C$.

Since g is ACL^p, it follows from Fubini's theorem that $g|U_t$ is ACL^p for almost every $t \in \mathbb{R}^1$ where $U_t = \{ x \in U : x_i = t \}$. By 4.1, for such t $D_i' g(x)$ exists in U_t except in a set of zero $(n-1)$-dimensional measure. By Fubini's theorem we conclude that $D_i' g(x)$ exists a.e.

Now we write ψ_{ik} for the function denoted above by ψ_k. Since $D_i' g(x)$ exists a.e. for all i, the functions

$$F_k = \sum_{i=1}^{n} \psi_{ik} ,$$

defined a.e. in U, converge a.e. to zero. By Egorov's theorem, for every $\eta > 0$ there exists a compact set $E \subset U$ with $m(U \setminus E) < \eta$ such that the functions F_k converge to zero uniformly in E. It follows that $D_i' g$ is continuous in E for all i.

It suffices to show that g is differentiable at every $z \in E$ which is a point of linear density in the direction of each coordinate axis. We may assume that $z = 0$. For every $\varepsilon \in]0, 1[$ there exists $\delta > 0$ such that (1) $Q = \{ x \in \mathbb{R}^n : |x_i| \leq \delta, i = 1, \ldots, n \} \subset U$, (2) $|D_i' g(x) - D_i' g(0)| < \varepsilon$ for $x \in Q \cap E$ and for all i, (3) $|F_k(x)| < \varepsilon$ for $x \in E$ and $1/k < \delta$, and (4) $m_1(J \cap E)(1 + \varepsilon) > m_1(J)$ for any line segment $J \subset Q$ lying in some coordinate axis and containing 0.

Let $h \in \mathbb{R}^n$ be such that $0 < |h| < \delta/2$. Then there exists an n-interval $R = \{ x \in \mathbb{R}^n : a_i \leq x_i \leq b_i, i = 1, \ldots, n \}$ such that $h_i - \varepsilon|h| \leq a_i < h_i < b_i \leq h_i + \varepsilon|h|$ and such that the points $a_i e_i$ and $b_i e_i$ belong to E for all $i = 1, \ldots, n$. Since g is monotone, the function $x \mapsto |g(x) - g(0) - g'(0)h|$ attains its maximum in R at some point $y \in \partial R$. Suppose $y = b_i e_i + k$ where $k \in \mathbb{R}_i^{n-1}$. Then

$$|g(h) - g(0) - g'(0)h| \leq |g(y) - g(0) - g'(0)h|$$
$$\leq |g(y) - g(b_i e_i) - D'_i g(b_i e_i)k| + |g(b_i e_i) - g(0) - D_i g(0)b_i|$$
$$+ |D'_i g(b_i e_i)k - D'_i g(0)k| + |g'(0)y - g'(0)h|$$
$$\leq n\varepsilon|h| + 2\varepsilon|h| + n\varepsilon|h| + n|g'(0)||\varepsilon||h|$$

because $|k| \leq n|h|$, $|b_i| \leq 2|h|$, and $|y - h| \leq n\varepsilon|h|$. The same estimate is obtained if $y = a_i e_i + k$, $k \in \mathbb{R}_i^{n-1}$. This shows that g is differentiable at 0 and the lemma is proved.   □

4.5. Proof of I.2.4. Let $f: G \to \mathbb{R}^n$ be a qr mapping. Each component f_j is $f^\sharp F_I$-extremal by 2.8 and so monotone by 3.3. We can apply 4.4 to $g = f_j$ and conclude that each f_j, therefore f, is differentiable a.e.   □

4.6. Remark. As mentioned at the beginning of this chapter, a K-qr mapping f in fact belongs to ACL^P for some $p = p(n, K) > n$, and so the differentiability a.e. of f would follow from 4.1. To prove this requires a lengthy argument which we have elected to by–pass here. A good reference is [BI2]. The history of this important result is as follows. In [B] B. Bojarski proved it for qc mappings in the case $n = 2$. For qc mappings in all dimensions it was proved by F.W. Gehring in [G7]. O. Martio gave in [M2] a partial result for qr mappings, but his bound for p in a compact set C depended on the multiplicity bound $\sup_{x \in C} i(x, f)$. T. Iwaniec proved the general result in [I1] and at the same time sharpened Gehring's theorem in [G7]. Recently K. Astala [As] proved that the optimal p for qc mappings in the plane is $2K/(K-1)$. For more comments on this subject we refer to [I5].

5. Discreteness and Openness of Quasiregular Mappings

In this section we shall complete the proof of Reshetnyak's theorem I.4.1, which states that a nonconstant qr mapping is both discrete and open. We start by proving that qr mappings are sense–preserving (Theorem I.4.5). Recall from I.4 that a continuous mapping $f: G \to \mathbb{R}^n$ of a domain in $\mathbb{R}^n$ is sense–preserving (weakly sense–preserving) if the topological index satisfies $\mu(y, f, D) > 0$ ($\mu(y, f, D) \geq 0$) for all domains $D \subset\subset G$ and all $y \in fD \smallsetminus f\partial D$.

5.1. Lemma. Let G be a domain in $\mathbb{R}^n$ and let $f: G \to \mathbb{R}^n$ be ACL^n with $J_f(x) \geq 0$ a.e. Then f is weakly sense–preserving.

Proof. Let $D \subset\subset G$ be a domain and let $y \in fD \smallsetminus f\partial D$. Let (f_j) be an approximating sequence for f as in I.1.4. Then $f_j \to f$ uniformly in $\overline{D}$ and $J_{f_j} \to J_f$ in $L^1(\overline{D})$. We may assume that there exists a connected neighborhood U of y such that $\overline{U} \cap f\partial D = \emptyset$ and $\overline{U} \cap f_j \partial D = \emptyset$ for all j.

Fix j. Set $S_j = \{x \in D : J_{f_j}(x) \leq 0\}$ and $V_j = \{z \in U : \mu(z, f_j, D) < 0\}$. Since $\mu(z, f_j, U)$ is constant in each component of $\mathbb{R}^n \setminus f_j \partial D$, we have $V_j = U$ or $V_j = \emptyset$. We claim that $V_j \subset f_j S_j$. Suppose there exists a point $z \in V_j \setminus f_j S_j$. The set $f_j^{-1}(z) \cap D$ is finite. Otherwise there would exist a limit point x_0 of $f_j^{-1}(z) \cap D$. Since $f_j^{-1}(z) \cap \partial D = \emptyset$, $x_0 \in f_j^{-1}(z) \cap D$. We have $J_{f_j}(x_0) > 0$ because $z \notin f_j S_j$, so f_j must be a homeomorphism in some neighborhood of x_0. But this contradicts the fact that x_0 is a limit point of $f_j^{-1}(z) \cap D$, and thus $f_j^{-1}(z) \cap D$ is finite. Write $f_j^{-1}(z) \cap D = \{x_1, \ldots, x_k\}$. There exist disjoint neighborhoods $U_{x_1}, \ldots, U_{x_k}$ in D of the points x_i such that the mappings $f_j | U_{x_i}$ are homeomorphisms. Then by D_4 – D_6 in I.4.4,

$$\mu(z, f_j, D) = \sum_{i=1}^{k} \mu(z, f_j, U_{x_i}) = k > 0.$$

On the other hand, $z \in V_j$ implies $\mu(z, f_j, D) < 0$, a contradiction. The alternative is to have $V_j \subset f_j S_j$.

Set $S_j' = \{x \in D : J_{f_j}(x) < 0\}$. Then

$$m(V_j) \leq m(f_j S_j) \leq \int_{S_j} |J_{f_j}| \, dm = \int_{S_j'} |J_{f_j}| \, dm$$

$$= \int_D |J_{f_j}| \, dm - \int_{D \setminus S_j'} |J_{f_j}| \, dm \leq \int_D \left(|J_{f_j}| - J_{f_j} \right) dm.$$

The last integral tends to zero as $j \to \infty$ because $J_f \geq 0$ a.e. and $J_{f_j} \to J_f$ in $L^1(\overline{D})$. It follows that for some j_0, $m(V_j) < m(U)$ and thus $V_j = \emptyset$ for all $j \geq j_0$. In particular, $\mu(y, f_j, D) \geq 0$ for all $j \geq j_0$. Since $f_j | \overline{D} \to f | \overline{D}$ uniformly, $\mu(y, f_j, D) \to \mu(y, f, D)$ (see the proof of I.4.14(a)), and the required inequality $\mu(y, f, D) \geq 0$ follows.   □

5.2. Proof of Theorem I.4.5. Let $f: G \to \mathbb{R}^n$ be a nonconstant qr mapping, $D \subset\subset G$ a domain, and let $y \in fD \setminus f\partial D$. We have to show that $\mu(y, f, D) > 0$. Let Y be the y-component of $\mathbb{R}^n \setminus f\partial D$ and set $V = D \cap f^{-1}Y$. Clearly $D \setminus f^{-1}(y) \neq \emptyset$. There is therefore a point $x_0 \in V \cap \partial f^{-1}(y)$. Let $U = B^n(x_0, r) \subset V$ and set $E = \{x \in U : J_f(x) > 0\}$. We claim that $m(E) > 0$. Suppose $m(E) = 0$. The quasiregularity of f yields $f'(x) = 0$ a.e. in U. Since f is ACL, this implies that $U \subset f^{-1}(y)$. This contradicts $x_0 \in \partial f^{-1}(y)$, hence $m(E) > 0$.

We now can use the fact that f is differentiable a.e. and find a point $z \in E$ where f is differentiable. Set $g(x) = f'(z)(x - z) + f(z)$. Let W be a ball $B^n(z, \rho) \subset\subset V$. The homotopy $h_t(x) = tg(x) + (1 - t)f(x)$, $t \in [0, 1]$, $x \in \overline{W}$, satisfies the condition $f(z) \in h_t W \setminus h_t \partial W$ for all $t \in [0, 1]$, provided ρ is sufficiently small. From D_5 and D_6 in I.4.4 we conclude then that $\mu(f(z), f, W) = \mu(f(z), g, W) = 1$. With 5.1 we obtain

$$\mu(y, f, D) = \mu(f(z), f, D)$$
$$= \mu(f(z), f, W) + \mu(f(z), f, D \smallsetminus \overline{W}) \geq 1 + 0 = 1 .$$

The theorem is proved.   □

5.3. Lemma. *Let* $f \colon G \to \mathbb{R}^n$ *be a nonconstant* K-qr *mapping and let* $y \in \mathbb{R}^n$. *For each point* $z \in f^{-1}(y)$ *there are arbitrarily small numbers* $r > 0$ *such that* $S^{n-1}(z, r) \cap f^{-1}(y) = \emptyset$.

Proof. We may assume $y = 0$. The open set $G_0 = G \smallsetminus f^{-1}(0)$ is nonempty because f is not constant. By 2.8, $u = \log |f|$ is an $f^{\sharp} F_I$-extremal in G_0.

We first show that $\operatorname{int} f^{-1}(0) = \emptyset$. Suppose this is not true, and let X be a component of $\operatorname{int} f^{-1}(0)$. There exists a point $x_0 \in \partial X \cap G$. Then $x_0 \in \partial G_0$. We choose $R > 0$ such that $\overline{B}^n(x_0, R) \subset G$ and $X \smallsetminus B^n(x_0, R) \neq \emptyset$. We next fix $t \in {]}0, R[$ such that $d_n K^{2/n} (\log R/t)^{-1} < \frac{1}{2}$, where d_n is the constant in 3.11. For $0 < \rho \leq R$ set $M_\rho = \max\{ u(x) : x \in G_0 \cap \overline{B}^n(x_0, \rho) \}$. Choose a point a in $G_0 \cap \overline{B}^n(x_0, t)$ with the property that $u(a) = M_t$. Let $-\infty < M < 2M_t - M_R \leq M_t$, and let U be the a-component of the open set $\{ x \in G_0 : u(x) > M \} \cap B^n(x_0, R)$. As an extremal u is monotone, hence $\overline{U} \cap S^{n-1}(x_0, R) \neq \emptyset$. The function $v = u - M$ satisfies in U all the conditions in Lemma 3.11, appealing to which we obtain

$$M_t - M = \sup_{U_t} v < \tfrac{1}{2} \sup_U v \leq \tfrac{1}{2}(M_R - M) .$$

But this gives $2M_t - M_R \leq M$, which contradicts the choice of M. Accordingly, $\operatorname{int} f^{-1}(0) = \emptyset$.

Let $z \in f^{-1}(0)$ and suppose that for some $R > 0$ it is true that $\overline{B}^n(z, R) \subset G$ and $f^{-1}(0) \cap S^{n-1}(z, t) \neq \emptyset$ for all $t \in {]}0, R[$. As $z \in \partial G_0 \cap G$, we can apply the very same argument as above to arrive at another contradiction. The lemma is proved.   □

5.4. Corollary. *Let* $f \colon G \to \mathbb{R}^n$ *be a nonconstant qr mapping. Then* f *is light, i.e., for every* $y \in \mathbb{R}^n$ *the set* $f^{-1}(y)$ *is totally disconnected.* To complete the proof of I.4.1 we shall show that a light and sense–preserving mapping is always discrete and open. A mapping $f \colon G \to \mathbb{R}^n$ is called *quasiopen* if the following holds. For every $y \in \mathbb{R}^n$, $y \in \operatorname{int} fV$ whenever E is a compact component of $f^{-1}(y)$ and V is a neighborhood of E in G.

5.5. Lemma. *If* $f \colon G \to \mathbb{R}^n$ *is continuous and sense–preserving, then* f *is quasiopen.*

Proof. Let $y \in \mathbb{R}^n$ and let V be a neighborhood of a compact component E of $f^{-1}(y)$. Then there exists a domain $D \subset\subset V$ such that $E \subset D$ and $\partial D \cap f^{-1}(y) = \emptyset$ (see the proof of [V2, Theorem 3.1]). If Y is the y-component of $\mathbb{R}^n \smallsetminus f\partial D$ and if $z \in Y$, then $\mu(z, f, D) = \mu(y, f, D) > 0$ because f is sense–preserving. Therefore $z \in fD$. We infer that $Y \subset fD \subset fV$. Since Y is open, $y \in \operatorname{int} fV$. The lemma is proved.   □

5.6. Lemma. *If $f\colon G \to \mathbb{R}^n$ is continuous, light and sense–preserving, then f is discrete and open.*

Proof. By 5.5, f is quasiopen. A light and quasiopen mapping is clearly open. Let $y \in \mathbb{R}^n$ and $x \in f^{-1}(y)$. Since f is light, there exists a domain $D \subset\subset G$ such that $x \in D$ and $y \in fD \smallsetminus f\partial D$. Set $k = \mu(y, f, D)$. Since f is sense–preserving, $k > 0$. We claim that $N(y, f, D) = \operatorname{card} f^{-1}(y) \cap D \le k$, which proves the discreteness of f. Suppose that there exist at least $k + 1$ points $x_1, \dots, x_{k+1}$ in $f^{-1}(y) \cap D$. Using lightness again we find disjoint domains $D_1, \dots, D_{k+1}$ in D with $x_i \in D_i$ and $f^{-1}(y) \cap \partial D_i = \emptyset$, $i = 1, \dots, k+1$. Let $D_{k+2}, D_{k+3}, \dots$ be the components of $D \smallsetminus (\overline{D}_1 \cup \dots \cup \overline{D}_{k+1})$. Since $f^{-1}(y) \cap D$ is compact, $f^{-1}(y) \cap D \subset D_1 \cup \dots \cup D_p$ for some p. We have $\mu(y, f, D_i) > 0$ for $1 \le i \le k + 1$ and $\mu(y, f, D_i) \ge 0$ for $i \ge k + 2$. But then we get the contradiction

$$k = \mu(y, f, D) = \sum_{i=1}^{p} \mu(y, f, D_i) \ge k + 1 .$$

Hence $N(y, f, D) \le k$. The lemma is proved. $\square$

5.7. Proof of Theorem I.4.1. Let f be a nonconstant qr mapping. By 5.4 f is light and by Theorem I.4.5, which was proved in 5.2, f is sense–preserving. The openness and discreteness of f follow from 5.6. $\square$

5.8. Notes. The basic ideas of this section come from Reshetnyak's articles [Re2] and [Re4]. The proof of 5.1 is a simplified version of [Re2, Theorem 5] due to O. Martio. The argument in 5.2 is from [Re2, Lemma 7], except that we have used the differentiability of f a.e. Reshetnyak obtains 5.3 as a corollary of a stronger result [Re2, Theorem 10], which says that a closed set of zero capacity has a preimage of zero capacity. Lemma 5.5 appears in [V2, Theorem 4.1 and Theorem 5.1]. Lemma 5.6 is essentially [Re2, Theorem 11].

6. Pullbacks of General Kernels

So far we have considered pullbacks under qr mappings only of the trivial kernel $F_I(y, k) = |k|^n$. Since we have now filled a number of gaps in our presentation in proving Theorems I.2.4, I.4.1, and I.4.5, we are in a position to use freely material from preceding chapters. In particular, we can take advantage of results in Chapter II when dealing with pullbacks of general kernels.

    Let G and G' be domains in $\mathbb{R}^n$, let $F\colon G' \times \mathbb{R}^n \to \mathbb{R}^1$ be a variational kernel, and let $f\colon G \to G'$ be a qr mapping. We define the pullback $f^{\sharp}F$ of F by the same formula (2.2) as earlier, i.e.,

$$(6.1) \qquad f^\sharp F(x,h) = \begin{cases} F\big(f(x), J_f(x)^{1/n} f'(x)^{-1*} h\big) & \text{if } x \in A, \\ |h|^n & \text{if } x \in G \smallsetminus A, \end{cases}$$

where A is the Borel set of points $x \in G$ at which all the partial derivatives of f exist and $J_f(x) > 0$. By II.7.4 we now know that $m(G \smallsetminus A) = 0$ if f is nonconstant.

We claim that $f^\sharp F$ is a variational kernel. We may assume that f is not constant. Using Lusin's theorem for the mappings f and f' and the fact from II.7.4(3) that $m(E) = 0$ is equivalent to $m(fE) = 0$ for any measurable $E \subset G$, we conclude that $f^\sharp F$ satisfies (A) in 1.3. The remaining conditions (B), (C), and (D) also follow readily from II.7.4. Let us check (C) in order to give estimates for the structure constants of $f^\sharp F$. For almost every $x \in G$ and all $h \in \mathbb{R}^n$

$$f^\sharp F(x,h) \le \beta(F) J_f(x) |f'(x)^{-1*} h|^n \le \beta(F) J_f(x) |f'(x)^{-1}|^n |h|^n$$
$$\le \beta(F) K_I(f) |h|^n ,$$

whence

$$(6.2) \qquad \beta(f^\sharp F) \le \beta(F) K_I(f) .$$

Similarly we obtain

$$(6.3) \qquad \alpha(f^\sharp F) \ge \alpha(F) K_O(f)^{-1} .$$

6.4. Example. For $F = F_\sigma$ as in Example 1.4 we obtain

$$f^\sharp F(x,h) = F_\tau(x,h) ,$$

where

$$(6.5) \qquad \tau(x) = \begin{cases} J_f(x)^{2/n} f'(x)^{-1} \sigma(f(x)) f'(x)^{-1*} & \text{if } x \in A, \\ I & \text{if } x \in G \smallsetminus A. \end{cases}$$

The next lemma shows that the ACL^n class is preserved under qr mappings. It is not generally true that a composition of two mappings in ACL^n is necessarily ACL^n. For $n = 2$ this is seen by taking $f(x) = (x_1, 0)$ and $g(x) = |x|^{-1/2} x$. Then both f and g are ACL^2, but $g \circ f$ is not.

6.6. Lemma. If $f \colon G \to G'$ is qr and $v \colon G' \to \mathbb{R}^1$ is ACL^n, then $u = v \circ f$ is ACL^n and

$$(6.7) \qquad \nabla u(x) = f'(x)^* \nabla v(f(x)) \quad a.e.$$

Proof. We may assume that f is not constant. To prove that u is ACL^n let (v_j) be an approximating sequence for v as in I.1.4. Each function $u_j = v_j \circ f$ is ACL^n and $\nabla u_j(x) = f'(x)^* \nabla v_j(f(x))$ a.e. Let $U \subset\subset G$. The transformation formula in I.4.14 gives

$$\int\limits_{U} |\nabla u_j - \nabla u_k|^n \, dm = \int\limits_{U} |f'(x)^* (\nabla v_j(f(x)) - \nabla v_k(f(x)))|^n \, dx$$

(6.8)
$$\le K_O(f) \int\limits_{U} |\nabla v_j \circ f - \nabla v_k \circ f|^n J_f \, dm$$

$$\le K_O(f) N(f, U) \int\limits_{fU} |\nabla v_j - \nabla v_k|^n \, dm \ .$$

Combined with the fact that $u_j \to u$ locally uniformly, (6.8) implies that $(u_j | U)$ is a Cauchy sequence in $W_n^1(U)$. Therefore $u_j | U \to u | U$ in $W_n^1(U)$ so we conclude that u is ACL^n.

To prove (6.7) let $\varphi \colon G \to \mathbb{R}^n$ be in $C_0^\infty(G)$. Estimating as in (6.8) we see that

(6.9) $$\lim_{j \to \infty} \int\limits_{G} f'(x)^* \nabla v_j(f(x)) \cdot \varphi(x) \, dx = \int\limits_{G} f'(x)^* \nabla v(f(x)) \cdot \varphi(x) \, dx \ .$$

Since $\nabla u_j(x) = f'(x)^* \nabla v_j(f(x))$ a.e. and $u_j \to u$ locally in W_n^1, we obtain with the aid of (6.9) the equality

$$\int\limits_{G} \nabla u \cdot \varphi \, dm = \int\limits_{G} f'(x)^* \nabla v(f(x)) \cdot \varphi(x) \, dx \ .$$

This certifies that $f'^*(\nabla v \circ f)$ is the weak gradient of u, from which (6.7) follows.   □

Let $f \colon G \to G'$ be a nonconstant qr mapping, let $F \colon G' \times \mathbb{R}^n \to \mathbb{R}^1$ be a variational kernel, and let $v \colon G' \to \mathbb{R}^n$ be a continuous F-extremal. Set $u = v \circ f$. If D is a normal domain for f, then we get by the integral transformation formula in I.4.14, by 6.6, and by (D) in 1.3 that

(6.10)
$$I_{f^\sharp F}(u, D) = \int\limits_{D} F\big(f(x), J_f^{1/n}(x) f'(x)^{-1*} \nabla u(x)\big) \, dx$$

$$= \int\limits_{D} J_f(x) F\big(f(x), \nabla v(f(x))\big) \, dx = N(f, D) I_F(v, fD) \ .$$

In case f is qc we thus have $I_{f^\sharp F}(u, D) = I_F(v, fD)$ for all domains $D \subset\subset G$, and it is easy to see that u is $f^\sharp F$-extremal. In fact, if $D \subset\subset G$ and if $w \in C(\overline{D}) \cap W_n^1(D)$ with $w | \partial D = u | \partial D$, we get in this case

$$I_{f^\sharp F}(u, D) = I_F(v, fD) \le I_F(w \circ f^{-1}, fD) = I_{f^\sharp F}(w \circ f^{-1} \circ f, D)$$
$$= I_{f^\sharp F}(w, D) \ ,$$

which shows that u is $f^\sharp F$-extremal.

We shall next show that u is $f^\sharp F$-extremal even in the general case.

6.11. Theorem. *Let* $f: G \to G'$ *be qr, let* $F: G' \times \mathbb{R}^n \to \mathbb{R}^1$ *be a variational kernel, and let* v *be a continuous F-extremal. Then* $u = v \circ f$ *is $f^{\sharp}F$-extremal.*

Proof. We may assume that f is not constant. Let U be a normal domain for f. By 1.17 and 1.12 it is enough to show

$$(6.12) \qquad \int_U \nabla_h f^{\sharp} F(\,\cdot\,, \nabla u) \cdot \nabla \varphi \, dm = 0$$

for all $\varphi \in C_0^{\infty}(U)$. We define a function $\psi: fU \to \mathbb{R}^1$ by

$$\psi(y) = \sum_{x \in f^{-1}(y) \cap U} i(x, f)\varphi(x) \,.$$

Clearly spt ψ is a compact subset of fU. We claim that ψ is ACL^n.

To prove that ψ is ACL we adapt the arguments in the proof of Lemma II.5.3, where the mapping $g_U(y) = m^{-1} \sum \{ i(x, f)x : x \in f^{-1}(y) \cap U \}$ was shown to be ACL. The proof of the continuity of ψ is similar to the proof in II.5.3. If $\beta: \overline{J}_0 \to \overline{J}$ and $\alpha_j: \overline{J}_0 \to U$, $j = 1, \ldots, m = N(f, U)$, are the paths that appear at the end of the proof of II.5.3, we can write

$$\psi|\overline{J} = \sum_{j=1}^{m} \varphi \circ \alpha_j \circ \beta^{-1} \,.$$

Since each $\alpha_j \circ \beta^{-1}$ is absolutely continuous, so is $\psi|\overline{J}$. Hence ψ is ACL.

To complete the proof of $\psi \in \mathrm{ACL}^n$ we use the argument in Lemma II.7.1. If $h_{ij}: Q_i \to D_{ij}$, $j = 1, \ldots, m$, $i = 1, 2, \ldots$, are the qc mappings in the proof of II.7.1, we have

$$(6.13) \qquad \psi|Q_i = \sum_{j=1}^{m} \varphi \circ h_{ij} \,.$$

This gives for almost every $y \in Q_i$,

$$|D_k \psi(y)|^n \le M m^{n-1} \sum_{j=1}^{m} |h'_{ij}(y)|^n \,,$$

where $M < \infty$ depends only on φ. The fact that $D_k \psi \in L^n(fU)$ then follows as in II.7.1.

Let $x \in U$ be a point where f is differentiable, $J_f(x) > 0$, and $f(x)$ is nonexceptional for F. For the moment we write $\xi(h) = f^{\sharp}F(x, h)$, $\eta(h) = F(f(x), h)$, $Ah = J_f(x)^{1/n} f'(x)^{-1*} h$. Then

$$\nabla_h f^{\sharp} F(x, h) \cdot k = \xi'(h)k = \eta'(Ah)Ak = \nabla_h F(f(x), Ah) \cdot Ak \,.$$

We shall here substitute $h = \nabla u(x)$ and $k = \nabla \varphi(x)$. From conditions (D) and (B) for F it follows that $\lambda^{n-1} \nabla_h F(f(x), a) = \nabla_h F(f(x), \lambda a)$ for all

$\lambda > 0$ and $a \in \mathbb{R}^n$. For almost every x we have $\nabla u(x) = f'(x)^* \nabla v(f(x))$ by 6.6. By these remarks we obtain for almost every $x \in U$,

$$\nabla_h f^\sharp F(x, \nabla u(x)) \cdot \nabla \varphi(x) = \nabla_h F(f(x), \nabla v(f(x))) J_f(x) \cdot f'(x)^{-1*} \nabla \varphi(x) .$$

From (6.13) it follows that

$$\nabla \psi(y) = \sum_{j=1}^{m} h'_{ij}(y)^* \nabla \varphi(h_{ij}(y))$$

for almost every $y \in Q_i$, and we get

$$\int_U \nabla_h f^\sharp F(\,\cdot\,, \nabla u) \cdot \nabla \varphi \, dm = \sum_{i,j} \int_{D_{ij}} \nabla_h f^\sharp F(\,\cdot\,, \nabla u) \cdot \nabla \varphi \, dm$$

$$= \sum_{i,j} \int_{Q_i} \nabla_h F(\,\cdot\,, \nabla v) \cdot h'_{ij}{}^* (\nabla \varphi \circ h_{ij}) \, dm$$

$$= \sum_{i} \int_{Q_i} \nabla_h F(\,\cdot\,, \nabla v) \cdot \nabla \psi \, dm = \int_{fU} \nabla_h F(\,\cdot\,, \nabla v) \cdot \nabla \psi \, dm .$$

Since $\psi \in W_{n,0}^1(fU)$, the last integral is zero by Remark 1.15. The theorem is proved. $\square$

6.14. Notes. In [Re8, Theorem 1] Yu.G. Reshetnyak proved Theorem 6.11 for the special kernel $F = F_\sigma$ given in 1.4. Theorem 6.11 is contained in [GLM1, 7.10] but the proof there differs from ours. For 6.6 see also [LF].

7. Further Properties of Extremals

Into this section we have collected some properties of extremals that are no less fundamental than those discussed in the preceding sections but were not needed at the time. For example, up to this point we have not presented a general existence result for extremals. Only the examples in Corollary 2.8 are presently available to us.

Throughout this section $F: U \times \mathbb{R}^n \to \mathbb{R}^1$ is a variational kernel with structure constants α and β. We start with the following comparison principle. The uniqueness of continuous extremals follows as a corollary.

7.1. Theorem. Let U be bounded, and let u and v be continuous functions in $\overline{U}$ that are F-extremals in U. If $u \geq v$ on ∂U, then $u \geq v$ in U.

Proof. Suppose that $u(x_0) < v(x_0)$ for some $x_0 \in U$, and let $0 < \varepsilon < v(x_0) - u(x_0)$. Let D be the component of $V = \{ x \in U : v(x) > u(x) + \varepsilon \}$ that contains x_0. Then $D \subset\subset U$, $w = u + \varepsilon = v$ on ∂D, and $v > w$ in D.

The function w is also an extremal, and $\eta = (v+w)/2$ coincides with both w and v on ∂D. Therefore $I_F(v, D) \leq I_F(\eta, D)$ and $I_F(w, D) \leq I_F(\eta, D)$. Let E be the subset of D where both $\nabla v(x)$ and $\nabla w(x)$ exist but are not equal. Then the strict convexity in (B) implies that

$$(7.2) \qquad F(x, \nabla\eta(x)) < \tfrac{1}{2}\big(F(x, \nabla v(x)) + F(x, \nabla w(x))\big)$$

a.e. in E. In $D \smallsetminus E$ (7.2) is replaced by equality a.e. Hence, if $m(E) > 0$, then $I_F(\eta, D) < \big(I_F(v, D) + I_F(w, D)\big)/2 \leq I_F(\eta, D)$. This shows $m(E) = 0$. It follows that $v - w$ is the constant 0 in D. But then we get the contradiction $v(x_0) > u(x_0) + \varepsilon = w(x_0) = v(x_0)$, and the theorem follows. $\qquad\square$

7.3. Corollary. *Let U, u, and v be as in 7.1. If $u = v$ on ∂U, then $u = v$ in U.*

Harnack's inequality relates the maximum and minimum of nonnegative extremals in compact subsets of U. Since we are dealing with the borderline case $p = n$, we obtain this result without difficulty thanks to a proof devised by S. Granlund [Gr].

7.4. Theorem. *Let u be a nonnegative continuous F-extremal in U. If $B^n(x_0, R) \subset U$ and if $V_r = B^n(x_0, r)$, $0 < r < R$, then*

$$(7.5) \qquad\qquad\qquad \sup_{V_r} u \leq a \inf_{V_r} u \,,$$

where $a = \exp\big(D_n\beta\alpha^{-1}(\log(R/r))^{-1}\big)$ and $D_n > 0$ depends only on n.

Proof. Let $\varepsilon > 0$. We apply 3.1 to $u + \varepsilon$ and the condenser $(A, C) = (B^n(x_0, R), \overline{B}^n(x_0, \rho))$ where $\rho = (Rr)^{1/2}$. If $v = \log(u + \varepsilon)$, 3.1(2) gives

$$(7.6) \qquad \int_{V_\rho} |\nabla v|^n \, dm \leq C_n\beta\alpha^{-1}\omega_{n-1}\Big(\log\frac{R}{\rho}\Big)^{1-n}.$$

By 3.3 $u + \varepsilon$ is monotone and so is v. Therefore we obtain with the help of 3.4 and (7.6)

$$\operatorname{osc}(v, V_r)^n \log\frac{\rho}{r} \leq \int_r^\rho \operatorname{osc}(v, V_t)^n t^{-1} \, dt = \int_r^\rho \operatorname{osc}(v, \partial V_t)^n t^{-1} \, dt$$

$$\leq A_n \int_{V_\rho} |\nabla v|^n \, dm \leq A_n C_n\beta\alpha^{-1}\omega_{n-1}\Big(\log\frac{R}{\rho}\Big)^{1-n}.$$

Hence

$$\log\Big(\sup_{V_r} u + \varepsilon\Big) - \log\Big(\inf_{V_r} u + \varepsilon\Big) \leq \lambda = 2(A_n C_n\beta\alpha^{-1}\omega_{n-1})^{1/n}\Big(\log\frac{R}{r}\Big)^{-1}.$$

Letting $\varepsilon \to 0$ gives $\sup_{V_r} u \leq e^\lambda \inf_{V_r} u$. The theorem is proved. $\qquad\square$

As a final topic in this section we take up the existence of extremals. We need the concept of weak convergence. Let X be a normed linear space with coefficients in $\mathbb{R}^1$. By X' we denote the dual of X, the linear space of bounded linear functionals $h\colon X \to \mathbb{R}^1$ with the norm $\|h\| = \sup\{\, h(x) : \|x\| \leq 1\,\}$. We say that the sequence (x_j) in X *converges weakly* to an element $x \in X$ if $h(x_j) \to h(x)$ for all $h \in X'$.

7.7. Lemma. *If X is a reflexive and separable Banach space and (x_j) is a bounded sequence in X, then there is a subsequence (x_{j_k}) that converges weakly to an element x of X.*

The proof of 7.7 can be found in [Y, p. 126]. The lemma remains true without the assumption of separability, but the proof is more difficult (see [Y, p. 141]).

7.8. Duals of Sobolev spaces. Let U be open in $\mathbb{R}^n$, let $1 < p < \infty$, and let $q = p/(p-1)$. The function spaces $W_p^1(U)$ and $W_{p,0}^1(U)$ (of real valued functions) are indicated here for short by W and W_0. In this context we let L_m^p be the spaces of mappings $g\colon U \to \mathbb{R}^m$ in $L^p(U)$ and let L^p be L_1^p.

In order to describe the duals of W and W_0 we define the mapping $P\colon W \to L_{n+1}^p$ by $P(u) = (u, D_1 u, \dots, D_n u)$. Then P is clearly a quasi–isometric linear isomorphism of W onto PW, the latter being a closed subspace of L_{n+1}^p because W is complete. If $h \in W'$, then h factors as $h = h_0 \circ P$, where $h_0 \in (PW)'$. By the Hahn–Banach theorem h_0 has an extension $h_1 \in (L_{n+1}^p)'$. Accordingly, there exists $v \in L_{n+1}^q$ such that $h_1 = L_v$, where L_v is defined by

$$L_v(w) = \sum_{i=1}^{n+1} \langle w_i, v_i \rangle\,, \quad \langle w_i, v_i \rangle = \int_U w_i v_i\, dm\,.$$

On the other hand, if $v \in L_{n+1}^q$, then the restriction $L_v|PW$ defines via the isomorphism $P\colon W \to PW$ an element in W'. By considering the restriction $P|W_0$ instead of P, we can make the corresponding observations for W_0'.

From the above it follows that a sequence (u_j) in W converges weakly to an element u in W if and only if for all $v \in L_{n+1}^q$, $\langle u_j, v_1 \rangle \to \langle u, v_1 \rangle$ and $\langle D_i u_j, v_{i+1} \rangle \to \langle D_i u, v_{i+1} \rangle$, $i = 1, \dots, n$.

As noted PW and PW_0 are closed subspaces of L_{n+1}^p. Since L_{n+1}^p is separable and reflexive (here we need $1 < p < \infty$), the same is true of PW and PW_0, and consequently of W and W_0. We can therefore apply 7.7 to the case where X is W or W_0 to obtain the following result.

7.9. Proposition. *Let X be either $W_p^1(U)$ or $W_{p,0}^1(U)$, $1 < p < \infty$. If (u_j) is a bounded sequence in X, then there is a subsequence (u_{j_k}) that converges weakly to an element u in X. In particular, ∇u_{j_k} converges weakly to ∇u in $L^p(U)$. If, in addition, $u_j \to v$ in $L^p(U)$, then $v = u$ and so belongs to X.*

For the existence of extremals we need the following lower semicontinuity property for variational integrals. We shall give a simple proof due to P. Lindqvist [Li1].

7.10. Proposition. Let u_i, $i = 0, 1, \ldots$, be functions in $W_n^1(U)$ such that $\nabla u_i \to \nabla u_0$ weakly in $L^n(U)$. Then

$$(7.11) \qquad\qquad I_F(u_0, U) \le \liminf_{i \to \infty} I_F(u_i, U) .$$

In the proof of 7.10 we shall make use of the following easy case ($p = 2$) of the Banach–Saks theorem.

7.12. Lemma. Let v_i, $i = 0, 1, \ldots$, be functions in $L^2(A)$ such that $v_i \to v_0$ weakly in $L^2(A)$. Then there exists a subsequence (v_{i_j}) for which

$$\lim_{k \to \infty} \left\| \frac{1}{k}(v_{i_1} + \ldots + v_{i_k}) - v_0 \right\|_2 = 0 .$$

Proof. As before we write

$$\langle g, h \rangle = \int_A gh \, dm , \quad g, h \in L^2(A) .$$

Set $w_i = v_i - v_0$. The weak converge of (v_i) implies that we can inductively choose a sequence $i_1 < i_2 < \ldots$ such that $|\langle w_{i_j}, w_{i_{k+1}} \rangle| \le 1/k$ for $1 \le j \le k$, $k = 1, 2, \ldots$. Furthermore, as a weakly convergent sequence is bounded in norm, $\|w_i\|_2^2 \le M$ for all i and for some M. Then

$$\left\| \frac{1}{k}(w_{i_1} + \ldots + w_{i_k}) \right\|_2^2 \le \frac{1}{k^2} \sum_{j=1}^{k} \|w_{i_j}\|_2^2 + \frac{2}{k^2} \sum_{l=1}^{k-1} \sum_{j=1}^{l} |\langle w_{i_j}, w_{i_{l+1}} \rangle|$$

$$\le \frac{1}{k^2} \left(kM + 2\left(1 + 2 \cdot \frac{1}{2} + \ldots + (k-1)\frac{1}{k-1}\right)\right) < \frac{M+2}{k} ,$$

which implies the lemma. □

Proof of 7.10. We may assume that the limit

$$\lim_{i \to \infty} I_F(u_i, U)$$

exists and is finite. Since $I_F(u_0, U_m) \to I_F(u_0, U)$ whenever $U_1 \subset U_2 \subset \ldots$ is an exhaustion of U by open sets $U_m \subset\subset U$, it suffices to prove (7.11) for such a set $U_m = V$.

Because $m(V) < \infty$ we have $L^2(V) \subset L^{n/(n-1)}(V)$, hence $\nabla u_i \to \nabla u_0$ weakly in $L^2(V)$. Lemma 7.12 then delivers a subsequence, again denoted by (u_i), such that $\|(\nabla u_1 + \ldots + \nabla u_k)/k - \nabla u_0\|_{2,V} \to 0$. Let $\varepsilon > 0$. By (A) in 1.3 and Egorov's theorem there exists a compact set $E \subset V$ with

$m(V \setminus E) < \varepsilon$ such that $F|E \times \mathbb{R}^n$ is continuous, and such that for some further subsequence, still denoted by (u_i), $(\nabla u_1 + \ldots + \nabla u_k)/k \to \nabla u_0$ uniformly in E. By (B) in 1.3 we have a.e.

$$F\Big(\cdot, \frac{1}{k}(\nabla u_1 + \ldots + \nabla u_k)\Big) \leq \frac{1}{k}\Big(F(\cdot, \nabla u_1) + \ldots + F(\cdot, \nabla u_k)\Big).$$

Fatou's lemma then yields

$$I_F(u_0, E) \leq \liminf_{k \to \infty} I_F\Big(\frac{1}{k}(u_1 + \ldots + u_k), E\Big)$$
$$\leq \liminf_{k \to \infty} \frac{1}{k}\Big(I_F(u_1, V) + \ldots + I_F(u_k, V)\Big) = \lim_{i \to \infty} I(u_i, V).$$

Since $u_0 \in W_n^1(U)$ and $I_F(u_0, V) \leq \beta \|\nabla u_0\|_n^n$, the result follows from the absolute continuity property of integrals.   □

We shall now prove the existence result for extremals in the following form.

7.13. Theorem. *Let U be bounded and let $v \in W_n^1(U)$. Then there exists an F-extremal in U with boundary values v.*

Proof. Let (u_i) be a sequence in $W_n^1(U)$ such that $u_i - v \in W_{n,0}^1(U)$ and

$$\lim_{i \to \infty} I_F(u_i, U) = I_0 = \inf_{w \in \mathcal{F}_v} I_F(w, U)$$

where $\mathcal{F}_v = \{w : w - v \in W_{n,0}^1(U)\}$. From (C) in 1.3 it follows that $\|\nabla u_i\|_n \leq M$ for some M. Set $w_i = u_i - v$. Let $B^n(r)$ be a ball which contains U. Then $w_i \in W_{n,0}^1(B^n(r))$ and by the Poincaré inequality III.2.1, $\|w_i\|_n \leq r\|\nabla w_i\|_n \leq r(M + \|\nabla v\|_n)$. With $\|\nabla w_i\|_n \leq M + \|\nabla v\|_n$ we have thus $\|w_i\|_{1,n} \leq M'$ for some M'. By 7.9 there exists a subsequence (w_{i_j}) of (w_i) which converges weakly to an element $w_0 \in W_{n,0}^1(U)$ and $\nabla w_{i_j} \to \nabla w_0$ weakly in $L^n(U)$. If we set $u_0 = w_0 + v$, then $\nabla u_{i_j} \to \nabla u_0$ weakly in $L^n(U)$. By 7.10, $I_F(u_0, U) \leq I_0$ and u_0 is the required extremal.   □

Next we study the comparison principle with general boundary values. The following lemma is useful.

7.14. Lemma. *Let $u \in W_{n,0}^1(U)$. Then $u^+ \in W_{n,0}^1(U)$. If $w \in W_n^1(U)$ and if $0 \leq w \leq u^+$, then also $w \in W_{n,0}^1(U)$.*

Proof. Let $\varphi_i \in C_0^\infty(U)$, $i = 1, 2, \ldots$, be such that $\varphi_i \to u$ in $W_n^1(U)$. Then $g_i = \max(\varphi_i, 0)$ has compact support, $g_i \to u^+$ in $L^n(U)$, and (g_i) is bounded in $W_n^1(U)$. By 7.9, $g_i \to u^+$ in $W_n^1(U)$. Since each g_i can be approximated in $W_n^1(U)$ by functions in $C_0^\infty(U)$, we get $u^+ \in W_{n,0}^1(U)$. To prove the second statement we may assume $0 \leq u$. Let $\psi_i \in C^\infty(U)$, $i = 1, 2, \ldots$, be such that $\psi_i \to w$ in $W_n^1(U)$. We may assume $\psi_i \geq 0$.

Then $h_i = \min(\psi_i, \varphi_i)$ has compact support, $h_i \to w$ in $L^n(U)$, and (h_i) is bounded in $W_n^1(U)$. As in the first part we conclude that $w \in W_{n,0}^1(U)$. $\square$

Let $v_1, v_2 \in W_n^1(U)$ be such that $v_1 \geq v_2$ on the boundary of U, by which we mean $\xi = \min(v_1 - v_2, 0) \in W_{n,0}^1(U)$. Let u_i be an F-extremal in U with boundary values v_i, $i = 1, 2$. We claim that $\eta = \min(u_1 - u_2, 0) \in W_{n,0}^1(U)$. Write $u_i = v_i + h_i$, $h_i \in W_{n,0}^1(U)$, $i = 1, 2$, and observe $\xi + \min(h_1 - h_2, 0) \leq \eta \leq 0$. From 7.14 we conclude first that $\xi + \min(h_1 - h_2, 0) \in W_{n,0}^1(U)$ and then that $\eta \in W_{n,0}^1(U)$. By 1.15 we have

$$
\begin{aligned}
0 &= \int_U \left(\nabla_h F(\,\cdot\,, \nabla u_1) - \nabla_h F(\,\cdot\,, \nabla u_2) \right) \cdot \nabla \eta \, dm \\
&= \int_E \left(\nabla_h F(\,\cdot\,, \nabla u_1) - \nabla_h F(\,\cdot\,, \nabla u_2) \right) \cdot (\nabla u_1 - \nabla u_2) \, dm,
\end{aligned}
$$

(7.15)

where $E = \{ x \in U : u_1(x) < u_2(x) \}$. By 1.8(4) and (7.15), $\nabla u_1 - \nabla u_2 = 0$ a.e. in E, hence $\nabla \eta = 0$ a.e. in U. This implies $\eta = 0$ a.e. We have proved the following comparison principle, which is an extension of 7.1.

7.16. Theorem. Let $v_1, v_2 \in W_n^1(U)$ be such that $\min(v_1 - v_2, 0) \in W_{n,0}^1(U)$ and let u_i be an F-extremal with boundary values v_i, $i = 1, 2$. Then $u_1 \geq u_2$ a.e. In particular, the boundary values determine the F-extremal uniquely in $W_n^1(U)$. Moreover, the maximum principle holds, i.e. if $v_1 \leq M$, then $u_1 \leq M$ a.e.

7.17. Notes. The proof of 7.1 is taken from [GLM1, 4.18]. In the literature one finds many proofs of the results 7.4, 7.10, and 7.13 under various assumptions (see [Mor2], [Se]). In our situation, (i.e., in the borderline case $p = n$) simple proofs are possible, as shown by S. Granlund [Gr] and P. Lindqvist [Li1]. For the special case $F = F_\sigma$ (see 1.4) a still simpler proof of 7.13 can be given (see [BI2]). In Chapter VII we shall complete the result of 7.13. First in VII.3.6 we show that F-extremals can be taken continuous and then in VII.4.12 we study the continuity up to the boundary.

8. The Limit Theorem

The limit of a locally uniformly convergent sequence of K-qr mappings is K-qr. This result was proved by Yu.G. Reshetnyak in [Re5]. His proof involved a lengthy discussion of differential forms with norms. Following the presentation by P. Lindqvist in [Li2] we shall give a proof which is a simplified version of Reshetnyak's original argument. To put a finishing touch on the section we apply the limit theorem to prove that nonconstant qr mappings with sufficiently small dilatation are local homeomorphisms provided $n \geq 3$.

8.1. Lemma. Let $g: G \to \mathbb{R}^n$ be ACL^n, let $\zeta \in C_0^\infty(G)$, and let $h = (g_1, \ldots, g_{i-1}, \zeta, g_{i+1}, \ldots, g_n)$. Then

$$(8.2) \qquad \int_G \zeta J_g \, dm = - \int_G g_i J_h \, dm \,.$$

Proof. By the approximation result I.1.4 we may assume that g is C^∞. We may also assume $i = 1$. Write $g_\zeta = (\zeta g_1, g_2, \ldots, g_n)$. Let ω be the $(n-1)$-form $\zeta g_1 dg_2 \wedge \ldots \wedge dg_n$. Then $d\omega = J_{g_\zeta} dx_1 \wedge \ldots \wedge dx_n$. If Q is a closed n-interval in G, the elementary Stokes theorem gives

$$(8.3) \qquad \int_Q J_{g_\zeta} \, dm = \int_Q d\omega = \int_{\partial Q} \omega \,.$$

If $Q_1, \ldots, Q_k$ are closed cubes in G with disjoint interiors and $\mathrm{spt}\, \zeta \subset A = Q_1 \cup \ldots \cup Q_k$, then by summing equalities of type (8.3) we obtain

$$\int_G J_{g_\zeta} \, dm = \int_A J_{g_\zeta} \, dm = \int_{\partial A} \omega = 0 \,.$$

Since $J_{g_\zeta} = \zeta J_g + g_1 J_h$, (8.2) follows. □

The following lemma could be proved like (3.15) for the component functions $u = f_i$, but we give an alternate proof.

8.4. Lemma. Let $f: G \to \mathbb{R}^n$ be qr. Then

$$(8.5) \qquad \int_G \psi^n J_f \, dm \leq n^n K_O(f)^{n-1} \int_G |f|^n |\nabla \psi|^n \, dm$$

for every nonnegative $\psi \in C_0^\infty(G)$.

Proof. By choosing $\zeta = \psi^n$ and $i = 1$ in 8.1 we get

$$\int_G \psi^n J_f \, dm = -n \int_G \psi^{n-1} f_1 J_{f_\psi} \, dm$$

where $f_\psi = (\psi, f_2, \ldots, f_n)$. We have $J_{f_\psi}(x) \leq |\nabla \psi(x)||f'(x)|^{n-1}$ a.e. Then Hölder's inequality and the quasiregularity of f yield

$$\int_G \psi^n J_f \, dm \leq n \left(\int_G |f|^n |\nabla \psi|^n dm \right)^{1/n} \left(\int_G \psi^n K_O(f) J_f \, dm \right)^{(n-1)/n},$$

from which (8.5) follows. □

8.6. Theorem. *Let $f_j: G \to \mathbb{R}^n$, $j = 1, 2, \ldots$, be a sequence of K-qr mappings converging locally uniformly to a mapping f. Then f is qr and*

$$(8.7) \qquad K_O(f) \leq \liminf_{j \to \infty} K_O(f_j) , \quad K_I(f) \leq \liminf_{j \to \infty} K_I(f_j) .$$

We shall first prove weak convergence of the Jacobians in the following form.

8.8. Lemma. *The limit mapping f is ACL^n and*

$$(8.9) \qquad \lim_{j \to \infty} \int_G \psi^n J_{f_j} \, dm = \int_G \psi^n J_f \, dm \quad \text{for all } \psi \in C_0^\infty(G) .$$

Proof. Let $V \subset\subset G$ be a ball and let $\psi \in C_0^\infty(G)$ be such that $0 \leq \psi \leq 1$ and $\psi|V = 1$. Then 8.4 and $K(f_j) \leq K$ imply the existence of $M_V < \infty$ such that $\|f_j'\|_{n,V} \leq M_V$ for all j. By 7.9, $f|V \in W_n^1(V)$. As f is also continuous, it is ACL^n.

To prove (8.9) we shall show by induction on $i = 1, \ldots, n+1$ that

$$(8.10) \qquad \begin{aligned} &\lim_{j \to \infty} \int_G \varphi^{i-1} J(f_{j1}, \ldots, f_{j,i-1}, g_i, \ldots, g_n) \, dm \\ &\qquad = \int_G \varphi^{i-1} J(f_1, \ldots, f_{i-1}, g_i, \ldots, g_n) \, dm \end{aligned}$$

for any $\varphi \in C_0(G)$ and any ACL^n mapping $g: G \to \mathbb{R}^n$. Here we use the notation $J(h)$ for the Jacobian determinant of a mapping h. It is clear that (8.10) implies

$$(8.11) \qquad \begin{aligned} &\lim_{j \to \infty} \int_G \varphi^{i-1} f_{ji} J(f_{j1}, \ldots, f_{j,i-1}, g_i, \ldots, g_n) \, dm \\ &\qquad = \int_G \varphi^{i-1} f_i J(f_1, \ldots, f_{i-1}, g_i, \ldots, g_n) \, dm . \end{aligned}$$

Equality (8.10) is trivially true for $i = 1$. Suppose (8.10) is true for i. If $\varphi \in C_0^\infty(G)$, 8.1 and (8.11) inform us that

$$\begin{aligned} &\lim_{j \to \infty} \int_G \varphi^i J(f_{j1}, \ldots, f_{ji}, g_{i+1}, \ldots, g_n) \, dm \\ &= - \lim_{j \to \infty} i \int_G \varphi^{i-1} f_{ji} J(f_{j1}, \ldots, f_{j,i-1}, \varphi, g_{i+1}, \ldots, g_n) \, dm \\ &= -i \int_G \varphi^{i-1} f_i J(f_1, \ldots, f_{i-1}, \varphi, g_{i+1}, \ldots, g_n) \, dm \\ &= \int_G \varphi^i J(f_1, \ldots, f_i, g_{i+1}, \ldots, g_n) \, dm . \end{aligned}$$

Finally, by uniformly approximating any member φ of $\mathcal{C}_0(G)$ by functions in $\mathcal{C}_0^\infty(G)$ we complete the induction step. The lemma is proved. $\square$

Proof of 8.6. By 8.8, f is ACL^n. Let $V \subset\subset G$ be a ball, let $\psi \in \mathcal{C}_0^\infty(G)$ be such that $0 \leq \psi \leq 1$ and $\psi|V = 1$, and let M_V be as in the proof of 8.8. It follows from 7.10 that for $i = 1, \ldots, n$

$$(8.12) \qquad \int_V |\nabla f_i|^n \, dm \leq \liminf_{j \to \infty} \int_V |\nabla f_{ji}|^n \, dm .$$

To see this we may assume that the right hand side is obtained as a limit by a subsequence (g_k) of (f_j). Since $\|f'_j\|_{n,V} \leq M_V$ for all j, there exists by 7.9 a subsequence (f_{j_k}) of (g_k) such that $\nabla f_{j_k i} \to \nabla f_i$ weakly in $L^n(V)$, so (8.12) follows from 7.10.

Inequality (8.12) and Lemma 8.8 imply that

$$\int_V |f'|^n \, dm \leq d \liminf_{j \to \infty} \int_V |f'_j|^n \, dm \leq d K \lim_{j \to \infty} \int_G \psi^n J_{f_j} \, dm$$

$$= d K \int_G \psi^n J_f \, dm$$

for some constant $d > 0$. We deduce from this the inequality

$$\int_V |f'|^n \, dm \leq d K \int_V J_f \, dm .$$

By Lebesgue's theorem, $|f'(x)|^n \leq d K J_f(x)$ a.e. We have established the quasiregularity of f.

It remains to prove the dilatation inequalities (8.7). If f is constant, these are obvious . Suppose that f is nonconstant. Its branch set B_f is of measure zero. Hence it suffices to check the inequalities in $G \setminus B_f$. Let $x_0 \in G \setminus B_f$ and let U be a neighborhood of x_0 such that $f|\overline{U}$ is injective. Since $f_j \to f$ uniformly in $\overline{U}$, there exists j_0 and a ball $W = B^n(f(x_0), r) \subset\subset fU$ such that $1 = \mu(f(x_0), f, U) = \mu(y, f_j, U)$ for $y \in W$ and $j \geq j_0$. Let U_0 be a neighborhood of x_0 such that for some j_1, $f_j \overline{U}_0 \subset W$ for all $j \geq j_1$. Then $f_j|U_0$ is injective for all $j \geq j_1$ because $\mu(y, f_j, U) = 1$ for all $y \in f_j U_0$. The result then follows by an appeal to the corresponding result for qc mappings (see [V4, 37.2]). $\square$

We shall now apply Theorem 8.6 to prove that for $n \geq 3$ nonconstant qr mappings with suitably small dilatation are locally homeomorphic. We need a convergence result for the local indices.

8.13. Lemma. *Let $f_j\colon G \to \mathbb{R}^n$ be a sequence of sense–preserving discrete open mappings which converge to a discrete open mapping $f\colon G \to \mathbb{R}^n$ locally uniformly. Then*

$$i(x_0, f) \geq \limsup_{j \to \infty} i(x_0, f_j)$$

for every $x_0 \in G$.

Proof. The argument is similar to the one at the end of the proof of 8.6. We choose $r > 0$ such that $D = U(x_0, f, r)$ is a normal neighborhood of x_0 with respect to f. Then $i(x_0, f) = \mu(f(x_0), f, D)$. Since $f_j \to f$ uniformly on $\overline{D}$, there exists j_0 such that $\mu(f_j(x_0), f_j, D) = \mu(y_0, f, D) = i(x_0, f)$ for $j \geq j_0$. On the other hand, $\mu(f_j(x_0), f_j, D) = \sum\{\, i(x, f_j) : x \in D \cap f_j^{-1}(f_j(x_0)) \,\}$. Thus $\mu(f_j(x_0), f_j, D) \geq i(x_0, f_j)$, and the lemma follows. $\qquad\square$

8.14. Theorem. *For every $n \geq 3$ there exists $K > 1$ such that every nonconstant K-qr mapping $f\colon G \to \mathbb{R}^n$ is a local homeomorphism.*

Proof. Suppose the theorem is not true in some dimension $n \geq 3$. Then there exists a sequence of nonconstant K_j-qr mappings $f_j\colon G_j \to \mathbb{R}^n$ such that $K_j \leq 2$ and $B_{f_j} \neq \emptyset$ for all j, where $K_j \to 1$. By III.5.8 there exists $x_j \in B_{f_j}$ with $i(x_j, f_j) \leq 18$. From II.4.3 we then get $H(x_j, f_j) < C$ where C depends only on n. Performing similarity mappings we may assume with notation from II.4 that (1) $x_j = 0 = f_j(x_j)$, (2) $U(0, f_j, r)$ is a normal neighborhood of 0 for $0 < r \leq 1$, (3) $l^*(0, f_j, 1) = 1$, (4) $L(0, f_j, t) < Cl(0, f_j, t)$ for $0 < t \leq 1$. Condition (3) implies $\overline{B}^n \subset G_j$.

Since $f_j B^n \subset B^n$, it follows from III.2.7 that the restrictions $g_j = f_j | B^n$ form a normal family. Passing to a subsequence, we may therefore assume that $g_j \to g$ locally uniformly in B^n. By 8.6, $g\colon B^n \to \mathbb{R}^n$ is 1-qr, and hence is either a Möbius transformation or a constant by I.2.5. Since $0 \in B_{g_j}$, it follows from 8.13 and $g_j(0) = 0$ that $gB^n = \{0\}$.

Set $l_j = l(0, f_j, 1)$. Since $L(0, f_j, 1) = 1$, (4) implies that $l_j \geq 1/C$. Let E_j be the condenser $\bigl(U(0, f_j, l_j), \overline{U}(0, f_j, l_j/2)\bigr)$. By II.4.1(6), $L^*(0, f_j, l_j) = 1$. Hence we obtain the estimate $\operatorname{cap} E_j \geq \omega_{n-1}\bigl(\log 1/l_j^*\bigr)^{1-n}$ with $l_j^* = l^*(0, f_j, l_j/2)$. Since E_j is a normal condenser, II.10.9 gives

$$\operatorname{cap} E_j \leq K_j i(0, f_j) \operatorname{cap} f E_j \leq 36\omega_{n-1}(\log 2)^{1-n}.$$

We obtain $l_j^* \leq C_1 < 1$ where C_1 depends only on n. But the diameter $d(g_j \overline{B}^n(C_1))$ has the lower bound $l_j/2 \geq \frac{1}{2}C$, which contradicts the fact that $g_j \to 0$ uniformly on $\overline{B}^n(C_1)$. The theorem is proved. $\qquad\square$

8.15. Notes. The results 8.13 and 8.14 are from [MRV3]. Theorem 8.14 was also proved by V.M. Gol'dshtein [Go]. T. Iwaniec has in [I3] given a simple proof of 8.14 that relies on analytical aspects of the theory. The proof of 8.14 does not give any explicit lower bound for K. The same holds true for [Go] and [I3].

Chapter VII. Boundary Behavior

Many results of the boundary behavior of the theory of planar analytic functions have their counterparts for qr mappings for all dimensions $n \geq 2$. In the early stage of the development of the theory some of these counterparts were established by O. Martio and S. Rickman [MR1]. Later M. Vuorinen continued this line of research in a number of articles, see 2.3 and 7.4. The tool in [MR1] and mostly in Vuorinen's articles is the method of extremal length. S. Granlund, P. Lindqvist, and O. Martio introduced for a given variational kernel F in [GLM2] a substitute for the harmonic measure, called F-harmonic measure. This potential theoretic notion has turned out to be a useful tool when studying the properties of qr mappings near the boundary.

To this chapter we have selected some typical known results on boundary behavior of qr mappings. The question how large sets are removable for bounded qr mappings stayed open for a long time, but was recently solved in [IM] and [I4], see 1.18. In Section 1 we give the proof of the result that a set of zero n-capacity is removable (which is weaker than the optimal one). One of the main open problems for qr mappings of a ball is the question whether such a map, when bounded, must have some radial limits. In Section 2 we obtain radial limits almost everywhere for maps with a bound of the volume growth, i.e. growth of $A(r)$ (see IV.1). Section 3 is devoted to the proof that F-extremals are essentially continuous. Section 4 deals with the study of the continuity of F-extremals up to the boundary and in Section 5 we introduce the F-harmonic measure. In Section 6 we apply estimates of the F-harmonic measure to prove Phragmén–Lindelöf type theorems. We end the chapter by a discussion of Lindelöf's theorem for qr mappings in a ball.

1. Removability

In III.2.10 we proved that an isolated singularity is removable for a qm mapping if the mapping omits a set of positive capacity. Recall that the term capacity means n-capacity. In this section we first extend this result and show that sets of zero n-capacity are removable. Next we consider the connection between p-capacity and Hausdorff dimension. We end the section with a discussion of results on removability of sets of zero p-capacity, $p < n$.

1.1. Theorem. *Let G be a domain in $\overline{\mathbb{R}}^n$, let F be closed in G, and let $\operatorname{cap}_n F = 0$. If $f: G \setminus F \to \overline{\mathbb{R}}^n$ is a qm mapping such that $\operatorname{cap}_n(\overline{\mathbb{R}}^n \setminus f(G \setminus F)) > 0$, then f can be extended to a qm mapping $f^*: G \to \overline{\mathbb{R}}^n$. Moreover $K_I(f^*) = K_I(f)$, $K_O(f^*) = K_O(f)$.*

Proof. By III.2.9, f can be extended continuously to a map $f^*: G \to \overline{\mathbb{R}}^n$. We claim that f^* is qm. To prove this we may assume that G is a bounded domain in $\mathbb{R}^n$, that $|f^*| < 1$, and that $F = G \cap F_0$, where F_0 is compact and $\operatorname{cap}_n F_0 = 0$. By III.2.3, $m_n(F) = 0$. With the help of III.2.2 we find functions $\varphi_j \in C_0^\infty(\mathbb{R}^n)$, $j = 1, 2, \ldots$, with the properties

(1) $0 \le \varphi_j \le 1$,
(2) $\varphi_j | U_j = 1$ for some neighborhood U_j of F,
(3) $\varphi_j(x) \to 0$ for a.e. $x \in \mathbb{R}^n$,
(4) $\lim_{j \to \infty} \|\nabla \varphi_j\|_n = 0$.

Let $\psi \in C_0^\infty(G)$ be nonnegative and set $\psi_j = (1 - \varphi_j)\psi$. Then $\operatorname{spt} \psi_j \subset G \setminus F$. By VI.2.8 each component function f_k is an $F_1 = f^\sharp F_I$-extremal. By VI.3.14,

$$(1.2) \qquad \|\psi_j \nabla f_k\|_n \le C_1 \|f_k \nabla \psi_j\|_n \le C_1 \|\nabla \psi_j\|_n,$$

where C_1 depends only on n and $K(f)$. Write $M = \max |\psi|$. We have

$$(1.3) \qquad \begin{aligned} \|\nabla(\psi_j f_k)\|_n &\le \|\psi_j \nabla f_k\|_n + \|\nabla \psi_j\|_n \\ &\le (C_1 + 1)\|\nabla \psi_j\|_n \\ &\le (C_1 + 1)(\|\nabla \psi\|_n + M\|\nabla \varphi_j\|_n). \end{aligned}$$

By (4) and (1.3), the norm $\|\nabla(\psi_j f_k)\|_n$ stays bounded. By (1) and (3), $\psi_j f_k$ converges to ψf_k in $L^n(G)$. Hence by VI.7.9, $\psi_j f_k$ converges weakly in $W_n^1(G)$ to ψf_k and $\psi f_k \in W_n^1(G)$. We conclude that $f^* \in \operatorname{ACL}^n(G)$. Since $m_n(F) = 0$, f^* is qr and $K_O(f^*) = K_O(f)$, $K_I(f^*) = K_I(f)$. □

1.4. Remark. Theorem 1.1 was originally proved in [MRV2, 4.1] by showing that f^* is Hölder continuous and using the fact that the Hausdorff dimension $\dim_{\mathcal{H}} F$ is zero (which will be proved in 1.16), after which it follows by application of I.(4.12). The proof arrangement above has the advantage that with some extra methods it extends to sets F of p-capacity zero for some $p \in \,]0, n[$. We shall discuss this in 1.18. Inequality (1.2) is a so-called Caccioppoli type estimate and one of the main tasks to obtain stronger results is to obtain (1.2) for some $p < n$.

We now turn to the question of the connection of Hausdorff measure and p-capacity. Recall from I.1.1 the notation Λ_h for the h-Hausdorff measure for a function h. We first give a condition which implies $\operatorname{cap}_p F = 0$. We can present a simple proof thanks to the article [V6] by J. Väisälä.

1.5. Theorem. *Let* $1 < p \leq n$ *and for* $0 < r < 1$ *set*

$$h(r) = \begin{cases} r^{n-p}, & \text{if } 1 < p < n; \\ (\log \frac{1}{r})^{1-n}, & \text{if } p = n. \end{cases}$$

If $C \subset \mathbb{R}^n$ *is compact with* $\Lambda_h C < \infty$, *then* $\operatorname{cap}_p C = 0$.

Proof. For $0 < r < 1$ let E_r be the condenser $(B^n(C, r), C)$ where $B^n(C, r) = \{y \in \mathbb{R}^n : d(C, y) < r\}$. We first show that if

$$a = \sup_{0 < r < 1} \operatorname{cap}_p E_r = \lim_{r \to 0} \operatorname{cap}_p E_r$$

is finite, then $\operatorname{cap}_p C = 0$. By II.10.2, $\mathsf{M}_p(\Gamma_r) = \operatorname{cap}_p E_r$ where $\Gamma_r = \Gamma_{E_r}$. For $0 < s < r$ let $\Gamma_{s,r}$ be the family $\Delta(\partial B^n(C, s), \partial B^n(C, r); B^n(C, r) \setminus \overline{B}^n(C, s))$. Suppose $\operatorname{cap}_p C > 0$. By II.1.6(3),

$$(1.6) \qquad \mathsf{M}_p(\Gamma_r)^{1/(1-p)} \geq \mathsf{M}_p(\Gamma_s)^{1/(1-p)} + \mathsf{M}_p(\Gamma_{s,r})^{1/(1-p)}.$$

As $s \to 0$, $\mathsf{M}_p(\Gamma_s) \to a > 0$ and $\mathsf{M}_p(\Gamma_{s,r}) = \operatorname{cap}_p(B^n(C, r), \overline{B}^n(C, s)) \to \operatorname{cap}_p E_r = \mathsf{M}_p(\Gamma_r)$. Inequality (1.6) implies $a = \infty$, which proves our claim.

To proceed we first let $p = n$. It is enough to show that $\mathsf{M}_n(\Gamma_r)$ stays bounded for small $r > 0$. Let $0 < r < 1$ and let the balls $B^n(x_i, r_i)$, $i = 1, 2, \ldots$, with $r_i < r^2$ cover C such that

$$(1.7) \qquad \sum_i \left(\log \frac{1}{r_i}\right)^{1-n} \leq \Lambda_h(C) + 1.$$

Let $\Gamma_i = \Delta(S^{n-1}(x_i, r_i), S^{n-1}(x_i, r); B^n(x_i, r) \setminus \overline{B}^n(x_i, r_i))$. Then Γ_r is minorized by $\cup_i \Gamma_i$ and hence

$$(1.8) \qquad \begin{aligned} \mathsf{M}_n(\Gamma_r) &\leq \sum_i \mathsf{M}_n(\Gamma_i) = \omega_{n-1} \sum_i \left(\log \frac{r}{r_i}\right)^{1-n} \\ &\leq 2^{n-1}\omega_{n-1} \sum_i \left(\log \frac{1}{r_i}\right)^{1-n} \leq 2^{n-1}\omega_{n-1}(\Lambda_h(C) + 1). \end{aligned}$$

Then let $1 < p < n$ and write $\beta = (n-p)/(p-1)$. Now let $0 < r^\beta < \frac{1}{2}$ and cover C by balls as before with (1.7) replaced by

$$(1.9) \qquad \sum_i r_i^{n-p} \leq \Lambda_h(C) + 1.$$

By II.1.7(b) and $r_i^{\beta/2} < r^\beta < \frac{1}{2}$ we get

$$\mathsf{M}_p(\Gamma_i) = \omega_{n-1} \left(\frac{\beta}{r_i^{-\beta} - r^{-\beta}}\right)^{p-1} \leq a_0 r_i^{n-p},$$

where a_0 depends only on n and p. The boundedness of $\mathsf{M}_p(\Gamma_r)$ follows as in (1.8). The theorem is proved. $\qquad \square$

1.10. Corollary. *If* $1 < p < n$ *and* $F \subset \mathbb{R}^n$ *is a Borel set with* $\dim_{\mathcal{H}} F < n - p$, *then* $\text{cap}_p F = 0$.

To prove a converse statement to Theorem 1.5 we need some preliminary lemmas. For $\alpha > 0$ and $A \subset \mathbb{R}^n$ we let $\gamma_\alpha(A)$ be the infimum of the sums $\sum_j |B_j|^{\alpha/n}$ when A is covered by a countable number of balls B_j. Here $|B_j|$ is the n-measure of B_j. Our purpose is to estimate the outer measure γ_α, $0 < \alpha \le n$, by means of p-capacity, $p > n - \alpha$.

Let $V \subset \mathbb{R}^n$ be an open ball and $v \in L^p(V)$, $1 \le p < \infty$. For $\alpha > 0$ we define the Morrey type *maximal function* of v by

$$(1.11) \qquad M_{\alpha,p}v(x) = \sup_B \left(\frac{1}{|B|^{\alpha/n}} \int_B |v|^p dm \right)^{1/p},$$

where B runs over open balls with $x \in B \subset V$. This maximal function can be related to γ_α as follows.

1.12. Lemma. *If* v *is as above and* $\alpha > 0$, *then*

$$\gamma_\alpha(V_\lambda) \le \frac{5^\alpha}{\lambda^p} \|v\|_{p,V}^p$$

for every $\lambda > 0$ *where* $V_\lambda = \{ x \in V : M_{\alpha,p}v(x) > \lambda \}$.

Proof. For every $x \in V_\lambda$ there exists a ball B_x such that $x \in B_x \subset V$ and $\|v\|_{p,B_x}^p > \lambda^p |B_x|^{\alpha/n}$. By a well–known Vitali type covering lemma (see [St, 1.7, p. 9]) there exists a countable subset $\{ B_j : j = 1, 2, \ldots \}$ of $\{ B_x : x \in V_\lambda \}$ such that the balls B_j are disjoint and $V_\lambda \subset \bigcup_j B_j^*$ where B_j^* is the ball concentric with B_j and $d(B_j^*) = 5d(B_j)$. We obtain

$$\gamma_\alpha(V_\lambda) \le \sum_j |B_j^*|^{\alpha/n} \le \frac{5^\alpha}{\lambda^p} \sum_j \|v\|_{p,B_j}^p \le \frac{5^\alpha}{\lambda^p} \|v\|_{p,V}^p \,,$$

and the lemma is proved. $\qquad\qquad\square$

1.13. Lemma. *Let* V *be an open ball in* $\mathbb{R}^n$, *let* $u \in C_0^1(V)$, *and let* $1 < p < \infty$, $0 \le n - \alpha < p$. *Then there exists a constant* $C_0 < \infty$ *depending only on* n, p, *and* α *such that*

$$|u| \le C_0 d(V)^{1-n/p+\alpha/p} M_{\alpha,p} \nabla u \,.$$

Proof. We may assume $V = B^n(R)$. For $x \in V$ we write

$$u(x) = -\int_0^\infty \left(\frac{d}{dt} \fint_{B^n} u(x+ty)\,dy \right) dt = -\int_0^\infty \left(\fint_{B^n} y \cdot \nabla u(x+ty)\,dy \right) dt \,.$$

Hence, by Hölder's inequality,

$$|u(x)| \leq \int_0^\infty \left(\fint_{B^n} |\nabla u(x+ty)|\, dy \right) dt = \int_0^\infty \left(\fint_{B^n(x,t)} |\nabla u|\, dm \right) dt$$

$$\leq \int_0^R \left(\fint_{B^n(x,t)} |\nabla u|^p dm \right)^{\frac{1}{p}} dt + \left(\int_V |\nabla u|\, dm \right) \left(\int_R^\infty \frac{dt}{\Omega_n t^n} \right)$$

$$\leq \int_0^R \left(\frac{1}{|B^n(x,t)|^{\alpha/n}} \int_{B^n(x,t)} |\nabla u|^p dm \right)^{\frac{1}{p}} |B^n(x,t)|^{\frac{1}{p}(\frac{\alpha}{n}-1)} dt$$

$$+ \left(\frac{1}{|V|^{\alpha/n}} \int_V |\nabla u|^p dm \right)^{\frac{1}{p}} |V|^{\frac{1}{p}(\frac{\alpha}{n}-1)+1} \Omega_n^{-1}(n-1)^{-1} R^{1-n}$$

$$\leq \left(\frac{p}{\alpha-n+p} + \frac{1}{n-1} \right) \Omega_n^{\frac{1}{p}(\frac{\alpha}{n}-1)} R^{1-\frac{n}{p}+\frac{\alpha}{p}} M_{\alpha,p} \nabla u(x) \,.$$

The lemma is proved. $\qquad\qquad\qquad\qquad\qquad\qquad\qquad\qquad\qquad\qquad\qquad\qquad\square$

1.14. Lemma. *With assumptions in Lemma 1.13 we have*

$$\gamma_\alpha\{ x \in V : |u(x)| > \lambda \} \leq \frac{5^\alpha C_0^p d(V)^{p+\alpha-n}}{\lambda^p} \int_V |\nabla u|^p dm$$

for all $\lambda > 0$.

Proof. By 1.12 and 1.13 we obtain

$$\gamma_\alpha\{ x \in V : |u(x)| > \lambda \} \leq \gamma_\alpha\{ x \in V : M_{\alpha,p}\nabla u(x) > \lambda C_0^{-1} d(V)^{\frac{n}{p}-\frac{\alpha}{p}-1} \}$$

$$\leq 5^\alpha C_0^p \lambda^{-p} d(V)^{p+\alpha-n} \int_V |\nabla u|^p dm \,. \qquad\qquad\square$$

1.15. Theorem. *Let* $1 < p \leq n$, $\alpha > n-p$, *and let* $C \subset \mathbb{R}^n$ *be compact with* $\mathrm{cap}_p\, C = 0$. *Then* $\mathcal{H}^\alpha(C) = 0$.

Proof. Let the ball $V = B^n(R)$ contain C and let $u \in C_0^\infty(V)$ be a function with $u|C = 1$ and $0 \leq u \leq 1$. Then by 1.14,

$$\gamma_\alpha\{ x \in V : u(x) > \tfrac{1}{2} \} \leq 5^\alpha C_0^p 2^p d(V)^{p+\alpha-n} \int_V |\nabla u|^p dm \,.$$

This shows that $\gamma_\alpha(C) = 0$. Let $\delta, \varepsilon > 0$. Since $\gamma_\alpha(C) = 0$, we can cover C by balls $B^n(x_i, r_i)$, $i = 1, 2, \ldots$, such that $\sum_i \lambda_\alpha(2r_i)^\alpha < \min(\lambda_\alpha \delta^\alpha, \varepsilon)$, whence $2r_i < \delta$ for all i. It follows that $\mathcal{H}_\delta^\alpha(C) < \varepsilon$, hence $\mathcal{H}^\alpha(C) = 0$. $\quad\square$

1.16. Corollary. *If $1 < p \le n$ and $F \subset \mathbb{R}^n$ is a Borel set with $\mathrm{cap}_p F = 0$, then $\dim_{\mathcal{H}} F \le n - p$.*

1.17. Remark. For $p = n$ a result sharper than the one given in 1.15 is true, namely, $\mathrm{cap}_n C = 0$ implies $\Lambda_h C = 0$ where $h(r) = (\log(1/r))^{1-n-\varepsilon}$, $\varepsilon > 0$ (see [Wa2] for details).

1.18. Removability of sets with positive Hausdorff dimension. Let G be a domain in $\mathbb{R}^n$ and let $F \subset G$ be closed in G such that $G \smallsetminus F$ is a domain. For $n = 2$ the classical theorem by P. Painlevé and A.S. Besicovitch (see [CL, p. 5]) says that a bounded analytic function $f \colon G \smallsetminus F \to \mathbb{R}^2$ extends to an analytic function to G if $\mathcal{H}^1(F) = 0$. If $f \colon G \smallsetminus F \to \mathbb{R}^2$ is K-qr and bounded, it extends to a K-qr mapping of G if $\mathcal{H}^{1/K}(F) = 0$. This follows from the representation $f = \varphi \circ h$, where h is K-qc and φ analytic, and from the fact that h is locally Hölder continuous with exponent $1/K$.

Until very recently no general result for removing sets of positive Hausdorff dimension in connection with qr mappings for $n \ge 3$ was proved. Now the following is known.

1.19. Theorem. *There exists $\alpha = \alpha(n, K) \in {]0, n[}$ such that every bounded K-qr mapping $f \colon G \smallsetminus F \to \mathbb{R}^n$ with $\dim_{\mathcal{H}} F < \alpha$ has an extension to a K-qr mapping of G.*

This theorem was for even dimensions proved by T. Iwaniec and G. Martin in [IM] and for all dimensions by Iwaniec in [I4] (see also [IS] and [Le3]). We shall not prove 1.19. The proof is based on the use of Hodge theory on differential forms in connection with quasiregular mappings. The result in 1.19 is achieved from the following two steps. Following Iwaniec and Martin let us call a mapping $g \colon D \to \mathbb{R}^n$ in $W^1_{p,\mathrm{loc}}(D)$, where D is a domain in $\mathbb{R}^n$, *weakly* qr if for some $K \in [1, \infty[$,

(1) $J_g(x) \ge 0$ a.e.,
(2) $|Dg(x)| \le K \inf_{|h|=1} |Dg(x)h|$ a.e.

One step is to show that there exists $p = p(n, K) < n$ such that a weakly quasiregular mapping $g \in W^1_{p,\mathrm{loc}}(D)$ satisfies a Caccioppoli type estimate like (1.2) with the exponent p instead of n. Another step is to show that such a g is in fact in $W^1_{n,\mathrm{loc}}(D)$, and so g coincides outside a zero set with a qr mapping (see Theorem 3.9). Now suppose that with this p we have $\mathrm{cap}_p F = 0$ in 1.19. As in the proof of 1.1 one shows that $f \in W^1_{p,\mathrm{loc}}(G)$, which means that f is weakly qr and hence its continuous representative is qr. The needed connection between Hausdorff dimension and p-capacity is given in 1.10. In even dimensions linearity in the proof of 1.19 can be used to a certain extent. An essential tool in [IM] is a singular integral operator which is a counterpart of the Beurling–Ahlfors operator in the plane. Ideas in this direction originate from the article [DS] by S.K. Donaldson and D.P. Sullivan.

P. Järvi and M. Vuorinen proved in [JV] a removability theorem where F is a special Cantor set of positive Hausdorff dimension and $f \colon G \smallsetminus F \to \mathbb{R}^n$ is a K-qr mapping omitting a finite number $m = m(n, K)$ of points. The proof is based on the Picard–Schottky type theorem IV.3.3. This extends a theorem on analytic functions in the plane by L. Carleson [Car]. Given any positive integer p there exists by Theorem IV.2.2 a nonconstant $K(p)$-qr map of $\mathbb{R}^3$ into itself omitting p points. Since ∞ is an essential singularity for the map in IV.2.2, this shows for $n = 3$ that the bound $m(n, K)$ must tend to ∞ as $K \to \infty$.

Without proof we state a converse to Theorem 1.19 for $n = 3$.

1.20. Theorem[R17]. *For any $\alpha > 0$ there exists a Cantor set $F \subset \mathbb{R}^3$ with $\dim_{\mathcal{H}} F \leq \alpha$ and a bounded $K(\alpha)$-qr map $f: \mathbb{R}^3 \setminus F \to \mathbb{R}^3$ that does not extend continuously to any point of F.*

1.21. Notes. The proof of Theorem 1.1 follows the lines in [IM]. (Related results were given in [Mik2].) For $p = n$ Theorem 1.15 was proved by Yu.G. Reshetnyak [Re2, p. 646] (see too [Re7]). The proof arrangement for 1.15 is taken from [BI2]. For the history of Theorems 1.5 and 1.15, see [Wa1], [Wa2], [V6]. P. Koskela and O. Martio [KM] have studied the removability question with the additional assumption that the map is Hölder continuous in G. In [KM] they also improve slightly Theorem 1.1. Substantial progress in the removability question is obtained in [IM], [I4], [IS], and [L3] (cf. 1.19).

2. Asymptotic and Radial Limits

One of the main open questions in boundary behavior of qr mappings of the unit ball in space is the existence of radial limits if the mapping is bounded. Fatou's theorem says that a bounded analytic function of the unit disk has radial limits at almost every boundary point. For 2-dimensional qr mappings we get radial limits in general only in a smaller set, the reason being that a qc mapping of the unit disk onto itself can be singular on the boundary with respect to the linear measure. In space we know existence of radial limits only in cases were further restrictions are posed on the mapping. One such result is presented in Theorem 2.7. By a normal family argument we prove that radial limit implies the existence of angular (or conical) limit (Theorem 2.1).

Results on the existence of asymptotic limits have been studied rather successfully in various situations. For qr mappings of a ball omitting a set of positive capacity we get asymptotic limits on the boundary in a set which is dense with respect to capacity. We shall return to questions of existence of asymptotic limits in Section 7. We close the section by proving a counterpart of a theorem by F. and M. Riesz.

If $y \in S^{n-1}$ and $0 < \varphi < \pi/2$, we set

$$K(y, \varphi) = \{ x \in \mathbb{R}^n : y \cdot (y - x) > |y - x| \cos \varphi \} .$$

2.1. Theorem. Let $f\colon B^n \to \overline{\mathbb{R}}^n$ be a qm mapping with $\operatorname{cap} \mathbf{C} fB^n > 0$ and let $y \in \partial B^n$. Suppose that the radial limit

$$b = \lim_{t \to 1} f(ty)$$

exists. Then also the angular limit

(2.2)
$$\lim_{\substack{x \to y \\ x \in K(y,\varphi)}} f(x)$$

exists and equals b for every $\varphi \in \,]0, \pi/2[$.

Proof. Let $0 < \varphi < \varphi_0 < \pi/2$ and set $K_0 = K(y, \varphi_0) \cap B^n(y, 1)$. Suppose that the limit (2.2) does not exist. Then there is a sequence (x_i) of points in $K(y, \varphi)$ converging to y with $f(x_i) \to b_1 \neq b$. We may assume that $K_i = K(y, \varphi_0) \cap B^n(y, 2|y - x_i|) \subset B^n$ for all i. Let h_i be a similarity map of K_0 onto K_i. By III.2.7 and Ascoli's theorem the family of maps $g_i = f \circ h_i$ is normal. Hence by VI.8.6 there exists a subsequence of (g_i) converging to a qm map g. There exists a point $z \in \overline{K}(y, \varphi) \cap S^{n-1}(y, \frac{1}{2}) \subset K_0$ which is a limit of a subsequence of $(h_i^{-1}(x_i))$. Then $g(z) = b_1$. But g maps the whole segment $[0, y] \cap K_0$ on the point b. This is possible only if g is constant. This contradiction proves the theorem. $\square$

2.3. Remark. M. Vuorinen has in [Vu4, 3.1] proved a variant of 2.1 for bounded mappings where the assumption of a radial limit is replaced by the existence of a limit

$$b = \lim_{\substack{x \to y \\ x \in E}} f(x) \,,$$

where E is contained in a cone $K(y, \psi)$ and the *lower capacity density* $\operatorname{cap} \underline{\operatorname{dens}}(E, 0) > 0$. Here

$$\operatorname{cap} \underline{\operatorname{dens}}(E, z) = \liminf_{\rho \to 0} \gamma(z, E, \rho)$$

and

$$\gamma(z, E, \rho) = \mathsf{M}\big(\Delta(S^{n-1}(z, 2r), \overline{B}^n(z, r) \cap E; \overline{B}^n(2r))\big) \,.$$

The proof makes use of a 2-constants theorem which we will present in 6.18. See too [Vu3, 4.7 and 5.3].

If $f\colon G \to \overline{\mathbb{R}}^n$ is a qm mapping and $b \in \partial G$, we say that f has an *asymptotic limit* at b if $c = \lim_{t \to 1} f(\gamma(t))$ exists for some path $\gamma\colon [0, 1[\to G$ with $\gamma(t) \to b$ as $t \to 1$. We recall from V.9.2 that c is also called an *asymptotic value*.

2.4. Theorem. *Let* $f: B^n \to \overline{\mathbb{R}}^n$ *be a K-qm mapping with* $\mathrm{cap}(\overline{\mathbb{R}}^n \smallsetminus fB^n) > 0$. *If* $E \subset S^{n-1}$ *is the set of points at which* f *has some asymptotic limit, then* $\mathrm{cap}(E \cap B^n(y, \varepsilon)) > 0$ *for all* $y \in S^{n-1}$ *and* $\varepsilon > 0$.

Proof. Suppose there exists $y \in S^{n-1}$ and $\varepsilon > 0$ such that $\mathrm{cap}(E \cap B^n(y, \varepsilon)) = 0$. Then there exists a path $\beta: [0, 1[\to B^n$ such that $\beta(t) \to c \in S^{n-1} \cap B^n(y, \varepsilon/2)$ as $t \to 1$ and $\lim_{t \to 1} f(\beta(t))$ does not exist. We find a sequence $s_1 < s_2 < \ldots$ of positive numbers such that $s_i \to 1$ and

$$\lim_{k \to \infty} f(\beta(s_{2k})) = u \neq \lim_{k \to \infty} f(\beta(s_{2k+1})) = v .$$

We may assume that the spherical diameter $q(F_k)$ of $F_k = f(\beta[s_{2k}, s_{2k+1}])$ exceeds a number $a > 0$ for all k. Let Γ_k be the family of rectifiable paths $\alpha: [0, 1] \to \overline{\mathbb{R}}^n$ with $\alpha(0) \in F_k$ and $\alpha(1) \in \overline{\mathbb{R}}^n \smallsetminus fB^n$. Then by III.2.6 there is $\delta > 0$ such that $\mathsf{M}(\Gamma_k) \geq \delta$ for all k. We define $\rho \in\,]0, \varepsilon[$ by the formula

(2.5) $$\frac{K\omega_{n-1}}{(\log \frac{\varepsilon}{2\rho})^{n-1}} = \frac{\delta}{2} .$$

Let Γ_k^* be the family of maximal rectifiable liftings of the paths in Γ_k starting at some point in $\beta[s_{2k}, s_{2k+1}]$. Let m be the first k such that $\beta[s_{2k}, s_{2k+1}] \subset B^n(c, \rho)$ and let Γ_m' be the subfamily of paths in Γ_m^* that end outside $B^n(y, \varepsilon)$. Then $\mathsf{M}(\Gamma_m') \leq \omega_{n-1}(\log(\varepsilon/2\rho))^{1-n} = \delta/2K$ by (2.5). But

$$\delta \leq \mathsf{M}(\Gamma_m) \leq \mathsf{M}(f\Gamma_m^*) \leq K\,\mathsf{M}(\Gamma_m^*) ,$$

and hence $\mathsf{M}(\Gamma_m^* \smallsetminus \Gamma_m') \geq \delta/2K$. The set A of endpoints of the paths in $\Gamma_m^* \smallsetminus \Gamma_m'$ is contained in $B^n(y, \varepsilon) \cap S^{n-1}$ and $\mathrm{cap}\, A > 0$. But f has a limit along each path in $\Gamma_m^* \smallsetminus \Gamma_m'$, and the theorem follows. $\square$

For an isolated boundary point we get the following counterpart to a theorem by F. Iversen.

2.6. Theorem. *Let* $f: G \to \overline{\mathbb{R}}^n$ *be a K-qm mapping and let* $b \in \partial G$ *be an isolated essential singularity of* f. *Then every point in* $\overline{\mathbb{R}}^n \smallsetminus fG$ *is an asymptotic value of* f.

Proof. Let $z \in \overline{\mathbb{R}}^n \smallsetminus fG$. We may assume $z = 0$. Choose $r_0 > 0$ such that $\overline{B}^n(b, r_0) \subset G \cup \{b\}$ and set $U_0 = B^n(b, r_0) \smallsetminus \{b\}$. Since $0 \notin fG$, there exists $r' > 0$ such that $\overline{B}^n(r') \cap fS^{n-1}(b, r_0) = \emptyset$. By III.2.11 we can find a cap $C \subset S^{n-1}(r')$ and a component C^* of $f^{-1}C$ in U_0. For $y \in S^{n-1}$ let $\gamma_y: \,]0, r'] \to \overline{B}^n(r')$ be the path $\gamma_y(t) = ty$. For $r'y \in C$ we let γ_y^* be a maximal lifting of γ_y terminating in C^*. Each γ_y^* is a path $\gamma_y^*: \,]r_y, r'] \to U_0$ such that $\gamma_y^*(t) \to b$ as $t \to r_y$. If $r_y = 0$ for some y, then 0 is an asymptotic value of f. In fact, we show that $r_y = 0$ for almost all $r'y \in C$.

Let $E_i = \{y \in S^{n-1} : r'y \in C, r_y > 1/i\}$, $i = 1, 2, \ldots$. It is enough to show $\mathcal{H}^{n-1}(E_i) = 0$ for each i. Fix i and set $\Gamma_i = \{\gamma_y^* : y \in E_i\}$. Then $\mathsf{M}(\Gamma_i) = 0$, and so $\mathsf{M}(f\Gamma_i) = 0$ by II.8.1. If $\rho \in \mathcal{F}(f\Gamma_i)$, we have

$$\int \rho^n \, dm \geq \int\limits_{S^{n-1}} \left(\int\limits_{1/i}^{r'} t^{n-1} \rho(ty) \, dt \right) dy \geq \frac{1}{i^{n-1}} \mathcal{H}^{n-1}(F_\rho) \,,$$

where

$$F_\rho = \{\, y \in S^{n-1} : \int\limits_{1/i}^{r'} \rho(ty) \, dt \geq 1 \,\} \,.$$

Then $E_i \subset F_\rho$. Since $\mathsf{M}(f\Gamma_i) = 0$, $\mathcal{H}^{n-1}(F_\rho)$ can be made arbitrarily small, and so $\mathcal{H}^{n-1}(E_i) = 0$. $\qquad\square$

Next we prove a Fatou–type theorem which states that radial limits exist almost everywhere if a suitable bound is put on the volume growth, i.e., on $A(r)$ (see II.1).

2.7. Theorem. *Let* $f \colon B^n \to \overline{\mathbb{R}}^n$ *be a* K-qm *mapping such that*

$$(2.8) \qquad \int\limits_0^1 A(r)(1-r)^{n-2} \left(\log \frac{1}{1-r} \right)^{n+\delta} dr < \infty$$

for some $\delta > 0$. *Then* f *has radial limits a.e. in* S^{n-1}.

Proof. We may assume that f is not constant. For $y \in S^{n-1}$ and for any positive integer k we let $\gamma_{y,k}$ be the path $\gamma_y|[r_k, r_{k+1}]$, where $\gamma_y(t) = ty$ and $r_k = 1 - 2^{-k}$. Set

$$E_k = \{\, y \in S^{n-1} : \sigma(f|\gamma_{y,k}|) \geq k^{-1-\delta/n} \,\}$$

where we recall the notation σ for the spherical metric. Let $\Gamma_k = \{\, \gamma_{y,k} : y \in E_k \,\}$. The constant function $\rho_k = k^{1+\delta/n}$ of $\overline{\mathbb{R}}^n$ is admissible (with respect to the spherical metric) for the family $f\Gamma_k$. Inequality II.(2.6) applied to the spherical measure then gives

$$(2.9) \qquad \begin{aligned} \mathsf{M}(\Gamma_k) &\leq K\,k^{n+\delta} \int\limits_{\overline{\mathbf{R}}^n} n(r_{k+1}, z) \, d\sigma(z) \\ &= K\,k^{n+\delta} 2^{-n} \omega_n A(r_{k+1}) \,. \end{aligned}$$

On the other hand,

$$(2.10) \qquad \mathsf{M}(\Gamma_k) = \mathcal{H}^{n-1}(E_k) \left(\log \frac{r_{k+1}}{r_k} \right)^{1-n} \geq \frac{\mathcal{H}^{n-1}(E_k)}{(1-r_k)^{n-1}} \,.$$

By using the relationships $1 - r_k = 2^j(1 - r_{k+j}) = 2^j(r_{k+j} - r_{k+j-1})$ and $k = (\log(1 - r_k)^{-1})/\log 2$ we obtain from (2.9) and (2.10) the estimate

$$\sum_{k \geq m} \mathcal{H}^{n-1}(E_k)$$

$$\leq cK \sum_{k \geq m} A(r_{k+1}) \left(\log \frac{1}{1 - r_{k+1}} \right)^{n+\delta} (1 - r_{k+2})^{n-2} (r_{k+2} - r_{k+1})$$

$$\leq cK \int_{r_{m+1}}^{1} A(r) \left(\log \frac{1}{1 - r} \right)^{n+\delta} (1 - r)^{n-2} dr$$

where c is a constant depending only on n. By (2.8) the integral above tends to 0 as $m \to \infty$. If m is any positive integer and $y \in S^{n-1} \setminus \bigcup_{k \geq m} E_k$, the radial limit exists at y by the definition of the sets E_k. The theorem is proved. $\qquad \square$

For a meromorphic function f of the disk radial limits exist a.e. if (see [N2, p. 204])

$$(2.11) \qquad \int_0^1 A(r)\, dr < \infty .$$

This condition is fullfilled for example if f is bounded. If q is an integer exceeding the number $C(n, K)$ in V.1.9 and f is a K-qm mapping of B^n omitting q points in $\overline{\mathbb{R}}^n$, then by V.1.9, $A(r) = O((1 - r)^{1-n})$. At the moment no better bound for the growth of $A(r)$ is known if the qr mapping is for example bounded. On the other hand, as mentioned in the beginning of this section, for $n = 2$ radial limits do not necessarily exist a.e. for bounded qr mappings of a disk.

With a somewhat similar technique we obtain in Theorem 2.12 a counterpart to the theorem of F. and M. Riesz, which for a bounded analytic function f of the disk says that f cannot have a constant value of radial limits in a set of positive measure on the unit circle without being constant. The statement is true more generally for meromorphic functions satisfying (2.11) (see [N2, p. 205]). Our condition (2.13) is formulated in terms of the multiplicity function $N(f, \overline{B}^n(r)) = \sup_y \operatorname{card}(f^{-1}(y) \cap \overline{B}^n(r))$ instead of $A(r)$.

2.12. Theorem. *Let* $f \colon B^n \to \overline{\mathbb{R}}^n$ *be a K-qm mapping such that*

$$(2.13) \qquad \int_0^1 N(f, \overline{B}^n(r))(1 - r)^{n-2} \left(\log \frac{1}{1 - r} \right)^{n-1+\delta} dr < \infty$$

for some $\delta > 0$. *Suppose there exists a measurable set* $E \subset S^{n-1}$ *with* $\mathcal{H}^{n-1}(E) > 0$ *such that the radial limit* $\lim_{t \to 1} f(ty)$ *exists and equals a constant* b *for all* $y \in E$. *Then* f *is constant.*

Proof. Suppose f is nonconstant. Fix $\delta > 0$ such that (2.13) is true and let the integral in (2.13) be $I(\delta)$. As in the proof of 2.7 we write $r_k = 1 - 2^{-k}$

and $\gamma_{y,k} = \gamma_y|[r_k, r_{k+1}]$ for $y \in S^{n-1}$ and $k = 1, 2, \ldots$. We may assume $b = 0$. If $\rho > 0$, we write

$$E(k, \rho) = E \cap \pi(f^{-1}\overline{B}^n(\rho) \cap S^{n-1}(r_k))$$

where $\pi: B^n \smallsetminus \{0\} \to S^{n-1}$ is the radial projection.

We will define inductively a sequence $\rho_1 > \rho_2 > \ldots$ of positive numbers as follows. Since f is nonconstant, we can by I.4.9 choose ρ_1 such that $\mathcal{H}^{n-1}(E(1, \rho_1)) < \mathcal{H}^{n-1}(E)/4$. We choose ρ_k, $k \geq 2$, such that

$$(2.14) \qquad \left(\log \frac{\rho_k}{\rho_{k+1}}\right)^{1-n} = c_1 \left(\log \frac{1}{1 - r_{k+1}}\right)^{n-1+\delta}, \quad k \geq 1,$$

where c_1 is a positive constant independent of k and is chosen later. We write

$$F_k = E(k+1, \rho_{k+1}) \smallsetminus \bigcup_{i=1}^{k} E(i, \rho_i)$$

and

$$\Gamma_k = \{\gamma_{y,k} : y \in F_k\}.$$

By the definition of the set F_k every path in $f\Gamma_k$ joins the sets $\mathbb{R}^n \smallsetminus B^n(\rho_k)$ and $\overline{B}^n(\rho_{k+1})$, hence

$$2^{k(n-1)}\mathcal{H}^{n-1}(F_k) \leq \mathsf{M}(\Gamma_k) \leq \omega_{n-1} K\, N(f, \overline{B}^n(r_{k+1}))\left(\log \frac{\rho_k}{\rho_{k+1}}\right)^{1-n}.$$

With (2.14) and the remarks in the proof of 2.7 we then obtain

$$\sum_{k \geq 1} \mathcal{H}^{n-1}(F_k)$$

$$\leq c_2 c_1 \sum_{k \geq 1} N(f, \overline{B}^n(r_{k+1}))(1 - r_{k+2})^{n-2}\left(\log \frac{1}{1 - r_{k+1}}\right)^{n-1+\delta}(r_{k+2} - r_{k+1})$$

$$\leq c_2 c_1 \int_0^1 N(f, \overline{B}^n(r))(1 - r)^{n-2}\left(\log \frac{1}{1 - r}\right)^{n-1+\delta} dr$$

$$= c_2 c_1 I(\delta)$$

where $c_2 > 0$ depends only on n. We now choose c_1 so that $c_2 c_1 I(\delta) = \mathcal{H}^{n-1}(E)/4$. Then

$$\mathcal{H}^{n-1}\left(\bigcup_{k=1}^{\infty} E(k, \rho)\right) \leq \tfrac{1}{2}\mathcal{H}^{n-1}(E)$$

and there is a point $y \in E \smallsetminus \bigcup_{k \geq 1} E(k, \rho_k)$. On the other hand,

$$\lim_{k \to \infty} \log \frac{\rho_1}{\rho_k} = \sum_{k \geq 1} \log \frac{\rho_k}{\rho_{k+1}} = c_3 \sum_{k \geq 1} \frac{1}{(k+1)^{1+\delta/(n-1)}}$$

with some constant $c_3 < \infty$, hence $\lim_{k \to \infty} \rho_k > 0$. According to the definition of the sets $E(k, \rho_k)$ we have $f(r_k y) \notin \overline{B}^n(\rho_k)$. This contradicts the assumption that the radial limit at y is the origin.   □

2.15. Notes. Theorem 2.1 is from [MR1, 5.8]. Theorem 2.4 is a stronger formulation of [MR1, 5.11] where we claimed only that asymptotic limits exist in a dense set. However, the proof of 2.4 is essentially contained in [MR1]. The proof of 2.6 is essentially contained in [MRV3, 3.15]. Theorem 2.7 is a stronger form of [MR1, 5.15] in several respects. In [MR1, 5.15] the mappings are bounded, and instead of a bound on the growth of $A(r)$, a stronger condition on the maximal multiplicity is required. Theorem 2.12 is a stronger form of [MR1, 5.17].

3. Continuity Results and the Reflection Principle

In this section we shall prove that we can relax the continuity in the definition of quasiregularity (I.2.1) and still remain essentially in the same class of mappings. We need such a result in the proof of a reflection principle. In this connection it is also convenient to show that an F-extremal as defined in VI.1.10 is always essentially continuous. As a tool we need a form of Morrey's lemma.

3.1. Morrey's lemma. *Let D be a bounded domain in $\mathbb{R}^n$, $u \in W_1^1(D)$, and suppose there are constants $C > 0$ and $\mu \in \,]0, 1]$ such that*

$$(3.2) \qquad I_1(x, u, s) = \int\limits_{B(x,s)} |\nabla u|\, dm \leq C s^{n-1+\mu}$$

for all balls $\overline{B}^n(x, s) \subset D$. Then there is a continuous $v \colon D \to \mathbb{R}^1$ and $M(n, \mu) < \infty$ such that $v = u$ a.e. and

$$(3.3) \qquad |v(x) - v(y)| \leq M(n, \mu) C |x - y|^\mu$$

for $x, y \in V$ and any ball $V = B^n(\xi, r)$ such that $B^n(\xi, 2r) \subset D$.

Proof. We first assume $u \in C^1$. Let $V = B^n(\xi, r)$ be a ball such that $B^n(\xi, 2r) \subset D$. For $x, z \in \overline{V}$ write

$$u(z) - u(x) = \int\limits_0^{|z-x|} w \cdot \nabla u(x + sw)\, ds$$

where $w = (z - x)/|z - x|$. We integrate with respect to z over V and obtain

$$u_V - u(x) = \frac{1}{|V|} \int_V \int_0^{|z-x|} w \cdot \nabla u(x + sw)\, ds\, dz \ .$$

Now we set

$$v(s, w) = |\nabla u| \chi_V (x + sw) \ .$$

Then

$$|u_V - u(x)| \le \frac{1}{|V|} \int_0^\infty \int_{S^{n-1}} \int_0^{2r} v(s, w) \rho^{n-1} d\rho\, dw\, ds$$

(3.4)
$$= \frac{2^n r^n}{n|V|} \int_0^\infty \int_{S^{n-1}} v(s, w)\, dw\, ds$$

$$= \frac{2^n r^n}{n|V|} \int_V |z - x|^{1-n} |\nabla u(z)|\, dz \ .$$

We write $I_1(s) = I_1(x, u, s)$ and get from (3.2) the estimate

$$\int_V |z - x|^{1-n} |\nabla u(z)|\, dz \le \int_0^{2r} s^{1-n} \int_{S(x,s)} |\nabla u|\, d\mathcal{H}^{n-1}\, ds$$

$$= \int_0^{2r} s^{1-n} I_1'(s)\, ds = (2r)^{1-n} I_1(2r) + (n-1) \int_0^{2r} s^{-n} I_1(s)\, ds$$

$$\le C\left(1 + \frac{n-1}{\mu}\right) 2^\mu r^\mu \ .$$

Combining this with (3.4) we obtain an inequality

$$|u_V - u(x)| \le C b(n, \mu) r^\mu$$

which implies a Hölder inequality for u of the form (3.3).

To treat the general case let $D' \subset\subset D$ and let $\varphi \in \mathcal{C}_0^\infty(D)$ be such that $\varphi|(D' + B^n(\delta)) = 1$ where $\delta = \frac{1}{2} d(D', \partial D)$. Then by well–known estimates (see [Z2, 1.6.1])

$$\int_{B(x,s)} |\nabla u_\varepsilon|\, dm = \int_{B(x,s)} |\nabla(\varphi u)_\varepsilon|\, dm$$

$$\le \int_{B(x,s)} |\nabla(\varphi u)|\, dm = \int_{B(x,s)} |\nabla u|\, dm$$

for $B^n(x, s) \subset D'$ and $\varepsilon \le \delta$. Here u_ε is the regularization of u as defined in I.1. From the first part of the proof we conclude that each u_ε, $\varepsilon \le \delta$, satisfies (3.3) with D replaced by D'. Hence the family $\{u_\varepsilon : 0 < \varepsilon \le \delta\}$

is equicontinuous in D'. By the diagonal method we thus obtain a limit function $v: D \to \mathbb{R}^1$ of a sequence (u_{ε_j}) with $\varepsilon_j \to 0$ that satisfies (3.3) in D. Moreover, $v = u$ a.e. The lemma is proved.   □

In III.2.1 we quoted one form of Poincaré's inequality for functions in $W_{p,0}^1(B^n(r))$. Here we need the following form, the proof of which can be deduced from [GT, (7.45)].

3.5. Lemma. Let $D_r = B^n(2r) \setminus \overline{B}^n(r)$, let $1 \le p < \infty$, and let $u \in W_p^1(D_r)$. Then

$$\int\limits_{D_r} |u - u_{D_r}|^p dm \le d_0 r^p \int\limits_{D_r} |\nabla u|^p dm$$

where u_{D_r} is the average of u over D_r and d_0 is a constant depending only on n.

The essential continuity of F-extremals can now be given a simple proof. This is due to the fact that we are dealing with the borderline case $p = n$.

3.6. Theorem. Let $U \subset \mathbb{R}^n$ be open, let F be a variational kernel in U, and let $u \in W_{n,\mathrm{loc}}^1(U)$ be an F-extremal. Then u is essentially continuous, i.e., there exists a continuous function v in U such that $v = u$ a.e.

Proof. Let $x_0 \in U$ and let $r_0 > 0$ be such that $\overline{B}^n(x_0, 4r_0) \subset U$. It is enough to show the existence of v in $B^n(x_0, r_0)$. We are going to apply Morrey's lemma to $B^n(x_0, r_0)$. Fix $x \in B^n(x_0, r_0)$ and write $V_r = B^n(x, r)$ for $r > 0$. Let $0 < r \le r_0$ and let $\varphi \in \mathcal{C}_0^\infty(V_{2r})$ be such that $0 \le \varphi \le 0$, $\varphi|V_r = 1$, and $|\nabla \varphi| \le 2/r$. Then VI.(3.15) gives

$$I_n(r) = \int\limits_{V_r} |\nabla u|^n dm \le \frac{n^n \beta}{\alpha} \int\limits_{V_{2r}} |u - u_A|^n |\nabla \varphi|^n dm$$

$$\le \frac{2^n n^n \beta}{r^n \alpha} \int\limits_{V_{2r} \setminus V_r} |u - u_A|^n dm .$$

Using Poincaré's inequality 3.5 we obtain

$$I_n(r) \le \lambda \big(I_n(2r) - I_n(r) \big) ,$$

and so

(3.7) $$I_n(r) \le \frac{\lambda}{1 + \lambda} I_n(2r) ,$$

where $\lambda = \beta \alpha^{-1} d_1$ and d_1 depends only on n. By iteration of (3.7) we get an inequality

$$(3.8) \qquad I_n(s) \le \left(\frac{s}{r_0}\right)^\nu \int\limits_{B(x_0,4r_0)} |\nabla u|^n dm \,,$$

where

$$\nu = \frac{1}{\log 2} \log \frac{1+\lambda}{\lambda} > 0 \,.$$

Hölder's inequality gives

$$\int\limits_{B(x,s)} |\nabla u| \, dm \le I_n(s)^{1/n} \Omega_n^{(n-1)/n} s^{n-1} \,,$$

which together with (3.8) shows that the condition in Morrey's lemma 3.1 is satisfied for $D = B^n(x_0, r_0)$ with $\mu = \nu/n$. The theorem is proved. $\qquad\square$

In our definition of quasiregularity (I.2.1) we assume that the mapping f is in $\mathrm{ACL}^n(G)$. If we replace this condition by $f \in W^1_{n,\mathrm{loc}}(G)$, then f coincides up to a nullset with a quasiregular mapping (cf. the discussion in connection with Theorem 1.19). This result could be proved by applying Theorem 3.6 to coordinate functions f_j together with results in VI.2 formulated for Sobolev spaces. However, we prefer to give a proof where the condition in Morrey's lemma for each coordinate function is obtained directly without reference to extremals of variational integrals.

3.9. Theorem. *Let G be a domain in $\mathbb{R}^n$, $n \ge 2$, and let $f \colon G \to \mathbb{R}^n$ be a mapping such that*

(1) $f \in W^1_{n,\mathrm{loc}}(G)$, *and*
(2) *there exists K, $1 \le K < \infty$, such that*

$$|f'(x)|^n \le K J_f(x) \quad \text{a.e.}$$

Then there is a qr mapping $h \colon G \to \mathbb{R}^n$ such that $h = f$ a.e. and $K_O(h) \le K$.

Proof. We will show that each component f_i satisfies locally an inequality of the form (3.2). Let $D \subset\subset G$. Set $r_0 = \min(\frac{1}{2}d(D, \partial G), 1)$ and $D' = D + B^n(r_0)$. We choose a sequence $g_j \colon G \to \mathbb{R}^n$, $j = 1, 2, \ldots$, in $C^\infty(G)$ such that $g_j|D' \to f|D'$ in W^1_n. Let $x \in D$ and $V = B^n(x, r)$ where $0 < r \le r_x = d(x, \partial D')$. We write $M_j = \max_{x \in \partial V} g_{j1}(x)$. The elementary Stokes theorem gives

$$\int\limits_V J_{g_j} \, dm = \int\limits_{\partial V} g_{j1} \wedge dg_{j2} \wedge \ldots \wedge g_{jn}$$

$$(3.10) \qquad = \int\limits_{\partial V} (g_{j1} - M_j) \wedge dg_{j2} \wedge \ldots \wedge dg_{jn}$$

$$\le \mathrm{osc}(g_{j1}, \partial V) \int\limits_{\partial V} |g_j'|^{n-1} d\mathcal{H}^{n-1} \,.$$

By II.1.11,

$$\text{(3.11)} \qquad \operatorname{osc}(g_{j1}, \partial V)^n \le A_n r \int_{\partial V} |g_j'|^n d\mathcal{H}^{n-1} \ .$$

By Hölder's inequality we get from (3.10) and (3.11) the estimate

$$\int_V J_{g_j} \, dm$$

$$\text{(3.12)} \qquad \le A_n^{\frac{1}{n}} r^{\frac{1}{n}} \left(\int_{\partial V} |g_j'|^n d\mathcal{H}^{n-1} \right)^{\frac{1}{n}} \left(\int_{\partial V} |g_j'|^n d\mathcal{H}^{n-1} \right)^{\frac{n-1}{n}} \omega_{n-1}^{\frac{1}{n}} r^{\frac{n-1}{n}}$$

$$= C_n r \int_{\partial V} |g_j'|^n d\mathcal{H}^{n-1}$$

where $C_n = A_n^{1/n} \omega_{n-1}^{1/n}$. Since $g_j|D' \to f|D'$ in W_n^1,

$$\lim_{j \to \infty} \int_V J_{g_j} \, dm = \int_V J_f \, dm$$

and

$$\int_{S(x,r)} |g_j'|^n d\mathcal{H}^{n-1} \longrightarrow \int_{S(x,r)} |f'|^n d\mathcal{H}^{n-1}$$

in $L^1[0, r_x]$ as functions in r. It follows from (3.12) that

$$\int_{B(x,r)} J_f \, dm \le C_n r \int_{S(x,r)} |f'|^n d\mathcal{H}^{n-1}$$

for almost every $r \in [0, r_x]$. On the other hand, from (2) we then get

$$\text{(3.13)} \qquad H(r) = \int_{B(x,r)} |f'|^n dm \le K C_n r \int_{S(x,r)} |f'|^n d\mathcal{H}^{n-1}$$

for almost every $r \in [0, r_x]$. The second integral in (3.13) is $H'(r)$ a.e., hence

$$\int_r^{r_x} \frac{dr}{r} \le K C_n \int_{H(r)}^{H(r_x)} \frac{dH}{H} \ .$$

We may assume $H(r) > 0$ for $r \in \,]0, r_x]$. We obtain

$$\text{(3.14)} \qquad H(r) \le r^\alpha r_x^{-\alpha} H(r_x) \le \gamma(D, f) r^\alpha , \qquad 0 \le r \le r_x ,$$

where $\alpha = (K C_n)^{-1}$ and

$$\gamma(D, f) = r_0^{-\alpha} \int_{D'} |f'|^n dm \ .$$

Finally, Hölder's inequality and (3.14) give

$$\int_{B(x,r)} |\nabla f_i|\, dm \le \int_{B(x,r)} |f'|\, dm \le H(r)^{1/n} \Omega_n^{(n-1)/n} r^{n-1}$$

$$\le \gamma(D, f)^{1/n} \Omega_n^{(n-1)/n} r^{n-1+\mu}$$

where $\mu = \alpha/n$. This proves (3.2) in D for $u = f_i$.

Morrey's lemma 3.1 implies that there is a continuous representative h in the class of f. Then h is in ACL^n, the weak partial derivatives of h coincide with those of f a.e. Hence h is qr and $K_O(h) \le K$. The theorem is proved.   □

3.15. Remark. The application of Morrey's lemma 3.1 also gives a direct proof of the Hölder continuity of F-extremals and qr mappings. However, the Hölder exponent is not the optimal one as it is for qr mappings in III.1.11.

We state the reflection principle in the following form.

3.16. Theorem. *Let G be a domain in B^n and let $E \subset \partial G \cap S^{n-1}$ be a nonempty set which is open in S^{n-1}. Suppose $f\colon G \to B^n$ is a K-qr mapping which satisfies*

$$(3.17) \qquad\qquad \lim_{x \to y} |f(x)| = 1$$

for all $y \in E$. Then there is an extension of f to a K-qm mapping g of the domain $G' = G \cup E \cup \alpha G$ such that $g(x) = \alpha(f(\alpha(x)))$ for $x \in \alpha G$ where α is the reflection in S^{n-1}.

Proof. By VI.2.8 $u = \log |f|$ is $f^\sharp F_I$-extremal in $G_0 = G \smallsetminus f^{-1}(0)$. By (3.17) there is a neighborhood D of E such that $\alpha D = D$ and $D \cap B^n \subset G_0 \smallsetminus \{0\}$. Applying VI.3.8 with $v = 0$ we obtain for every $y \in E$ a neighborhood V of y such that $u|B^n \cap V \in W_n^1(B^n \cap V)$. We first define a mapping $h\colon G' \to \mathbb{R}^n$ by

$$h(x) = \begin{cases} f(x) & \text{if } x \in G, \\ 0 & \text{if } x \in E, \\ \alpha(f(\alpha(x))) & \text{if } x \in \alpha G. \end{cases}$$

In $D \cap B^n$ we have

$$|f'(x)|^n \le K J_f(x) \le K^2 \ell(f'(x))^n \le K^2 |\nabla u(x)|^n \quad \text{a.e.}$$

We conclude $h|D \in W_{n,\mathrm{loc}}^1(D)$. Furthermore, $|h'(x)|^n \le K J_h(x)$ a.e. By 3.9 there exists a qr mapping $g_D\colon D \to \mathbb{R}^n$ that coincides with $h|D$ a.e. We obtain the required mapping by setting $g|G' \smallsetminus E = h|G' \smallsetminus E$ and $g|E = g_D|E$.   □

3.18. Notes. Our proof of Morrey's lemma 3.1 is based on the presentation in [GT, 7.19] (see also [Mor2, Theorem 3.5.2, p. 79]). The proof of Theorem 3.6 was pointed out to the author by P. Lindqvist. The idea originates from the article [W] by K.-O. Widman. More general cases are considered in [Se]. Theorem 3.9 appears in [Re1, Theorem 1]. A special case of the reflection principle 3.16 was proved in [MR1, 5.2]. Theorem 3.16 was given by O. Martio in [M6] and we have followed his proof arrangement.

4. The Wiener Condition

In this section we shall consider the continuity of F-extremals up to the boundary. Let U be a bounded open set in $\mathbb{R}^n$ and let F be a variational kernel in U with structure constants α and β. If needed, we extend F to $\mathbb{R}^n \times \mathbb{R}^n$ by letting the extension, still denoted by F, to be the trivial kernel $|h|^n$ in $\mathbb{R}^n \smallsetminus U$. Suppose $v \in W_n^1(U) \cap \mathcal{C}(\overline{U})$. By VI.7.13, VI.7.16, and 3.6 there exists a unique F-extremal $u \in W_n^1(U) \cap \mathcal{C}(U)$ with boundary values v, i.e. $u - v \in W_{n,0}^1(U)$. Henceforth we mean by F-extremal always the continuous representative. It turns out that a necessary and sufficient condition for u to be continuous in $\overline{U}$ (and then necessarily $u|\partial U = v|\partial U$) for any such v is that U satisfies the *Wiener condition*

$$(4.1) \qquad \int_0^1 \frac{\gamma(t)^{1/(n-1)}}{t}\, dt = \infty$$

at each boundary point x_0 of U. Here $\gamma(t) = \mathrm{cap}\big(B^n(x_0, 2t), (\mathbb{R}^n \smallsetminus U) \cap \overline{B}^n(x_0, t)\big)$ for $t > 0$. We shall prove the sufficiency part in Theorem 4.12.

Let $C \subset U$ be compact. If φ is a function in $\mathcal{C}_0^\infty(U)$ such that $0 \leq \varphi \leq 1$ and $\varphi|V = 1$ for some neighborhood V of C, we let $u \in W_n^1(U \smallsetminus C) \cap \mathcal{C}(U \smallsetminus C)$ be the F-extremal with boundary values φ. The function u is independent of the choice of φ and $0 \leq u \leq 1$ by VI.7.16. We call u the *F-potential* and φ a *boundary function* of the condenser (U, C). The integral

$$(4.2) \qquad \mathrm{cap}_F(U, C) = I_F(u, U \smallsetminus C) = \int_{U \smallsetminus C} F(\,\cdot\,, \nabla u)\, dm$$

is called the *F-capacity* of (U, C). By VI.1.8(3),

$$\mathrm{cap}_F(U, C) = \frac{1}{n} \int_{U \smallsetminus C} \nabla_h F(\,\cdot\,, \nabla u) \cdot \nabla u\, dm\,.$$

If w is any function in $W_n^1(U \smallsetminus C)$ with $u - w \in W_{n,0}^1(U \smallsetminus C)$, then

$$\int_{U \smallsetminus C} \nabla_h F(\,\cdot\,, \nabla u) \cdot \nabla u\, dm = \int_{U \smallsetminus C} \nabla_h F(\,\cdot\,, \nabla u) \cdot \nabla w\, dm$$

because u is an F-extremal, and so

$$(4.3) \qquad \mathrm{cap}_F(U,C) = \frac{1}{n} \int\limits_{U \smallsetminus C} \nabla_h F(\,\cdot\,, \nabla u) \cdot \nabla w \, dm \,.$$

We also observe that

$$(4.4) \qquad \alpha \, \mathrm{cap}(U,C) \le \mathrm{cap}_F(U,C) \le \beta \, \mathrm{cap}(U,C) \,.$$

4.5. Lemma. *Let u be the F-potential of a condenser $(B^n(2r), C)$, where $C \subset B^n(r)$. Let $0 < a < 1$ and set $C_a = C \cup \{\, x \in B^n(2r) : u(x) \ge a \,\}$. Then $u(x) \to 0$ as $x \to S^{n-1}(2r)$, C_a is a compact subset of $V = B^n(2r)$, $v = \min(u,a)/a$ is the F-potential of (V, C_a), and*

$$(4.6) \qquad \mathrm{cap}_F(V, C_a) = a^{1-n} \, \mathrm{cap}_F(V, C) \,.$$

Proof. We may assume $r = 1$. We first show that $u(x) \to 0$ as $x \to \partial V$. Suppose $u(x_i) > \varepsilon > 0$ for a sequence x_i, $2 - 1/i < |x_i| < 2$, converging to a point $x_0 \in \partial V$. Fix $i \ge 3$. As an extremal u is monotone, hence $\{\, x \in V \smallsetminus C : u(x) \ge \varepsilon \,\}$ contains a continuum K_i in $A_i = \overline{B}^n(\rho_i) \smallsetminus B^n(3/2)$, $\rho_i = 2 - (2 - |x_i|)/i$, connecting x_i and a point $z_i \in \partial A_i$. We can approximate u by a function $\psi_i | (V \smallsetminus C)$ such that $\psi_i \in C_0^\infty(V)$, $\|\nabla u - \nabla \psi_i\|_{n, V \smallsetminus C} < 1/i$, $|u(x) - \psi_i(x)| < \varepsilon/2$ for $x \in K_i$. Suppose $z_i \in S^{n-1}(\rho_i)$. By applying VI.3.4 to ψ_i and to caps in the spheres $S^{n-1}(z_i, r)$, $2 - \rho_i < r < 2 - |x_i|$, we obtain an estimate $\|\nabla \psi_i\|_{n, V \smallsetminus C} \ge M_i$, where $M_i \to \infty$ as $i \to \infty$. We get a similar estimate if $z_i \in S^{n-1}(3/2)$ by choosing the spheres $S^{n-1}(x_i, r)$, $2 - |x_i| < r < |x_i| - \frac{3}{2}$. Since $\|\nabla u\|_{n, V \smallsetminus C}$ is finite, we get a contradiction. It now follows that C_a is a compact subset of V.

Let $\varphi, \varphi_a \in C_0^\infty(V)$ be boundary functions for (V,C) and (V, C_a) respectively, and let $\varphi_a | W_a = 1$ for some neighborhood W_a of C_a. For $i = 1, 2, \dots$ there exists $\psi_i \in C_0^\infty(V \smallsetminus C)$ such that $\|(\varphi + \psi_i) - u\|_{1, n, V \smallsetminus C} < 1/i$ and $|\varphi + \psi_i - u| < 1/i$ in $\{\, x \in V : d(x, C_a) > 1/i \,\}$. Set $w_i = \max(\varphi + \psi_i, u)$ and $D_i = \{\, x \in V \smallsetminus C : w_i(x) < a \,\}$. We observe that if D is any component of $V \smallsetminus C$ such that $\overline{D} \subset V$, then $u | D = 1$ and so $\overline{D} \subset C_a$. Applying Harnack's inequality VI.7.4 to the component D of $V \smallsetminus C$ with $\overline{D} \cap \partial V \ne \emptyset$ we find $b < a$ such that $u(x) < b$ for $x \in V \smallsetminus W_a$. It follows that $\partial D_i \smallsetminus \partial V \subset W_a$, $i \ge i_0$, for some i_0. Let $i \ge i_0$. The set $\overline{D}_i$ is a compact set in $\overline{V} \smallsetminus C$ and w_i is continuous up to the boundary of D_i. Let $\hat{w}_i \in C(\overline{V} \smallsetminus C_a) \cap W_n^1(V \smallsetminus C_a)$ be the function defined by

$$\hat{w}_i | D_i = w_i \,,$$
$$\hat{w}_i | (V \smallsetminus C_a \smallsetminus D_i) = a \,.$$

Then $\hat{w}_i - a\varphi_a \in W_{n,0}^1(V \smallsetminus C_a)$ (see the proof of VI.1.11). We have $\|u - \hat{w}_i\|_{1, n, D_i} < 1/i$. Since $m(V \smallsetminus C_a \smallsetminus D_i) \to 0$ as $i \to \infty$, we get

$\|\hat{w}_i - u\|_{1,n,V \setminus C_a \setminus D_i} \to 0$. It follows that $u - a\varphi_a \in W^1_{n,0}(V \setminus C_a)$, hence v is the F-potential of (V, C_a).

To prove (4.6) we make use of (4.3) and obtain

$$\mathrm{cap}_F(V, C_a) = \frac{1}{na^n} \int\limits_{V \setminus C} \nabla_h F(\,\cdot\,, \nabla u) \cdot \nabla\big(\min(u, a)\big)\, dm$$

$$= \frac{1}{na^{n-1}} \int\limits_{V \setminus C} \nabla_h F(\,\cdot\,, \nabla u) \cdot \nabla\Big(\frac{\min(u, a)}{a}\Big)\, dm$$

$$= a^{1-n}\, \mathrm{cap}_F(V, C)\,.\qquad\qquad\square$$

In the following we denote by $c_0, c_1, \ldots$ positive constants depending only on n and β/α.

4.7. Lemma. *Let* V, C, *and* u *be as in 4.5. Then*

$$u(x) \ge c_0\, \mathrm{cap}(V, C)^{1/(n-1)} \quad \text{for} \quad x \in \overline{B}^n(3r/2) \setminus C\,.$$

Proof. We may assume $\mathrm{cap}(V, C) > 0$. Set

$$M = \max_{S(3r/2)} u\,, \quad m = \min_{S(3r/2)} u\,.$$

By Harnack's inequality VI.7.4, $0 < m \le M < 1$ and $M \le c_1 m$. Let $\varepsilon > 0$ be such that $M + \varepsilon < 1$. With the notation in Lemma 4.5 we have $C_{M+\varepsilon} \subset B^n(3r/2)$ because $u(x) \to 0$ as $x \to \partial V$. We then get by (4.4) and 4.5 that

$$\alpha\, \mathrm{cap}(V, C) \le \mathrm{cap}_F(V, C) = (M + \varepsilon)^{n-1}\, \mathrm{cap}_F(V, C_{M+\varepsilon})$$

$$\le \beta c_1^{n-1}(m + \varepsilon)^{n-1}\, \mathrm{cap}\big(V, \overline{B}^n(3r/2)\big)\,.$$

By VI.7.16, $u(x) \ge m$ for $x \in B^n(3r/2) \setminus C$, and the lemma follows. $\square$

4.8. Lemma. *For any bounded condenser* (A, C) *in* $\mathbb{R}^n$ *we denote by* $\lambda(A, C)$ *the* F-*potential of* (A, C). *Let* $x_0 \in \mathbb{R}^n$, $\rho > 0$, *and for* $i = 0, 1, \ldots$ *set* $B_i = B^n(x_0, 2^{1-i}\rho)$. *Then*

$$(4.9) \quad w(x) \le \exp\Big(-c_0 \sum_{i=1}^{k} \gamma(x_0, \complement U, 2^{1-i}\rho)^{1/(n-1)}\Big) \quad \text{if } x \in B_k \cap U,\ k \ge 1\,,$$

where $w = 1 - \lambda(B_0, \overline{B}_1 \cap \complement U)$ *and*

$$(4.10) \quad \gamma(x_0, E, r) = \mathrm{cap}\big(B^n(x_0, 2r), E \cap \overline{B}^n(x_0, r)\big)\,, \quad E \text{ closed}, r > 0\,.$$

Proof. Write $v_i = \lambda(B_{i-1}, \overline{B}_i \cap \complement U)$, $a_i = \gamma(x_0, \complement U, 2^{1-i}\rho)$, $i \ge 1$. By 4.7,

$$(4.11) \qquad\qquad v_i(x) \ge \delta_i \ge 1 - e^{-\delta_i}\,, \quad x \in U \cap \overline{B}_i\,,$$

where $\delta_i = c_0 a_i^{1/(n-1)}$. Write $w_1 = w = 1 - v_1$ and $w_i = e^{\delta_{i-1}} w_{i-1}$, $i = 2, 3, \ldots$. We have $w_1 | U \cap \overline{B}_1 \le e^{-\delta_1}$, hence $w_2 | U \cap \overline{B}_1 \le 1$. We claim that $w_2 | U \cap B_1 \le (1 - v_2) | U \cap B_1$. Let φ and ψ be boundary functions for $(B_0, \overline{B}_1 \cap \complement U)$ and $(B_1, \overline{B}_2 \cap \complement U)$ respectively. There exist functions $\varphi_i \in C_0^\infty (B_0 \smallsetminus (\overline{B}_1 \cap \complement U))$ and $\psi_i \in C_0^\infty (B_1 \smallsetminus (\overline{B}_2 \cap \complement U))$, $i = 1, 2, \ldots$, such that $1 - \varphi + \varphi_i \to w_1$ and $1 - \psi + \psi_i \to 1 - v_2$ in W_n^1. Since $\min(e^{\delta_1}(1 - \varphi + \varphi_i), w_2) \le \max(1 - \psi + \psi_i, 0)$ near the boundary of $U \cap B_1$, we see that $\min(1 - v_2 - w_2, 0) \in W_{n,0}^1(U \cap B_1)$. Therefore $1 - v_2 \ge w_2$ in $U \cap B_1$. From (4.11) we get $w_2 | U \cap \overline{B}_2 \le e^{-\delta_2}$, hence $w_1 | U \cap \overline{B}_2 \le e^{-\delta_1 - \delta_2}$. Continuing similarly we obtain (4.9). $\qquad \square$

4.12. Theorem. *Let* U *be a bounded open set in* $\mathbb{R}^n$, *let* $v \in C(\overline{U}) \cap W_n^1(U)$, *and let* u *be the F-extremal in* U *with* $u - v \in W_{n,0}^1(U)$. *Let* $x_0 \in \partial U$ *and suppose* (4.1) *holds. Then*

$$(4.13) \qquad \lim_{x \to x_0} u(x) = v(x_0) .$$

In particular, if (4.1) *holds for all* $x_0 \in \partial U$, *then* $u \in C(\overline{U})$.

Proof. Let $\rho > 0$ and let w be the function in Lemma 4.8. Set $B_i = B^n(x_0, 2^{1-i}\rho)$ and

$$g = w \max_{\partial U} v + \max_{\partial U \cap \overline{B}_0} v .$$

We have $u \le \max_{\partial U} v$ and $w | \partial B_0 = 1$. Let $\varphi_i \in C_0^\infty(U)$ be functions such that $v + \varphi_i \to u$ in $W_n^1(U)$. Let $\varepsilon > 0$. Then $\min(v + \varphi_i, u) < g + \varepsilon$ near $\partial(U \cap B_0)$, hence $u \le g$ in $U \cap B_0$. Similarly $h \le u$ in $U \cap B_0$, where

$$h = w \min_{\partial U} v + \min_{\partial U \cap \overline{B}_0} v .$$

Then

$$(4.14) \qquad \mathrm{osc}(u, U \cap B_k) \le \mathrm{osc}(v, \partial U \cap B_0) + \mathrm{osc}(v, \partial U) \sup_{U \cap B_k} w .$$

Next we estimate the sum appearing in (4.9) by an integral. Write $r_i = 2^{1-i}\rho$ and $\gamma(r) = \gamma(x_0, \complement U, r)$. If $r_{i+2} \le r \le r_{i+1}$, $i \ge 0$, we get by the help of auxiliary qc maps (see [MSa1, 2.7]) that

$$\frac{1}{b_0} \gamma(r_{i+2}) \le \mathrm{cap}(B_i, \complement U \cap B_{i+2}) \le \gamma(r)$$

$$\le b_0 \, \mathrm{cap}\big(B^n(x_0, 4r), \complement U \cap \overline{B}^n(x_0, r)\big) \le b_0 \gamma(r_{i+1}) ,$$

where $b_0 > 0$ depends only on n. Hence

$$(4.15) \qquad \frac{1}{b_0^{\frac{1}{n-1}}} \log 2 \sum_{i=2}^{k} \gamma(r_i)^{\frac{1}{n-1}} \le \int_{r_k}^{r_1} \frac{\gamma(r)^{\frac{1}{n-1}}}{r} \, dr \le b_0^{\frac{1}{n-1}} \log 2 \sum_{i=1}^{k-1} \gamma(r_i)^{\frac{1}{n-1}} .$$

Substituting (4.9) into (4.14) and using (4.15) we get (4.13) from (4.1) and the continuity of v. The theorem is proved.   □

4.16. Remark. For later use we observe that the estimate (4.14) is valid for any $x_0 \in \mathbb{R}^n$ if $U \cap B_k \neq \emptyset$.

We call an open set $U \subset \mathbb{R}^n$ *regular* if U is bounded and if (4.1) is satisfied for all $x_0 \in \partial U$. Any open set can be exhausted by regular sets, for example by taking unions of cubes.

4.17. Notes. In a more general setting Theorem 4.12 was first proved by V. Maz'ya in [Ma], see also [R5, 4.19]. In our situation we have here presented a simpler proof and it was given to the author by T. Kilpeläinen. The necessity of the Wiener condition (4.1) was proved by P. Lindqvist and O. Martio [LM].

5. F-Harmonic Measure

In classical function theory the notion of harmonic measure of a closed set $C \subset \partial G$ with respect to a domain G plays an important role. In [GLM2] S. Granlund, P. Lindqvist, and O. Martio introduced an n-dimensional analogue of it in the setting of F-extremals. It is called F-harmonic measure. Due to the nonlinearity in the theory of F-extremals the F-harmonic measure fails for $n \geq 3$ to be additive with respect to the set C (and is not a measure on sets on the boundary) as in the case for ordinary harmonic measure. In spite of this disadvantage the concept of F-harmonic measure has turned out to be useful in various estimates.

5.1. Sub-(super-)F-extremals. To define the F-harmonic measure for general bounded open sets we need an analogue for subharmonic functions. Let U be an open set in $\mathbb{R}^n$ and F a variational kernel in U (see VI.1.3). An upper semicontinuous function $u \colon U \to [-\infty, \infty[$ is a *sub-F-extremal* if u satisfies the following F-comparison principle: If $D \subset\subset U$ is a domain and $v \in \mathcal{C}(\overline{D})$ is an F-extremal in D with $v \geq u$ on ∂D, then $v \geq u$ on D. A function $u \colon U \to {]-\infty, \infty]}$ is called a *super-F-extremal* if $-u$ is a sub-F-extremal. It follows directly from the definition that a sub-F-extremal u satisfies the maximum principle: If $D \subset\subset U$ is a domain, then

$$(5.2) \qquad \max_{\overline{D}} u = \max_{\partial D} u .$$

An example of a sub-F_I-extremal in $\mathbb{R}^n$ is $u = \log |y|$. To see this we observe that u is an F_I-extremal in $\mathbb{R}^n \smallsetminus \{0\}$, and if $D \subset \mathbb{R}^n$ is a relatively compact domain in $\mathbb{R}^n$ with $0 \in \overline{D}$ and if $v \in \mathcal{C}(\overline{D})$ is an F_I-extremal in D, then $v > u$ in a neighborhood of 0. If $f \colon G \to \mathbb{R}^n$ is a nonconstant qr map, we conclude from the discreteness of f and from VI.6.11 that $\log|f|$ is sub-$f^\sharp F_I$-extremal.

Next we assume that U is bounded. If $g: \partial U \to [-\infty, \infty]$ is any function, we define the *Perron upper class* $\mathcal{U}_g$ of g (with respect to F) as the set of super-F-extremals u in U which are bounded from below and which satisfy

$$(5.3) \qquad \liminf_{x \to y} u(x) \geq g(y), \quad y \in \partial U .$$

Then we set

$$(5.4) \qquad \overline{H}_g = \inf\{u : u \in \mathcal{U}_g\} .$$

Similarly we define the *lower class* $\mathcal{L}_g$ and $\underline{H}_g$ by means of sub-F-extremals.

Let C be a closed subset of ∂U, and let $\chi_C: \partial U \to [0,1]$ be the characteristic function of C. Then the function $\overline{H}_{\chi_C}$ is called the *F-harmonic measure* of C with respect to U and it is denoted by $\omega(C, U; F)$. Clearly $0 \leq \omega(C, U; F) \leq 1$. It is true that $\overline{H}_g$ in (5.4) is either an F-extremal, identically ∞, or identically $-\infty$, and that the first case occurs if g is bounded [GLM4, 2.2]. In particular, the F-harmonic measure $\omega(C, U; F)$ is an F-extremal. We shall prove this for regular open sets, i.e., bounded open sets that satisfy (4.1) on the boundary. We need some results on limits of sequences of extremals. Recall that F-extremals are continuous.

5.5. Proposition. *Let F be a variational kernel in an open set $U \subset \mathbb{R}^n$, let (u_i) be a sequence of F-extremals which converges locally uniformly to a function u in U. Then u is an F-extremal.*

Proof. Let $D \subset\subset U$ be a regular domain. We conclude from VI.3.1 and VI.7.9 that $u|\overline{D} \in \mathcal{C}(\overline{D}) \cap W_n^1(D)$. Let $\varepsilon > 0$. By 4.12 there exists an F-extremal $v \in \mathcal{C}(\overline{D}) \cap W_n^1(D)$ with $v|\partial D = u|\partial D$. The function $v + \varepsilon$ is also an F-extremal in D. For some i_0 we have $u_i < v + \varepsilon$ on ∂D if $i \geq i_0$. By VI.7.1, $u_i \leq v + \varepsilon$ and so $u \leq v + \varepsilon$ in D. This gives $u \leq v$ in D. Similarly $u \geq v$ in D. We conclude that u is an F-extremal in U. $\square$

5.6. Proposition. *Let (u_i) be an increasing sequence of F-extremals in a domain $U \subset \mathbb{R}^n$. Then either*

(1) $u_i \to u$ *locally uniformly in U and u is an F-extremal, or*
(2) $u_i(x) \to \infty$ *at every point $x \in U$.*

Proof. Suppose that

$$\lim_{i \to \infty} u_i(x_0) < \infty$$

for some point $x_0 \in U$. Let $x \in U$. Let $D \subset\subset U$ be a domain that contains x_0 and x, and let E be a continuum in D such that $x_0 \in E$ and $B^n(x, R) \subset E$ for some $R > 0$. Set

$$m = \inf_{x \in D} u_1(x) .$$

Each $u_i - m$ is a nonnegative F-extremal in D. Theorem VI.7.4 applied to $u_i - m$ in D gives a constant $M < \infty$ such that

$$\sup_{x \in E} u_i(x) \le M$$

for all i. In particular, $\operatorname{osc}(u_i, B^n(x, R)) \le M - m$ for all i. By VI.3.16 the family $\{ u_i : i = 1, 2, \ldots \}$ is equicontinuous. As (u_i) is increasing the convergence of (u_i) must be locally uniform in U. Proposition 5.5 completes the proof.   □

We now continue the discussion of the F-harmonic measure. Let U be a regular open set in $\mathbb{R}^n$ and $C \subset \partial U$ closed. Suppose (φ_i) is a decreasing sequence of functions in $\mathcal{C}(\overline{U}) \cap W_n^1(U)$ such that $0 \le \varphi_i \le 1$, $\varphi_i|C = 1$, and

$$\bigcap_i \operatorname{spt} \varphi_i = C .$$

By 4.12 there exists $u_i \in \mathcal{C}(\overline{U}) \cap W_n^1(U)$ which is an F-extremal in U and for which $u_i|\partial U = \varphi_i|\partial U$. Applying the result corresponding to 5.6 for a decreasing sequence of F-extremals in each component of U we conclude that (u_i) converges locally uniformly in U to an F-extremal u in U. We call (u_i) a *generating sequence* for (C, U).

5.7. Lemma. *With the assumptions above,* $u = \omega(C, U; F)$.

Proof. We write $\omega = \omega(C, U; F)$. Clearly each u_i is in the upper Perron class $\mathcal{U}_{\chi_C}$, so $\overline{H}_{\chi_C} = \omega \le u_i$ for all i. Hence $\omega \le u$. Suppose $v \in \mathcal{U}_{\chi_C}$. The function $v^*\colon \partial U \to [0, \infty]$ defined by

$$v^*(y) = \liminf_{x \to y} v(x), \quad y \in \partial U ,$$

is clearly lower semicontinuous. Let $\varepsilon > 0$. There exists i_ε such that $v^* + \varepsilon > u_i|\partial U$ for all $i \ge i_\varepsilon$. From this we conclude that for any sufficiently large open set $D \subset\subset U$ we have $v + 2\varepsilon > u_i$ on ∂D for all $i \ge i_\varepsilon$. Since $v + 2\varepsilon$ is a super-F-extremal and u_i is an F-extremal, we have $v + 2\varepsilon \ge u_i$ in D for $i \ge i_\varepsilon$. This proves $\omega \ge u$.   □

5.8. Corollary. *If* U *is a regular open set and* $C \subset \partial U$ *closed, then* $\omega = \omega(C, U; F)$ *is an* F-extremal. *Moreover,*

(5.9) $$\lim_{x \to y} \omega(x) = 0 \quad \text{if } y \in \partial U \smallsetminus C .$$

Proof. The first statement is a repetition of Lemma 5.7. If $y \in \partial U \smallsetminus C$ and if (u_i) is a generating sequence for (C, U), then $u_i|V \cap \partial U = 0$ for some neighborhood V of y and $i \ge i_0$ for some i_0. Since each u_i is continuous in $\overline{U}$ and $0 \le \omega \le u_i$, (5.9) follows.   □

5.10. Corollary. *If U and C are as in 5.8 and if v is a sub-F-extremal in U with the properties*

$$\limsup_{x \to y} v(x) \leq 1 , \quad y \in C ,$$

$$\limsup_{x \to y} v(x) \leq 0 , \quad y \in \partial U \smallsetminus C ,$$

then $v \leq \omega$ in U.

Proof. Let (u_i) be a generating sequence for (C, U). Fix i. For $\varepsilon > 0$ there exists a neighborhood V of ∂U such that $v \leq u_i + \varepsilon$ on $V \cap U$. Since v is sub-F-extremal and $u_i + \varepsilon$ F-extremal, $v \leq u_i + \varepsilon$ in U. The result now follows from 5.7. $\qquad\square$

5.11. Remarks. 1. As mentioned earlier, $\omega(C, U; F)$ is an F-extremal even if U is not regular [GLM4, 2.2].

2. In the case of a regular open set U we can in the definition of $\omega(C, U; F)$ replace the upper Perron class $\mathcal{U}_{\chi_C}$ by the set of functions $u \in \mathcal{C}(\overline{U}) \cap W_n^1(U)$ which are F-extremals in U and which satisfy $0 \leq u \leq 1$ and (5.3) with $g = \chi_C$. This follows from 5.7.

5.12. Notes. Propositions 5.5 and 5.6 are taken from [GLM1, 4.21, 4.22]. Lemma 5.7 appears in this form in [GLM4, 5.1]. General references for the F-harmonic measure are [GLM2], [GLM3], [GLM4], and [HM].

6. Phragmén–Lindelöf Type Theorems

In this section we study growth of qr mappings in unbounded domains. The results are obtained as corollaries from corresponding estimates for sub-F-extremals. E. Phragmén and E. Lindelöf proved in [PL] that if a subharmonic function u in the plane sector $G = \{ z : |\arg z| < \theta/2 \}$, $\theta \leq 2\pi$, satisfies $\limsup_{x \to y} u(x) \leq 0$ for all $y \in \partial G \cap \mathbb{R}^2$, then either $u \leq 0$ in G or

$$(6.1) \qquad M_u(r) = M(r) = \sup_{\substack{|x|=r \\ x \in G}} u(x)$$

satisfies

$$(6.2) \qquad \liminf_{r \to \infty} M(r) r^{-\pi/\theta} > 0 .$$

By a method of T. Carleman [Carl] for an unbounded planar domain G (6.2) can be replaced by

$$(6.3) \qquad \liminf_{r \to \infty} M(r) e^{-\pi P(1,r)} > 0 .$$

Here (for $n = 2$)

(6.4)
$$P(\rho, \sigma) = \int_{I(\rho,\sigma)} \frac{dt}{t\theta(t)^{1/(n-1)}} ,$$

$t^{n-1}\theta(t)$ is the $(n-1)$-measure of $G \cap S^{n-1}(t)$, and $I(\rho, \sigma) = \{ t \in [\rho, \sigma] : \partial G \cap S^{n-1}(t) \neq \emptyset \}$.

A counterpart of (6.3) will be established for sub-F-extremals in Theorem 6.9. Another growth relation is given in Theorem 6.7. Both results are based on estimates of F-harmonic measures.

Let G be an unbounded domain in $\mathbb{R}^n$, let F be a variational kernel in G with structure constants α and β, and set $G_r = G \cap B^n(r)$ for $r > 0$. We let $\omega(x, r)$ be the value of the F-harmonic measure $\omega(\overline{G}_r \cap S^{n-1}(r), G_r; F)$ at $x \in G_r$. Suppose G satisfies the Wiener condition (4.1) at each boundary point $x_0 \in \partial G \cap \mathbb{R}^n$. Then clearly G_r is a regular set for all $r > 0$. A Phragmén–Lindelöf principle can be formulated as follows.

6.5. Proposition. *Let* $u: G \to [-\infty, \infty[$ *be a sub-F-extremal in an unbounded domain* G *which satisfies* (4.1) *for all* $x_0 \in \partial G \cap \mathbb{R}^n$. *Suppose*

$$\limsup_{x \to y} u(x) \leq 0 \quad \text{for all } y \in \partial G \cap \mathbb{R}^n.$$

Then either $u \leq 0$ *or*

(6.6)
$$\liminf_{r \to \infty} M(r)\omega(x, r) > 0 \quad \text{for each } x \in G.$$

Proof. Suppose $u(z) > 0$ for some $z \in G$. Let $r \geq |z|$. Then $M(r) > 0$ by (5.2) and $u(z)/M(r) \leq \omega(z, r)$ by 5.10. If x is another point in G, we get by Harnack's inequality VI.7.4 an inequality $\omega(z, r) \leq A\omega(x, r)$ for sufficiently large r and with A independent of r, and (6.6) follows. $\square$

Recall the notation $\gamma(x, E, r)$ from (4.10). The next result gives a growth estimate in terms of $\liminf_{r \to \infty} \gamma(0, \mathbb{R}^n \smallsetminus G, r)$.

6.7. Theorem. *Let* G *be an unbounded domain with*

$$\delta = \liminf_{r \to \infty} \gamma(0, \mathbb{R}^n \smallsetminus G, r) > 0 .$$

Suppose that u *is a sub-F-extremal such that*

$$\limsup_{x \to y} u(x) \leq 0 \quad \text{for all } y \in \partial G \cap \mathbb{R}^n .$$

Then either $u \leq 0$ *or*

(6.8)
$$\liminf_{r \to \infty} M(r)r^{-\mu} > 0$$

where $\mu > 0$ *is a constant depending only on* n, β/α, *and* δ.

Proof. We may assume that G satisfies (4.1) at each $y \in \partial G \cap \mathbb{R}^n$. We shall use Proposition 6.5 and estimate $\omega(x, r)$ for a fixed $x \in G$. We choose $a > |x|$ such that $\gamma(0, \mathbb{R}^n \smallsetminus G, t) \geq \delta/2$ for $t \geq a$. Let $r > 4a$ and set $2\rho = r/2$. If (v_i) is a generating sequence for $(S^{n-1}(r) \cap \partial G_r, G_r)$, we have $v_i | \partial G_r \cap \bar{B}^n(2\rho) = 0$, $i \geq i_0$. From 4.8, (4.14), and Remark 4.16 we get

$$(6.9) \qquad \omega(x, r) \leq v_i(x) \leq \exp\left(-c_0 k (\delta/2)^{1/(n-1)}\right), \quad i \geq i_0,$$

where k is the maximal integer such that $2^{1-k}\rho > a$. Inequality (6.9) yields

$$\omega(x, r) \leq \left(\frac{4a}{r}\right)^\mu, \quad r > 4a,$$

where $\mu = (\log 2)^{-1} c_0 (\delta/2)^{1/(n-1)}$. From this 6.5 gives the claim. $\qquad \square$

6.10. Corollary. *Let G be as in 6.7 and let $f: G \to \mathbb{R}^n$ be a nonconstant K-qr mapping such that*

$$\limsup_{x \to y} |f(x)| \leq 1 \quad \text{for all } y \in \partial G \cap \mathbb{R}^n.$$

Then either $|f| < 1$ or

$$(6.11) \qquad \liminf_{r \to \infty} \log M_f(r) r^{-\mu} > 0$$

where $M_f(r) = M_{|f|}(r)$ and $\mu > 0$ is a constant depending only on n, K, and δ.

Proof. We apply 6.7 to $u = \log |f|$ which is a sub-F-extremal for $F = f^\sharp F_I$. In the first case $|f| \leq 1$ and in fact $|f| < 1$ because f is open. $\qquad \square$

With similar methods we also get the following result for entire qr mappings.

6.12. Theorem. *Let $f: \mathbb{R}^n \to \mathbb{R}^n$ be a nonconstant K-qr mapping. If there exists a closed set $E \subset \mathbb{R}^n$ such that*

$$(6.13) \qquad \delta = \liminf_{r \to \infty} \gamma(0, E, r) > 0$$

and

$$(6.14) \qquad a = \sup_{x \in E} |f(x)| < \infty,$$

then the lower order of f satisfies $\lambda_f \geq d(n, K, \delta) > 0$. In particular, if $\lambda_f < d(n, K, \delta)$, then

$$(6.15) \qquad \limsup_{r \to \infty} m_f(r) = \infty,$$

where $m_f(r) = \min_{|x|=r} |f(x)|$.

Proof. To prove the first statement let $z \in \mathbb{R}^n$ be a point such that $|f(z)| > a$. Then $z \notin E$ and if G is the z-component of $\mathbb{R}^n \setminus E$, then

$$\sup_{x \in G} |f(x)| = \infty .$$

Then G and $(f/a)|G$ satisfy the conditions in 6.10 and an estimate $\lambda_f \geq d(n, K, \delta) > 0$ follows. (Recall the characterization of λ_f in terms of $M_f(r)$ from V.8.18.) $\qquad\square$

6.16. Corollary. *If a K-qr mapping $f: \mathbb{R}^n \to \mathbb{R}^n$ has at least one asymptotic value $b \in \mathbb{R}^n$, then $\lambda_f \geq C_0(n, K) > 0$.*

6.17. Remark. For $n = 2$ there is an inequality $\lambda_f \geq C_0(n, K, k)$ where k is the number of finite asymptotic values and $C_0(n, K, k) \to \infty$ as $k \to \infty$. The sharp bound $k/2$ for $K = 1$ was obtained by L.V. Ahlfors [A1]. It is not known whether a similar statement holds for qr maps for $n \geq 3$.

We take the opportunity to give here the following 2-constants theorem for qr mappings whose proof follows from that of 6.7.

6.18. Theorem. *Let $f: G \to \mathbb{R}^n$ be K-qr, let $0 < m < M$, let U be a ball $B^n(z, s)$ such that $U \cap G \neq \emptyset$, and suppose*

(1) $|f(x)| \leq M$ if $x \in G \cap U$,
(2) $\limsup_{x \to y} |f(x)| \leq m$ if $y \in \partial G \cap U$,
(3) $\delta = \inf\{\gamma(z, \mathbb{R}^n \setminus G, r) : 0 < r \leq s/2\} > 0$.

Then

$$\log |f(x)| \leq \left(\frac{4r}{s}\right)^\mu \log M + \left(1 - \left(\frac{4r}{s}\right)^\mu\right) \log m$$

if $x \in B^n(z, r) \cap G$ and $0 < r \leq s/4$, where $\mu > 0$ depends only on n, K, and δ.

Proof. We may assume that G satisfies (4.1) at each $x_0 \in \partial G \cap \mathbb{R}^n$. Then $G \cap U$ is a regular set. We consider the $f^\sharp F_I$-harmonic measure

$$\omega = \omega\big(S^{n-1}(z, s) \cap \partial(U \cap G), U \cap G; f^\sharp F_I\big) .$$

We apply 5.10 to $v = \big(\log(M/m)\big)^{-1} \log(|f|/m)$ and get

$$\log |f(x)| \leq \omega(x) \log M + \big(1 - \omega(x)\big) \log m .$$

Using the argument in the proof of (6.9) we now obtain

$$\omega(x) \leq \left(\frac{4r}{s}\right)^\mu \quad \text{if } x \in B^n(z, r) \cap G \text{ and } 0 < r \leq s/4 ,$$

which then gives the claim. $\qquad\square$

The counterpart to (6.3) for sub-F-extremals is the following.

6.19. Theorem. *Let G be an unbounded domain in $\mathbb{R}^n$ and let u be a sub-F-extremal in G such that*

$$\limsup_{x \to y} u(x) \leq 0 \quad \text{for all } y \in \partial G \cap \mathbb{R}^n .$$

Then either $u \leq 0$ or

$$(6.20) \qquad \liminf_{r \to \infty} M(r) \exp\left(-b_1(\alpha/\beta)^{1/n} P(1,r)\right) > 0$$

where $b_1 > 0$ depends only on n and $P(1,r)$ is the integral defined in (6.4).

Proof. If G_1 is an unbounded subdomain of G, then $P_1(\rho_0, r) \geq P(\rho_0, r)$ for some $\rho_0 \geq 1$, where $P_1(\rho_0, r)$ is an integral (6.4) for G_1. Therefore we may assume that G satisfies (4.1) at each $y \in \partial G \cap \mathbb{R}^n$.

We claim that an estimate

$$(6.21) \qquad \omega(z,r) \leq 4 \exp\left(-b_1(\alpha/\beta)^{1/n} P(|z|,r)\right)$$

holds for $z \in G_r$. To prove (6.21) let $z \in G_r$ and let (v_i) be a generating sequence for $(\partial G_r \cap S^{n-1}(r), G_r)$. We may assume $P(|z|, r) > 0$. Let $0 < \varepsilon < 1$ and let $\rho < r$ be such that $P(|z|, \rho) > P(|z|, r)(1 - \varepsilon)$. For some i_0, $v_i = 0$ in $\partial G_r \cap \overline{B}^n(\rho)$ if $i \geq i_0$. Fix $i \geq i_0$ and write $v = v_i$. Let $|z| \leq r_1 < s < r_2 \leq \rho$ and set

$$Q(t) = \max_{G_r \cap S(t)} v(x) , \quad |z| \leq t < r .$$

By VI.3.6,

$$(6.22) \qquad \begin{aligned} \int_{I(r_1,s)} \frac{Q(t)^n}{t\,\theta(t)^{1/(n-1)}} dt &\leq \int_{I(r_1,s)} \frac{Q(t)^n}{\mathcal{H}^{n-1}(C_t)^{1/(n-1)}} dt \\ &\leq a_0 \int_{G_s} |\nabla v|^n dm , \end{aligned}$$

where a_0 depends only on n and C_t is a cap in $S^{n-1}(t) \cap G_r$ such that $Q(t) = \max_{C_t} v$ and $\overline{C}_t \cap \partial G_r \neq \emptyset$.

We claim that

$$(6.23) \qquad \int_{G_s} |\nabla v|^n dm \leq n^n \frac{\beta}{\alpha} \left(\int_s^{r_2} \frac{dt}{t\,Q(t)^{n/(n-1)}\theta(t)^{1/(n-1)}} \right)^{1-n}$$

is true. We may assume that the integral in the right hand side of (6.23) is positive. If ζ is in $W^1_{n,0}(B^n(r_2)) \cap C(\overline{B}^n(r_2))$ with $0 \leq \zeta \leq 1$ and $\zeta | \overline{B}^n(s) = 1$, inequality VI.(3.12) yields

$$(6.24) \qquad \int_{G_s} |\nabla v|^n dm \leq n^n \frac{\beta}{\alpha} \int_{G_{r_2}} |v|^n |\nabla \zeta|^n dm .$$

We choose ζ such that

$$1 - \zeta(x) = \left(\int\limits_s^{r_2} \frac{dt}{t\, Q(t)^{n/(n-1)}\theta(t)^{1/(n-1)}} \right)^{-1} \int\limits_s^{|x|} \frac{dt}{t\, Q(t)^{n/(n-1)}\theta(t)^{1/(n-1)}}$$

for $s < |x| \le r_2$ and $\zeta|\overline{B}^n(s) = 1$. Since ζ is a radial function, we may write $\tilde\zeta(|x|) = \zeta(x)$ and so (6.24) gives

$$\int\limits_{G_s} |\nabla v|^n dm \le n^n \frac{\beta}{\alpha} \int\limits_s^{r_2} Q(t)^n |\tilde\zeta'(t)|^n \theta(t) t^{n-1} dt \;,$$

from which (6.23) follows.

Clearly $Q(t_1) \le Q(t_2)$ for $|z| \le t_1 < t_2 < r$. Inequalities (6.22) and (6.23) therefore imply

(6.25) $$a_0 n^n \frac{\beta}{\alpha} \ge \left(\frac{Q(r_1)}{Q(r_2)} \right)^n P(r_1, s) P(s, r_2)^{n-1} \;.$$

In (6.25) we choose s so that $P(r_1, s) = P(s, r_2)$. Then (6.25) gives

(6.26) $$\frac{Q(r_1)}{Q(r_2)} \le c_0 P(r_1, r_2)^{-1} \;,$$

where $c_0 = 2na_0^{1/n}(\beta/\alpha)^{1/n}$. We shall derive (6.21) by a process of iteration from (6.26). If $P(|z|, r) \le c_0 e(1 - \varepsilon)^{-1}$, then the right hand side of (6.21) is greater than 1 with $b_1 = a_0^{-1/n}(2en)^{-1}(1 - \varepsilon)$. As a second case suppose $P(|z|, \rho) = P(|z|, r)(1 - \varepsilon) > c_0 e$, let $k \ge 2$ be the integer such that

(6.27) $$e(k - 1) \le \frac{1}{c_0} P(|z|, \rho) < ek \;,$$

and let $\rho_0 = |z| < \rho_1 < \ldots < \rho_k = \rho$ be numbers such that $P(\rho_{j-1}, \rho_j) = P(|z|, \rho)/k$, $j = 1, \ldots, k$. Then (6.26) yields

$$Q(|z|) \le \left(\frac{c_0 k}{P(|z|, \rho)} \right)^k \le \left(\frac{k}{(k-1)e} \right)^k$$

$$= \left(1 - \frac{1}{k} \right)^{-k} e^{-k} < 4 \exp\left(-\frac{1 - \varepsilon}{c_0 e} P(|z|, r) \right) \;.$$

Letting $\varepsilon \to 0$ we get (6.21) in both cases with $b_1 = a_0^{-1/n}(2en)^{-1}$. The theorem now follows from 6.5. $\square$

A corollary for a qr mapping $f \colon G \to \mathbb{R}^n$ is obtained from Theorem 6.19 similarly to 6.10 by applying 6.19 to $u = \log |f|$.

6.28. Notes. Corollary 6.10 and Theorem 6.12 are from [RV, 3.3, 3.5]. The proof arrangement in [RV] is slightly different and is based on iteration of a 2-constants theorem, which is similar to 6.18 and which was proved in [R5, 4.22]

with the method from Maz'ya's article [Ma]. The Phragmén–Lindelöf principle
6.5 in our setting appears in [GLM2, 3.10] although the Wiener condition (4.1)
was in [GLM2] replaced by a stronger assumption. Theorem 6.19 and its proof
are taken from [GLM3]. Results related to this section are also given in [BO]
and [Mik3].

7. Asymptotic Values

E. Lindelöf proved in [Lin] that if $f: B^2 \to B^2$ is analytic and has an asymptotic limit a along a path at a boundary point b, then f has the angular
limit a at b, and hence a is the only asymptotic value f can have at b.
The same is true for qr mappings for $n = 2$, which can be seen from the
decomposition $\varphi \circ h$ of such a mapping with h qc and φ analytic.

In this section we study the situation for qr mappings in dimensions $n \geq$
3. It turns out that Lindelöf's theorem is not true in the same form. Essentially
the reason for this is that an arc C on the boundary ∂B^n does not imply the
F-harmonic measure $\omega(C, B^n; F)$ to be positive. In fact, for $n \geq 3$ a bounded
qr mapping of B^n into itself can have an infinite number of asymptotic limits
along paths at a boundary point (Theorem 7.1).

On the other hand, if a qr mapping $f: B^n \to B^n$, $n \geq 2$, has a limit a
along the $(n-1)$-dimensional set $A = \{ y \in \partial B^n : y_n > 0 \}$ at the boundary
point e_1, then f has the angular limit a at e_1 (Theorem 7.3). In [R6] it
was shown for $n = 4$ that it is not enough to assume the existence of a limit
along the $(n-2)$-dimensional set $C = \{ y \in \partial B^n : y_{n-1} > 0,\ y_n = 0 \}$ at e_1,
in fact, again an infinite number of asymptotic limits at a boundary point can
exist along sets in B^n of this type.

We shall give the proof of the following theorem for $n = 3$; for a somewhat
different proof for the general case $n \geq 3$, see [R6].

7.1. Theorem. *For each $n \geq 3$ there exists a bounded qr mapping f of
B^n and a point b such that*

(1) f *has infinitely many asymptotic values at b,*
(2) f *has no angular limit at b.*

Proof. We identify $\mathbb{R}^2$ with $\mathbb{R}^2 \times \{0\} \subset \mathbb{R}^3$ and set

$$X = [0,1] \times \,]0,1] \subset \mathbb{R}^2 \,,$$
$$A^* = \{ x \in \mathbb{R}^3 : (x_1, x_2) \in A,\ -x_2 < x_3 < x_2 \} \quad \text{if } A \subset X \,.$$

We shall construct a qr map of $\operatorname{int} X^*$ omitting a neighborhood of ∞ and
with infinitely many asymptotic values at 0. Let L be the simplicial 2-

complex with underlying space X as shown in Figure 5 and let γ be the set of 2-simplices of L.

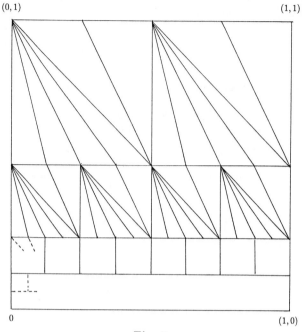

$(0,1)$ $(1,1)$

0 $(1,0)$

Fig. 5

We fix $\beta \geq \beta_0 > 1$ where the bound β_0 will be chosen sufficiently large later. Let ρ be the hyperbolic distance in the upper half plane of $\mathbb{R}^2$. Let $1 < \mu_1 < \nu_1$ and set

$$Y_0 = \{\, x \in X : x_2 \geq x_1^{\mu_1} \ \text{or} \ x_2 \leq x_1^{\nu_1} \,\},$$
$$Y_k = \{\, x \in X : \beta(k-1) < \rho(x, Y_0) \leq \beta k \,\}, \quad k = 1, 2, \ldots.$$

Let $c^1 \in B^2(1/2)$. As a first step we shall construct a map $f_1 \colon X^* \to \overline{B}^3(4)$, which is K-qr in $\operatorname{int} X^*$, and such that K and $f_1|Y_0^*$ are independent of c^1, and

(7.2)
$$\lim_{k \to \infty} \sup_{x \in Y_k^*} |f_1(x) - c^1| = 0.$$

Fix $C_0 \in \gamma$ and an affine map h_{C_0} onto the 2 simplex $T = \langle a, b, c \rangle \subset \mathbb{R}^2$ whose vertices are $a = e_1$, $b = -e_1/2 + \sqrt{3}e_2/2$, $c = -e_1/2 - \sqrt{3}e_2/2$. Then there exists a unique set of affine homeomorphisms $h_C \colon C \to T$ such that $h_C|C \cap D = h_D|C \cap D$, $C, D \in \gamma$. This follows because each interior vertex of L belongs to an even number of 2-simplices of L. We set $h_C^{-1}(a) = a_C$, $h_C^{-1}(b) = b_C$, $h^{-1}(c) = c_C$.

Let $w \colon T \to \overline{B}^2$ be the radial stretching, let $\psi_1 = w \times \operatorname{id}_I \colon T \times I \to \overline{B}^2 \times I$, $I = [-1, 1]$, and let $\varphi_1 \colon \overline{B}^2 \times I \to \overline{H}^3$ be the map $\varphi_1(r, \varphi, x_3) = (\rho, \varphi, \theta)$

defined by $\rho = e^{x_3}$, $\theta = \pi r/2$, in cylinder and spherical coordinates. Here $H^3 = \{x \in \mathbb{R}^3 : x_3 > 0\}$. For $C \in \gamma$ we define $g_C: \overline{C}^* \to T \times I$ by $g_C(x) = (h_C(x_1, x_2)_1, h_C(x_1, x_2)_2, x_3/x_2)$ and $F_C: \overline{C}^* \to \mathbb{R}^3$ by

$$F_C = \begin{cases} \varphi_1 \circ \psi_1 \circ g_C & \text{if } h_C \text{ is sense–preserving,} \\ v \circ \varphi_1 \circ \psi_1 \circ g_C & \text{if } h_C \text{ is sense–reversing.} \end{cases}$$

Here v is the reflection in ∂H^3. Let S_k be the similarity $S_k(x) = x/2^{k-1} + c^1$, $k = 1, 2, \ldots$. We still define some maps which enables us to move stepwise from S_k to S_{k+1}. To this end set

$$E_{\xi, \eta} = \varphi_1 \circ \psi_1(\langle \xi, \eta \rangle \times I),$$

where ξ and η are distinct points in $\{a, b, c\}$. For any ξ, η, ζ such that $\{\xi, \eta, \zeta\} = \{a, b, c\}$ we can define maps $\lambda_k(\xi, \eta)$ and $\lambda_k(\xi)$ of $R = \overline{B}^3(e) \smallsetminus B^3(1/e)$ into $\overline{B}^3(4)$, which are K_0-qc in $\operatorname{int} R$, K_0 independent of k, and such that

$$\lambda_k(\xi, \eta)|E_{\xi, \eta} = S_k|E_{\xi, \eta},$$
$$\lambda_k(\xi)|E_{\eta, \zeta} = S_{k+1}|E_{\eta, \zeta},$$
$$\lambda_k(\xi, \eta)|E_{\eta, \zeta} = \lambda_k(\eta)|E_{\eta, \zeta},$$
$$\lambda_k(\xi, \eta)|E_{\zeta, \xi} = \lambda_k(\xi)|E_{\zeta, \xi}.$$

For sufficiently large β_0, independent of $C \in \gamma$, we find for each $C \in \gamma$ a maximal integer k such that $|\operatorname{star} C| = \cup\{D \in \gamma : D \cap C \neq \emptyset\}$ is contained in $Y_k \cup Y_{k+1}$. We now are in a position to give the definition of the map f_1. If $C \in \gamma$ and k is the maximal integer k as above we set

$$f_1|C^* = \begin{cases} S_k \circ F_C & \text{if } a_C, b_C, c_C \in Y_k, \\ S_{k+1} \circ F_C & \text{if } a_C, b_C, c_C \in Y_{k+1}, \\ \lambda_k(a, b) \circ F_C & \text{if } a_C, b_C \in Y_k, \ c_C \in Y_{k+1}, \\ \lambda_k(a) \circ F_C & \text{if } a_C \in Y_k, \ b_C, c_C \in Y_{k+1}, \end{cases}$$

with the extension of the definition to all permutations of letters a, b, c. The given definition is by construction compatible on common boundaries $C^* \cap D^*$, $C, D \in \gamma$, and we obtain a map $f_1: X^* \to \overline{B}^3(4)$, which is K-qr in $\operatorname{int} X^*$ with K and $f_1|Y_0^*$ independent of c^1 and such that (7.2) holds. We observe that $f_1|\operatorname{int} X^*$ has c^1 as an asymptotic value at 0 since

$$\lim_{\substack{x \to 0 \\ x \in Z_1^*}} f_1(x) = c^1,$$

where $Z_1 = \{x \in X : x_2 = x_1^{\lambda_1}\}$ and λ_1 is some number in $]\mu_1, \nu_1[$, and $Z_1^* \cap \operatorname{int} X^*$ contains paths tending to 0.

Then let (c^1) be any sequence in $B^3(1/2)$ and let $1 < \lambda_1 < \lambda_2 < \ldots$. If $1 < \mu_1 < \lambda_1 < \nu_1 < \mu_2 < \lambda_2 < \nu_2 < \ldots$, we may deform f_1 in the sets Y^{j*}, $Y^j = \{x \in X : x_1^{\nu_j} < x_2 < x_1^{\mu_j}\}$, $j = 2, 3, \ldots$, by applying the construction from the first step to obtain a K-qr map $f_0: \operatorname{int} X^* \to B^3(4)$ such that

$$\lim_{\substack{x \to 0 \\ x \in Z_j^*}} f_0(x) = c^j ,$$

where $Z_j = \{ x \in X : x_2 = x_1^{\lambda_j} \}$. The required map satisfying (1) and (2) is then $f = f_0 \circ \Lambda$ where Λ is a suitable qc map of B^3 onto $\operatorname{int} X^*$. □

Positive results on the existence of an angular limit are obtained when a limit is assumed along a sufficiently large set on the boundary. These can be formulated in many ways for general domains. A simple such formulation for the unit ball was proved in [R6] and is the following.

7.3. Theorem. Let $n \geq 2$, let $A = \{ y \in \partial B^n : y_n > 0 \}$, and let $f \colon B^n \to B^n$ be a K-qr mapping such that for some $a \in \overline{B}^n$,

$$\lim_{\substack{y \to e_1 \\ y \in A}} \limsup_{x \to y} |f(x) - a| = 0 .$$

Then f has the angular limit a at e_1.

Proof. We may assume that $fB^n \subset B^n(1/2)$ and $a \in \overline{B}^n(1/2)$. We apply the 2-constants theorem 6.18 iteratively. Let us form a sequence $D_1, D_2, \ldots$ of subdomains of B^n inductively by setting

$$D_1 = B^n \cap \bigcup_{z \in A} B^n(z, \tfrac{1}{8} d(z, \partial B^n \smallsetminus A)) ,$$

$$D_{i+1} = B^n \cap \bigcup_{z \in D_i} B^n(z, \tfrac{}{8} d(z, \partial B^n \smallsetminus A)) .$$

Let E be a cone $K(e_1, \varphi)$, $0 < \varphi < \pi/2$. There exists a positive integer k and $t > 0$ such that $E \cap B^n(e_1, t) \subset D_k$. Given $\varepsilon > 0$ let $\eta > 0$ be such that $\eta < t$ and

$$\limsup_{x \to y} |f(x) - a| < \varepsilon \quad \text{if } y \in A \cap B^n(e_1, \eta) .$$

The 2-constants theorem 6.18 applied to balls $U = B^n(z, d(z, \partial B^n \smallsetminus A))$, $z \in A \cap B^n(e_1, \eta/2)$, and to the map $f - a$ with $8r = s = d(z, \partial B^n \smallsetminus A)$ gives

$$|f(x) - a| < e^{\beta \log \varepsilon} \quad \text{if } x \in D_1 \cap B^n(e_1, \eta/3)$$

where $\beta = 1 - (1/2)^\mu$ and $\mu > 0$ depends only on n and K. By repeated use of this we obtain

$$|f(x) - a| < e^{\beta^k \log \varepsilon} \quad \text{if } x \in D_k \cap B^n(e_1, \eta/3^k) .$$

Hence

$$\lim_{\substack{x \to e_1 \\ x \in E}} f(x) = a ,$$

and the theorem is proved. □

7.4. Remarks. 1. J. Heinonen and J. Rossi have in [HeR1, 2.1] extended Theorem 7.3 to the case of normal qm mappings.

2. For bounded K-qr mappings of $H = \{\, x \in \mathbb{R}^n : x_n > 0 \,\}$ M. Vuorinen gave in [Vu9, 3.3] the following result on the existence of an angular limit: Let f have a as an asymptotic value along a path $\gamma = (\gamma_1, \ldots, \gamma_n) \colon [0, 1[\to H$ with $\gamma(t) \to 0$ as $t \to 1$. There exists $\mu = \mu(n, K) > 0$ such that

$$(7.5) \qquad \lim_{t \to 1} \left(\frac{\gamma_n(t)}{P_n(\gamma(t))} \right)^{\mu} \log \frac{1}{|f(\gamma(t)) - a|} = \infty$$

implies f has the angular limit a at 0. Here P_n is the projection $\mathbb{R}^n \to \partial H$. The condition relates the tangentiality of γ to the speed of the approach of f to a along γ (see also [Vu4, 3.1] and Remark 2.3). For unbounded maps existence of angular limits are proved under various assumptions by Vuorinen in [Vu2, 4.3], [Vu3, 4.2, 5.2], [Vu4, 4.1], and [Vu7, 3.4, 4.2].

3. Further results on the existence of angular limits (or limits in general at a boundary point) have been proved for example in [Vu2, 4.1] and [GLM3, 4.21, 4.23, 4.27]. The result [GLM3, 4.27] is an interesting formulation by S. Granlund, P. Lindqvist, and O. Martio of a Lindelöf–type theorem for all dimensions, where they obtain an angular limit at a boundary point at which the boundary is locally a line segment.

Bibliography

[Ad]   R.A. ADAMS: *Sobolev Spaces.* – Academic Press, New York–San Francisco–London, 1975.

[Ag]   S. AGARD: *Angles and quasiconformal mappings in space.* – J. Analyse Math. 22 (1969) 177–200.

[AM]   S. AGARD and A. MARDEN: *A removable singularity theorem for local homeomorphisms.* – Indiana Math. J. 20 (1970) 455–461.

[A1]   L.V. AHLFORS: *Über die asymptotischen Werte der ganzen Funktionen endlichen Ordnung.* – Ann. Acad. Sci. Fenn. Ser. A 32;6 (1929) 1–15.

[A2]   L.V. AHLFORS: *Zur Theorie der Überlagerungsflächen.* – Acta Math. 65 (1935) 157–194.

[A3]   L.V. AHLFORS: *On quasiconformal mappings.* – J. Analyse Math. 3 (1954) 1–58, 207–208.

[A4]   L.V. AHLFORS: *Extension of quasiconformal mappings from two to three dimensions.* – Proc. Nat. Acad. Sci. U.S.A. 51 (1964) 768–771.

[A5]   L.V. AHLFORS: *Lectures on quasiconformal mappings.* – Van Nostrand Company, Toronto–New York–London, 1966.

[A6]   L.V. AHLFORS: *Conformal Invariants.* – McGraw–Hill, New York, 1973.

[ABe]   L.V. AHLFORS and L. BERS: *Riemann's mapping theorem for variable metrics.* – Ann. Math. 72 (1960) 385–404.

[AB]   L.V. AHLFORS and A. BEURLING: *Conformal invariants and function-theoretic null-sets.* – Acta Math. 83 (1950) 101–129.

[AVV]   G.D. ANDERSON, M.K. VAMANAMURTHY and M. VUORINEN: *Conformal invariants, quasiconformal maps, and special functions.* – Lecture Notes in Mathematics, vol. 1508, pp. 1–19. Springer, Berlin–Heidelberg–New York, 1992.

[As]   K. ASTALA: *Area distortion of quasiconformal mappings.* – (To appear).

[BA]   A. BEURLING and L.V. AHLFORS: *The boundary correspondence under quasiconformal mappings.* – Acta Math. 96 (1956) 125–142.

[B]   B. BOJARSKI: *Generalized solutions of a system of first order differential equations of elliptic type with discontinuous coefficients.* (Russian). – Mat. Sb. 43 (1957) 451–503.

[BI1]   B. BOJARSKI and T. IWANIEC: *Another approach to Liouville theorem.* – Math. Nachr. 107 (1982) 253–262.

[BI2]   B. BOJARSKI and T. IWANIEC: *Analytical foundations of the theory of quasiconformal mappings in R^n.* – Ann. Acad. Sci. Fenn. Ser. A I Math. 8 (1983) 257–324.

[BI3]   B. BOJARSKI and T. IWANIEC: *p-harmonic equation and quasiregular mappings.* – Banach Center Publ., vol. 19, pp. 25–38. Warsaw, 1987.

[BM]   V.A. BOTVINNIK and V.M. MIKLJUKOV: *A theorem of Phragmén–Lindelöf type for n-dimensional mappings with bounded distortion.* (Russian). – Sibirsk. Mat. Zh. 21 (1980) 232–235.

[Cald] A.P. CALDERÓN: *On the differentiability of absolutely continuous functions.* – Riv. Mat. Univ. Parma 2 (1951) 203–213.

[Ca]   E.D. CALLENDER: *Hölder continuity of n-dimensional quasiconformal mappings.* – Pacific J. Math. 19 (1960) 499–515.

[Carl] T. CARLEMAN: *Sur une inégalite différentielle dans la théorie des fonctions analytiques.* – C. R. Acad. Sci. Paris 196 (1933) 995–997.

[Car]   L. CARLESON: *A remark on Picard's theorem.* – Bull. Amer. Math. Soc. 67 (1961) 142–144.

[Ce]   L. CESARI: *Sulle funzioni assolutamente continue in due variabili.* – Ann. Scuola Norm. Sup. Pisa 10 (1941) 91–101.

[Ch1]   A.V. CHERNAVSKIĬ: *Finite–to–one open mappings of manifolds.* (Russian). – Mat. Sb. 65 (1964) 357–369.

[Ch2]   A.V. CHERNAVSKIĬ: *Remarks on the paper "On finite–to–one mappings of manifolds".* (Russian). – Mat. Sb. 66 (1965) 471–472.

[C]   P.T. CHURCH: *Differentiable open maps on manifolds.* – Trans. Amer. Math. Soc. 109 (1963) 87–100.

[CH]   P.T. CHURCH and E. HEMMINGSEN: *Light open maps on n-manifolds.* – Duke Math. J. 27 (1960) 527–536.

[CT]   P.T. CHURCH and J.G. TIMOURIAN: *Differentiable maps with small critical set or critical set image.* – Indiana Univ. Math. J. 27 (1978) 953–971.

[CL]   E.F. COLLINGWOOD and A.J. LOHWATER: *The Theory of Cluster Sets.* – Cambridge University Press, Cambridge, 1966.

[DS]   S.K. DONALDSON and D.P. SULLIVAN: *Quasiconformal 4-manifolds.* – Acta Math. 163 (1990) 181–252.

[D]   D. DRASIN: *The inverse problem of the Nevanlinna theory.* – Acta Math. 138 (1977) 83–151.

[EL]   A. EREMENKO and J.L. LEWIS: *Uniform limits of certain A-harmonic functions with applications to quasiregular mappings.* – Ann. Acad. Sci. Fenn. Ser. A I Math. 16 (1991) 361–375.

[F]   H. FEDERER: *Geometric Measure Theory.* – Springer, Berlin–Heidelberg–New York, 1969.

[Fl]   E. FLOYD: *Some characterizations of interior maps.* – Ann. Math. 51 (1950) 571–575.

[Fu]   B. FUGLEDE: *Extremal length and functional completion.* – Acta Math. 98 (1957) 171–219.

[GM]   D.B. GAULD and G.J. MARTIN: *Essential singularities of quasimeromorphic mappings.* – Math. Scand. (To appear).

[G1]   F.W. GEHRING: *The definitions and exceptional sets for quasiconformal mappings.* – Ann. Acad. Sci. Fenn. Ser. A I Math. 281 (1960) 1–28.

[G2]   F.W. GEHRING: *Symmetrization of rings in space.* – Trans. Amer. Math. Soc. 101 (1961) 499–519.

[G3]   F.W. GEHRING: *Extremal length definitions for the conformal capacity of rings in space.* – Michigan Math. J. 9 (1962) 137–150.

[G4]   F.W. GEHRING: *Rings and quasiconformal mappings in space.* – Trans. Amer. Math. Soc. 103 (1962) 353–393.

[G5]   F.W. GEHRING: *The Carathéodory convergence theorem for quasiconformal mappings in space.* – Ann. Acad. Sci. Fenn. Ser. A I Math. 336/11 (1963) 1–21.

[G6]   F.W. GEHRING: *Lipschitz mappings and the p-capacity of rings in n-space.* – Advances in the Theory of Riemann Surfaces. Ann. Math. Stud., vol. 66, pp. 175–193. Princeton Univ. Press, Princeton, NJ, 1971.

[G7]   F.W. GEHRING: *The L^p-integrability of the partial derivatives of quasiconformal mappings.* – Acta Math. 130 (1973) 265–277.

[GL]   F.W. GEHRING and O. LEHTO: *On the total differentiability of functions of a complex variable.* – Ann. Acad. Sci. Fenn. Ser. A I Math. 272 (1959) 1–9.

[GV1]   F.W. GEHRING and J. VÄISÄLÄ: *The coefficients of quasiconformality of domains in space.* – Acta Math. 114 (1965) 1–70.

[GV2]   F.W. GEHRING and J. VÄISÄLÄ: *Hausdorff dimension and quasiconformal mappings.* – J. London Math. Soc. 6(2) (1973) 504–512.

[GT]   D. GILBARG and N.S. TRUDINGER: *Elliptic Partial Differential Equations of Second Order.* – Grundlehren der mathematischen Wissenschaften, vol. 222. Springer, Berlin–Heidelberg–New York, 1977.

[Go]   V.M. GOL'DSHTEIN: *On the behavior of mappings with bounded distortion with distortion coefficient close to unity.* – Sibirsk. Mat. Zh. 12 (1971) 1250–1258.

[Gr]   S. GRANLUND: *Harnack's inequality in the borderline case.* – Ann. Acad. Sci. Fenn. Ser. A I Math. 5 (1980) 159–163.

[GLM1]   S. GRANLUND, P. LINDQVIST, and O. MARTIO: *Conformally invariant variational integrals.* – Trans. Amer. Math. Soc. 277 (1983) 43–73.

[GLM2]   S. GRANLUND, P. LINDQVIST, and O. MARTIO: *F-harmonic measure in space.* – Ann. Acad. Sci. Fenn. Ser. A I Math. 7 (1982) 233–247.

[GLM3]   S. GRANLUND, P. LINDQVIST, and O. MARTIO: *Phragmén–Lindelöf's and Lindelöf's theorem.* – Ark. Mat. 23 (1985) 103–128.

[GLM4]   S. GRANLUND, P. LINDQVIST, and O. MARTIO: *Note on the PWB–method in the nonlinear case.* – Pacific J. Math. 125 (1986) 381–395.

[Gre]   W.H. GREUB: *Linear Algebra.* – Grundlehren der mathematischen Wissenschaften, vol. 97, 3rd edn. Springer, Berlin–Heidelberg, 1967.

[Gro1]   M. GROMOV: *Hyperbolic manifolds, groups and actions.* – Proceedings of the 1978 Stony Brook Conference on Riemann Surfaces and Related Topics. Ann. Math. Stud., vol. 97, pp. 183–213. Princeton Univ. Press, Princeton, NJ, 1981.

[Gro2]   M. GROMOV: *Structures métriques pour les variétés riemanniennes. Rédigé par J. Lafontaine et P. Pansu.* – Cedic, Paris, 1981.

[Grö]   H. GRÖTZSCH: *Über die Verzerrung bei schlichten nichtkonformen Abbildungen und über eine damit zusammenhängende Erweiterung des Picardschen Satzes.* – Ber. Verh. Sächs. Akad. Wiss. Leipzig 80 (1928) 503–507.

[HK1]   J. HEINONEN and T. KILPELÄINEN: *On the Wiener criterion and quasilinear obstacle problems.* – Trans. Amer. Math. Soc. 310 (1988) 239–255.

[HK2]   J. HEINONEN and T. KILPELÄINEN: *A-superharmonic functions and supersolutions of degenerate elliptic equations.* – Ark. Mat. 26 (1988) 87–105.

[HKM]   J. HEINONEN, T. KILPELÄINEN, and O. MARTIO: *Nonlinear Potential Theory.* – Oxford University Press, Oxford, 1993.

[HKR]   J. HEINONEN, T. KILPELÄINEN, and J. ROSSI: *The growth of A-harmonic functions and quasiregular mappings along asymptotic paths.* – Indiana Univ. Math. J. 38 (1989) 581–601.

[HM]   J. HEINONEN and O. MARTIO: *Estimates for F-harmonic measures and Øksendal's theorem for quasiconformal mappings.* – Indiana Univ. Math. J. 36 (1987) 659–683.

[HeR1]   J. HEINONEN and J. ROSSI: *Lindelöf's theorem for normal quasimeromorphic mappings.* – Michigan Math. J. 37 (1990) 219–226.

[HeR2]   J. HEINONEN and J. ROSSI: *Remarks on the value distribution of quasimeromorphic mappings.* – Complex Variables (To appear).

[Hil]   H.M. HILDEN: *Three-fold branched coverings of S^3.* – Amer. J. Math. 98 (1976) 989–997.

[Hin]   A. HINKKANEN: *On the averages of the counting function of a meromorphic function.* – Ann. Acad. Sci. Fenn. Ser. A I Math. Diss. 26 (1980) 1–31.

[H1]   I. HOLOPAINEN: *Nonlinear potential theory and quasiregular mappings on Riemannian manifolds.* – Ann. Acad. Sci. Fenn. Ser. A I Math. Diss. 74 (1990).

[H2]   I. HOLOPAINEN: *Positive solutions of quasilinear elliptic equations on Riemannian manifolds.* – Proc. London Math. Soc. (3) 65 (1992) 651–672.

[HR1]   I. HOLOPAINEN and S. RICKMAN: *Classification of Riemannian manifolds in nonlinear potential theory.* – Potential Analysis (To appear).

[HR2]   I. HOLOPAINEN and S. RICKMAN: *A Picard type theorem for quasiregular mappings of R^n into n-manifolds with many ends.* – Rev. Mat. Iberoamericana 8 (1992) 131–148.

[HR3]   I. HOLOPAINEN and S. RICKMAN: *Quasiregular mappings of the Heisenberg group.* – Math. Ann. 294 (1992) 625–643.

[Hu]   S.-T. HU: *Homotopy Theory.* – Academic Press, New York–London, 1959.

[HW]   W. HUREWICZ and H. WALLMAN: *Dimension Theory.* – Princeton Univ. Press, Princeton, NJ, 1948.

[I1]   T. IWANIEC: *On L^p-integrability in PDE's and quasiregular mappings for large exponents.* – Ann. Acad. Sci. Fenn. Ser. A I Math. 7 (1982) 301–322.

[I2]   T. IWANIEC: *Some aspects of partial differential equations and quasiregular mappings.* – Proc. Internat. Congr. Math. (Warsaw, 1983), vol. 2, pp. 1193–1208. PWN, Warsaw, 1984.

[I3]   T. IWANIEC: *Stability property of Möbius mappings.* – Proc. Amer. Math. Soc. 100 (1987) 61–69.

[I4]   T. IWANIEC: *p-harmonic tensors and quasiregular mappings.* – Ann. Math. (To appear).

[I5]   T. IWANIEC: *L^p-theory of quasiregular mappings.* – Lecture Notes in Mathematics, vol. 1508, pp. 39–64. Springer, Berlin–Heidelberg–New York, 1992.

[IM]   T. IWANIEC and G. MARTIN: *Quasiregular mappings in even dimensions.* – Acta Math. (To appear).

[IS]   T. IWANIEC and C. SBORDONE: *Weak minima of variational integrals.* – (To appear).

[J]   P. JÄRVI: *On the behavior of quasiregular mappings in the neighborhood of an isolated singularity.* – Ann. Acad. Sci. Fenn. Ser. A I Math. 15 (1990) 341–353.

[JV]   P. JÄRVI and M. VUORINEN: *Self–similar Cantor sets and quasiregular mappings.* – J. reine angew. Math. 424 (1992) 31–45.

[Jo]   J. JORMAKKA: *The existence of quasiregular mappings from R^3 to closed orientable 3-manifolds.* – Ann. Acad. Sci. Fenn. Ser. A I Math. Diss. 69 (1988) 1–40.

[KR1]   A. KORÁNYI and H.M. REIMANN: *Quasiconformal mappings on the Heisenberg group.* – Invent. math. 80 (1985) 309–338.

[KR2]   A. KORÁNYI and H.M. REIMANN: *Foundations of the theory of quasiconformal mappings on the Heisenberg group.* – Invent. math. (To appear).

[KM]   P. KOSKELA and O. MARTIO: *Removability theorems for quasiregular mappings.* – Ann. Acad. Sci. Fenn. Ser. A I Math. 15 (1990) 381–399.

[La]   M.A. LAVRENT'EV: *Sur un critère différentiel des transformations homéomorphes des domaines a trois dimensions.* – Dokl. Akad. Nauk SSSR 22 (1938) 241–242.

[L]   H. LEBESGUE: *Sur de problème de Dirichlet.* – Rend. Circ. Mat. Palermo 24 (1907) 371–402.

[LV1]   O. LEHTO and K.I. VIRTANEN: *Boundary behaviour and normal meromorphic functions.* – Acta Math. 97 (1957) 47–65.

[LV2]   O. LEHTO and K.I. VIRTANEN: *Quasiconformal Mappings in the Plane.* – Grundlehren der mathematischen Wissenschaften, vol. 126, 2nd edn. Springer, Berlin–Heidelberg–New York, 1973.

[LF]   J. LELONG–FERRAND: *Étude d'une classe d'applications liées à des homo-morphismes d'algèbres de fonctions, et generalisant les quasi conformes.* – Duke Math. J. 40 (1973) 163–186.

[Le1]   J.L. LEWIS: *On a conditional theorem of Littlewood for quasiregular entire functions.* – J. Analyse Math. (To appear).

[Le2]   J.L. LEWIS: *Picard's theorem and Rickman's theorem by way of Harnack's inequality.* – Proc. Amer. Math. Soc. (To appear).

[Le3]   J.L. LEWIS: *On very weak solutions of certain elliptic systems.* – Preprint.

[Li1]   P. LINDQVIST: *A new proof of the lower–semicontinuity of certain convex variational integrals in Sobolev spaces.* – Helsinki University of Technology, Report–HTTK–MAT–A 97 (1977) 1–10.

[Li2]   P. LINDQVIST: *On the quasiregularity of a limit mapping.* – Ann. Acad. Sci. Fenn. Ser. A I Math. 11 (1986) 155–159.

[LM]   P. LINDQVIST and O. MARTIO: *Two theorems of N. Wiener for solutions of quasilinear elliptic equations.* – Acta Math. 155 (1985) 153–171.

[Lo]   C. LOEWNER: *On the conformal capacity in space.* – J. Math. Mech. 8 (1959) 411–414.

[MarR] A. MARDEN and S. RICKMAN: *Holomorphic mappings with bounded distor-tion.* – Proc. Amer. Math. Soc. 46 (1974) 226–228.

[M1]   O. MARTIO: *A capacity inequality for quasiregular mappings.* – Ann. Acad. Sci. Fenn. Ser. A I Math. 474 (1970) 1–18.

[M2]   O. MARTIO: *On the integrability of the derivative of a quasiregular mapping.* – Math. Scand. 35 (1974) 43–48.

[M3]   O. MARTIO: *Equicontinuity theorem with an application to variational in-tegrals.* – Duke Math. J. 42 (1975) 569–581.

[M4]   O. MARTIO: *On k-periodic quasiregular mappings in R^n .* – Ann. Acad. Sci. Fenn. Ser. A I Math. 1 (1975) 207–220.

[M5]   O. MARTIO: *Capacity and measure densities.* – Ann. Acad. Sci. Fenn. Ser. A I Math. 4 (1978/1979) 109–118.

[M6]   O. MARTIO: *Reflection principle for elliptic partial differential equations and quasiregular mappings.* – Ann. Acad. Sci. Fenn. Ser. A I Math. 6 (1981) 179–188.

[M7]   O. MARTIO: *Counterexamples for unique continuation.* – Manuscripta Math. 60 (1988) 21–47.

[M8]   O. MARTIO: *F-harmonic measures, quasihyperbolic distance and Milloux's problem.* – Ann. Acad. Sci. Fenn. Ser. A I Math. 12 (1987) 151–162.

[M9]   O. MARTIO: *Sets of zero elliptic harmonic measures.* – Ann. Acad. Sci. Fenn. Ser. A I Math. 14 (1989) 47–55.

[M10]   O. MARTIO: *Partial differential equations and quasiregular mappings.* – Lecture Notes in Mathematics, vol. 1508, pp. 65–79. Springer, Berlin–Heidelberg–New York, 1992.

[MR1]   O. MARTIO and S. RICKMAN: *Boundary behavior of quasiregular mappings.* – Ann. Acad. Sci. Fenn. Ser. A I Math. 507 (1972) 1–17.

[MR2]   O. MARTIO and S. RICKMAN: *Measure properties of the branch set and its image of quasiregular mappings.* – Ann. Acad. Sci. Fenn. Ser. A I Math. 541 (1973) 1–16.

[MRV1] O. MARTIO, S. RICKMAN, and J. VÄISÄLÄ: *Definitions for quasiregular mappings.* – Ann. Acad. Sci. Fenn. Ser. A I Math. 448 (1969) 1–40.

[MRV2] O. MARTIO, S. RICKMAN, and J. VÄISÄLÄ: *Distortion and singularities of quasiregular mappings.* – Ann. Acad. Sci. Fenn. Ser. A I Math. 465 (1970) 1–13.

[MRV3] O. MARTIO, S. RICKMAN, and J. VÄISÄLÄ: *Topological and metric proper-ties of quasiregular mappings.* – Ann. Acad. Sci. Fenn. Ser. A I Math. 488 (1971) 1–31.

[MSa1]   O. MARTIO and J. SARVAS: *Density conditions in the n-capacity.* – Indiana Univ. Math. J. 26 (1977) 761–776.

[MSa2]   O. MARTIO and J. SARVAS: *Injectivity theorems in plane and space.* – Ann. Acad. Sci. Fenn. Ser. A I Math. 4 (1978/79) 383–401.

[MS1]   O. MARTIO and U. SREBRO: *Periodic quasimeromorphic mappings.* – J. Analyse Math. 28 (1975) 20–40.

[MS2]   O. MARTIO and U. SREBRO: *Automorphic quasimeromorphic mappings in* R^n. – Acta Math. 135 (1975) 221–247.

[MS3]   O. MARTIO and U. SREBRO: *On the existence of automorphic quasi-meromorphic mappings in* R^n. – Ann. Acad. Sci. Fenn. Ser. A I Math. 3 (1977) 123–130.

[MS4]   O. MARTIO and U. SREBRO: *Universal radius of injectivity for locally quasiconformal mappings.* – Israel J. Math. 29 (1978) 17–23.

[MS5]   O. MARTIO and U. SREBRO: *On the local behavior of quasiregular maps and branched covering maps.* – J. Analyse Math. 36 (1979) 198–212.

[MV1]   O. MARTIO and J. VÄISÄLÄ: *Elliptic equations and maps of bounded length distortion.* – Math. Ann. 282 (1988) 423–443.

[MV2]   O. MARTIO and J. VÄISÄLÄ: *Global* L^p-integrability of the derivative of a *quasiconformal mapping.* – Complex Variables 9 (1988) 309–319.

[MZ]   O. MARTIO and W.P. ZIEMER: *Lusin's condition (N) and mappings with non-negative Jacobians.* – Michigan J. Math. (To appear).

[MaR]   P. MATTILA and S. RICKMAN: *Averages of the counting function of a quasiregular mapping.* – Acta Math. 143 (1979) 273–305.

[Ma]   V.G. MAZ'YA: *On the continuity at a boundary point of the solutions of quasilinear equations.* (Russian). – Vestnik Leningrad. Univ. 25, no. 13 (1970) 42–55.

[Mik1]   V.M. MIKLYUKOV: *Oriented quasiconformal mappings in space.* (Russian). – Dokl. Akad. Nauk SSSR 182 (1968) 266–267.

[Mik2]   V.M. MIKLYUKOV: *Removable singularities of quasiconformal mappings in space.* (Russian). – Dokl. Akad. Nauk SSSR 188 (1969) 525–527.

[Mik3]   V.M. MIKLYUKOV: *On a boundary property of n-dimensional mappings of bounded distortion.* (Russian). – Mat. Zametki 11 (1972) 159–164.

[Mik4]   V.M. MIKLYUKOV: *Asymptotic properties of subsolutions of quasilinear equations of elliptic type and mappings with bounded distortion.* (Russian). – Mat. Sb. 11 (1980) 42–66.

[Mil1]   J. MILES: *A note on Ahlfors' theory of covering surfaces.* – Proc. Amer. Math. Soc. 21 (1969) 30–32.

[Mil2]   J. MILES: *On the counting function for the a-values of a meromorphic function.* – Trans. Amer. Math. Soc. 147 (1970) 203–222.

[Mil3]   J. MILES: *Bounds on the ratio* $n(r,a)/S(r)$ *for meromorphic functions.* – Trans. Amer. Math. Soc. 162 (1971) 383–393.

[Mi1]   R. MINIOWITZ: *Distortion theorems for quasiregular mappings.* – Ann. Acad. Sci. Fenn. Ser. A I Math. 4 (1978/1979) 63–74.

[Mi2]   R. MINIOWITZ: *Spherically mean p-valent quasiregular mappings.* – Israel J. Math. 38 (1981) 199–208.

[Mi3]   R. MINIOWITZ: *Normal families of quasimeromorphic mappings.* – Proc. Amer. Math. Soc. 84 (1982) 35–43.

[Mi4]   R. MINIOWITZ: *A volume-area inequality.* – J. London Math. Soc. (2) 25 (1982) 88–98.

[Mor1]   C.B. MORREY: *On the solution of quasilinear elliptic partial differential equations.* – Trans. Amer. Math. Soc. 43 (1938) 126–166.

[Mor2]   C.B. MORREY: *Multiple Integrals in the Calculus of Variations.* – Grundlehren der mathematischen Wissenschaften, vol. 130. Springer, Berlin–Heidelberg–New York, 1966.

[Mos]   J. MOSER: *On Harnack's theorem for elliptic differential equations.* –
        Comm. Pure and Appl. Math. 14 (1961) 577–591.

[Mo]   G.D. MOSTOW: *Quasiconformal mappings in n-space and the rigidity of
        hyperbolic space forms.* – Inst. Hautes Études Sci. Publ. Math. 34 (1968)
        53–104.

[N1]   R. NEVANLINNA: *Zur Theorie der meromorphen Funktionen.* – Acta Math.
        46 (1925) 1–99.

[N2]   R. NEVANLINNA: *Analytic Functions.* – Grundlehren der mathematischen
        Wissenschaften, vol. 162. Springer, Berlin–Heidelberg–New York, 1970.

[O]   M. OHTSUKA: *Dirichlet Problem, Extremal Length, and Prime Ends.* – Van
        Nostrand, Reinhold, 1970.

[Ø1]   B. ØKSENDAL: *Dirichlet forms, quasiregular functions and Brownian mo-
        tion.* – Invent. math. 91 (1988) 273–297.

[Ø2]   B. ØKSENDAL: *Weighted Sobolev inequalities and harmonic measure asso-
        ciated to quasiregular functions.* – Comm. Partial Differential Equations 15
        (1990) 1447–1459.

[Pa1]   P. PANSU: *Quasiconformal mappings and manifolds of negative curva-
        ture.* – Curvature and Topology of Riemann Manifolds. Proc. 17th Intern.
        Taniguchi symposium held in Katata, Japan, Aug. 26–31, 1985, Lecture
        Notes in Mathematics, vol. 1201. Springer, Berlin–Heidelberg–New York,
        1986.

[Pa2]   P. PANSU: *An isoperimetric inequality on the Heisenberg group.* – Proceed-
        ings of "Differential Geometry on Homogeneous Spaces", Torino, 1983,
        159–174.

[Pel]   K. PELTONEN: *On the existence of quasiregular mappings.* – Ann. Acad.
        Sci. Fenn. Ser. A I Math. Diss. 85 (1992) 1–48.

[Pe1]   M. PESONEN: *Poletskiĭ's inequality, condition (N), and the integral trans-
        formation formula for quasiregular mappings.* (Finnish). – University of
        Helsinki, Lic. thesis, 1980 (Unpublished).

[Pe2]   M. PESONEN: *A path family approach to Ahlfors's value distribution theory.*
        – Ann. Acad. Sci. Fenn. Ser. A I Math. Diss. 39 (1982) 1–32.

[Pe3]   M. PESONEN: *Simplified proofs of some basic theorems for quasiregular
        mappings.* – Ann. Acad. Sci. Fenn. Ser. A I Math. 8 (1983) 247–250.

[PL]   E. PHRAGMÉN and E. LINDELÖF: *Sur une extension d'un principe classique
        de l'analyse et sur quelques propriétés des fonctions monogènes dans le
        voisinage d'un point singulier.* – Acta Math. 31 (1908) 381–406.

[Pi]   E. PICARD: *Sur une propriété des fonctions entières.* – C. R. Acad. Sci.
        Paris 88 (1879) 1024–1027.

[P1]   E.A. POLETSKIĬ: *The modulus method for non–homeomorphic quasiconfor-
        mal mappings.* (Russian). – Mat. Sb. 83 (1970) 261–272.

[P2]   E.A. POLETSKIĬ: *On the removal of singularities of quasiconformal map-
        pings.* (Russian). – Mat. Sb. 92 (134) (1973) 261–273.

[RR]   T. RADO and P.V. REICHELDERFER: *Continuous Transformations in Anal-
        ysis.* – Grundlehren der math. Wissenschaften, vol. 75. Springer, Berlin–
        Göttingen–Heidelberg, 1955.

[Re1]   YU.G. RESHETNYAK: *Bounds on moduli of continuity for certain mappings.*
        (Russian). – Sibirsk. Mat. Zh. 7 (1966) 1106–1114.

[Re2]   YU.G. RESHETNYAK: *Space mappings with bounded distortion.* (Russian).
        – Sibirsk. Mat. Zh. 8 (1967) 629–659.

[Re3]   YU.G. RESHETNYAK: *The Liouville theorem with minimal regularity con-
        ditions.* (Russian). – Sibirsk. Mat. Zh. 8 (1967) 835–840.

[Re4]   YU.G. RESHETNYAK: *On the condition of the boundedness of index for
        mappings with bounded distortion.* (Russian). – Sibirsk. Mat. Zh. 9 (1968)
        368–374.

[Re5]   YU.G. RESHETNYAK: *Mappings with bounded distortion as extremals of Dirichlet type integrals.* (Russian). – Sibirsk. Mat. Zh. 9 (1968) 652–666.

[Re6]   YU.G. RESHETNYAK: *Stability theorems for mappings with bounded distortion.* (Russian). – Sibirsk. Mat. Zh. 9 (1968) 667–684.

[Re7]   YU.G. RESHETNYAK: *The concept of capacity in the theory of functions with bounded distortion.* (Russian). – Sibirsk. Mat. Zh. 10 (1969) 1109–1138.

[Re8]   YU.G. RESHETNYAK: *On extremal properties of mappings with bounded distortion.* (Russian). – Sibirsk. Mat. Zh. 10 (1969) 1300–1310.

[Re9]   YU.G. RESHETNYAK: *The local structure of mappings with bounded distortion.* (Russian). – Sibirsk. Mat. Zh. 10 (1969) 1311-1313.

[Re10]   YU.G. RESHETNYAK: *On the branch set of mappings with bounded distortion.* (Russian). – Sibirsk. Mat. Zh. 11 (1970) 1333-1339.

[Re11]   YU.G. RESHETNYAK: *Space Mappings with Bounded Distortion.* – Transl. of Math. Monographs, vol. 73. Amer. Math. Soc., Providence, R.I., 1989.

[R1]   S. RICKMAN: *Path lifting for discrete open mappings.* – Duke Math. J. 40 (1973) 187–191.

[R2]   S. RICKMAN: *A path lifting construction for discrete open mappings with application to quasimeromorphic mappings.* – Duke Math. J. 42 (1975) 797–809.

[R3]   S. RICKMAN: *A quasimeromorphic mapping with given deficiencies in dimension three.* – Sympos. Math., vol. XVIII, pp. 535–549. Academic Press, London, 1976.

[R4]   S. RICKMAN: *On the value distribution of quasimeromorphic maps.* – Ann. Acad. Sci. Fenn. Ser. A I Math. 2 (1976) 447–466.

[R5]   S. RICKMAN: *On the number of omitted values of entire quasiregular mappings.* – J. Analyse Math. 37 (1980) 100–117.

[R6]   S. RICKMAN: *Asymptotic values and angular limits of quasiregular mappings of a ball.* – Ann. Acad. Sci. Fenn. Ser. A I Math. 5 (1980) 185–196.

[R7]   S. RICKMAN: *A defect relation for quasimeromorphic mappings.* – Ann. Math. 114 (1981) 165–191.

[R8]   S. RICKMAN: *Value distribution of quasimeromorphic mappings.* – Ann. Acad. Sci. Fenn. Ser. A I Math. 7 (1982) 81–88.

[R9]   S. RICKMAN: *Value distribution of quasiregular mappings.* – Proc. Value Distribution Theory, Joensuu 1981. Lecture Notes in Mathematics, vol. 981, pp. 220–245. Springer, Berlin–Heidelberg–New York, 1983.

[R10]   S. RICKMAN: *Quasiregular mappings and metrics on the n-sphere with punctures.* – Comment. Math. Helv. 59 (1984) 134–148.

[R11]   S. RICKMAN: *The analogue of Picard's theorem for quasiregular mappings in dimension three.* – Acta Math. 154 (1985) 195–242.

[R12]   S. RICKMAN: *Sets with large local index of quasiregular mappings in dimension three.* – Ann. Acad. Sci. Fenn. Ser. A I Math. 10 (1985) 493–498.

[R13]   S. RICKMAN: *Existence of quasiregular mappings.* – Proceedings of the Workshop on Holomorphic Functions and Moduli I. Math. Sci. Res. Inst. Publ., vol. 10, Berkeley, pp. 179–185. Springer, New York, 1988.

[R14]   S. RICKMAN: *Topics in the theory of quasiregular mappings.* – Conformal Geometry. Aspects of Math., vol. E 12, pp. 147–189. Friedr. Vieweg & Sohn, Braunschweig, 1988.

[R15]   S. RICKMAN: *Quasiconformal mappings.* – Ann. Acad. Sci. Fenn. Ser. A I Math. 13 (1988) 371–385.

[R16]   S. RICKMAN: *Defect relation and its realization for quasiregular mappings.* – Preprint 20, University of Helsinki, 1992.

[R17]   S. RICKMAN: *Nonremovable Cantor sets for bounded quasiregular mappings.* – Preprint 42, Institut Mittag–Leffler, 1989/1990.

[R18]   S. RICKMAN: *Picard's theorem and defect relation for quasiregular map-pings.* – Lecture Notes in Mathematics, vol. 1508, pp. 93–103. Springer, Berlin–Heidelberg–New York, 1992.

[RS]   S. RICKMAN and U. SREBRO: *Remarks on the local index of quasiregular mappings.* – J. Analyse Math. 46 (1986) 246–250.

[RV]   S. RICKMAN and M. VUORINEN: *On the order of quasiregular mappings.* – Ann. Acad. Sci. Fenn. Ser. A I Math. 7 (1982) 221–231.

[Ru]   W. RUDIN: *Real and Complex Analysis.* – McGraw–Hill, New York, 1970.

[S]   S. SAKS: *Theory of the Integral.* – Dover Publications, New York, 1964.

[Sa1]   J. SARVAS: *Symmetrization of condensers in n-space.* – Ann. Acad. Sci. Fenn. Ser. A I Math. 522 (1972) 1–44.

[Sa2]   J. SARVAS: *On the local behavior of quasiregular mappings.* – Ann. Acad. Sci. Fenn. Ser. A I Math. 1 (1975) 221–226.

[Sa3]   J. SARVAS: *The Hausdorff dimension of the branch set of a quasiregular mapping.* – Ann. Acad. Sci. Fenn. Ser. A I Math. 1 (1975) 297–307.

[Sa4]   J. SARVAS: *Coefficient of injectivity for quasiregular mappings.* – Duke Math. J. 43 (1976) 147–158.

[Sas]   S. SASTRY: *Upper bounds for value distribution of quasimeromorphic maps.* – Thesis, Purdue University, 1992.

[Se]   J. SERRIN: *Local behavior of solutions of quasilinear equations.* – Acta Math. 111 (1964) 247–302.

[Sr1]   U. SREBRO: *Analogues of the elliptic modular function in R^3.* – Symposium on Complex Analysis, Canterbury, 1973. London Math. Soc. Lecture Notes Series, vol. 12, pp. 125–128. Cambridge Univ. Press, Cambridge, 1974.

[Sr2]   U. SREBRO: *Quasiregular mappings.* – Advances in Complex Analysis, Lecture Notes in Mathematics, vol. 503, pp. 148–163. Springer, Berlin–Heidelberg–New York, 1976.

[St]   E. M. STEIN: *Singular Integrals and Differentiability Properties of Functions.* – Princeton Univ. Press, Princeton, NJ, 1970.

[Sto1]   S. STOÏLOW: *Sur les transformations continues et la topologie des fonctions analytiques.* – Ann. Sci. École Norm. Sup. 45 (1928) 347–382.

[Sto2]   S. STOÏLOW: *Leçons sur les principes topologiques de la théorie des fonctions analytiques.* – Gauthier-Villars, Paris, 1938.

[To]   S. TOPPILA: *On the counting function for the a-values of a meromorphic function.* – Ann. Acad. Sci. Fenn. Ser. A I Math. 2 (1976) 565–572.

[T]   P. TUKIA: *Automorphic quasimeromorphic mappings for torsionless hyperbolic groups.* – Ann. Acad. Sci. Fenn. Ser. A I Math. 10 (1985) 545–560.

[V1]   J. VÄISÄLÄ: *Two new characterizations for quasiconformality.* – Ann. Acad. Sci. Fenn. Ser. A I Math. 362 (1965) 1–12.

[V2]   J. VÄISÄLÄ: *Minimal mappings in euclidean spaces.* – Ann. Acad. Sci. Fenn. Ser. A I Math. 366 (1965) 1–22.

[V3]   J. VÄISÄLÄ: *Discrete open mappings on manifolds.* – Ann. Acad. Sci. Fenn. Ser. A I Math. 392 (1966) 1–10.

[V4]   J. VÄISÄLÄ: *Lectures on n-Dimensional Quasiconformal Mappings.* – Lecture Notes in Mathematics, vol. 229. Springer, Berlin–Heidelberg–New York, 1971.

[V5]   J. VÄISÄLÄ: *Modulus and capacity inequalities for quasiregular mappings.* – Ann. Acad. Sci. Fenn. Ser. A I Math. 509 (1972) 1–14.

[V6]   J. VÄISÄLÄ: *Capacity and measure.* – Michigan Math. J. 22 (1975) 1–3.

[V7]   J. VÄISÄLÄ: *A survey of quasiregular maps in R^n.* – Proc. Internat. Congr. Math. (Helsinki, 1978), vol. 2, pp. 685–691. Acad. Sci. Fennica, Helsinki, 1980.

[V8]   J. VÄISÄLÄ: *Quasimöbius maps.* – J. Analyse Math. 44 (1984/85) 218–234.

[Vu1]   M. VUORINEN: *Exceptional sets and boundary behavior of quasiregular mappings in n-space.* – Ann. Acad. Sci. Fenn. Ser. A I Math. Diss. 11 (1976) 1–44.

[Vu2]   M. VUORINEN: *On the Iversen–Tsuji theorem for quasiregular mappings.* – Math. Scand. 41 (1977) 90–98.

[Vu3]   M. VUORINEN: *On the boundary behavior of locally K-quasiconformal mappings in space.* – Ann. Acad. Sci. Fenn. Ser. A I Math. 5 (1980) 79–95.

[Vu4]   M. VUORINEN: *Capacity densities and angular limits of quasiregular mappings.* – Trans. Amer. Math. Soc. 263 (1981) 343–354.

[Vu5]   M. VUORINEN: *Lindelöf-type theorems for quasiconformal and quasiregular mappings.* – Proc. Complex Analysis Semester, Banach Center Publ., vol. 11, pp. 353–362. Warsaw, 1983.

[Vu6]   M. VUORINEN: *Some inequalities for the moduli of curve families.* – Michigan Math. J. 30 (1983) 369–380.

[Vu7]   M. VUORINEN: *On functions with a finite or locally bounded Dirichlet integral.* – Ann. Acad. Sci. Fenn. Ser. A I Math. 9 (1984) 177–194.

[Vu8]   M. VUORINEN: *Conformal invariants and quasiregular mappings.* – J. Analyse Math. 45 (1985) 69–115.

[Vu9]   M. VUORINEN: *Koebe arcs and quasiregular mappings.* – Math. Z. 190 (1985) 95–106.

[Vu10]   M. VUORINEN: *On the distortion of n-dimensional quasiconformal mappings.* – Proc. Amer. Math. Soc. 96 (1986) 275–283.

[Vu11]   M. VUORINEN: *On quasiregular mappings and domains with a complete conformal metric.* – Math. Z. 194 (1987) 459–470.

[Vu12]   M. VUORINEN: *Conformal Geometry and Quasiregular Mappings.* – Lecture Notes in Mathematics, vol. 1319. Springer, Berlin–Heidelberg, 1988.

[Vu13]   M. VUORINEN: *On Picard's theorem for quasiregular mappings.* – Proc. Amer. Math. Soc. 107 (1989) 383–394.

[Wa1]   H. WALLIN: *A connection between α-capacity and L^p-classes of differentiable functions.* – Ark. Mat. 5 (1964) 331–341.

[Wa2]   H. WALLIN: *Metrical characterization of conformal capacity zero.* – J. Math. Anal. Appl. 58 (1977) 298–311.

[Wh]   G.T. WHYBURN: *Analytic Topology.* – Amer. Math. Soc. Colloquium Publications, New York, 1942.

[W]   K.–O. WIDMAN: *Hölder continuity of solutions of elliptic systems.* – Manuscripta Math. 5 (1971) 299–308.

[Y]   K. YOSHIDA: *Functional Analysis.* – Grundlehren der mathematischen Wissenschaften, vol. 123. Springer, Berlin–Heidelberg–New York, 1971.

[Z1]   W.P. ZIEMER: *Extremal length and p-capacity.* – Michigan Math. J. 16 (1969) 43–51.

[Z2]   W.P. ZIEMER: *Weakly Differentiable Functions.* – Graduate Texts in Mathematics, vol. 120. Springer, Berlin–Heidelberg–New York, 1989.

[Zo1]   V.A. ZORICH: *The theorem of M. A. Lavrent'ev on quasiconformal mappings in space.* (Russian). – Mat. Sb. 74 (1967) 417–433.

[Zo2]   V.A. ZORICH: *Isolated singularities of mappings with bounded distortion.* (Russian). – Mat. Sb. 81 (1970) 634–638.

[Zo3]   V.A. ZORICH: *Global homeomorphism theorem for space quasiconformal mappings, its development and related open problems.* – Lecture Notes in Mathematics, vol. 1508, pp. 132–148. Springer, Berlin–Heidelberg–New York, 1992.

List of Symbols

Index